Advances in Stability Theory at the End of the 20th Century

Stability and Control: Theory, Methods and Applications

A series of books and monographs on the theory of stability and control

Edited by A.A. Martynyuk
Institute of Mechanics, Kiev, Ukraine
V. Lakshmikantham,
Florida Institute of Technology, USA

Volume 1
Theory of Integro-Differential Equations
V. Lakshmikantham and M. Rama Mohana Rao

Volume 2
Stability Analysis: Nonlinear Mechanics Equations
A.A. Martynyuk

Volume 3
Stability of Motion of Nonautonomous Systems (Method of Limiting Equations)
J. Kato, A.A. Martynyuk and A.A. Shestakov

Volume 4
Control Theory and its Applications
E.O. Roxin

Volume 5
Advances in Nonlinear Dynamics
Edited by S. Sivasundaram and A.A. Martynyuk

Volume 6
Solving Differential Problems by Multistep Initial and Boundary Value Methods
L. Brugnano and D. Trigiante

Volume 7
Dynamics of Machines with Variable Mass
L. Cveticanin

Volume 8
Optimization of Linear Control Systems: Analytical Methods and Computational Algorithms
F.A. Aliev and V.B. Larin

Volume 9
Dynamics and Control
Edited by G. Leitmann, F.E. Udwadia and A.V. Kryazhimskii

Volume 10
Volterra Equations and Applications
Edited by C. Corduneanu and I.W. Sandberg

Volume 11
Nonlinear Problems in Aviation and Aerospace
Edited by S. Sivasundaram

Volume 12
Stabilization of Programmed Motion
E.Ya. Smirnov

Volume 13
Advances in Stability Theory at the End of the 20th Century
Edited by A.A. Martynyuk

This book is part of a series. The publisher will accept continuation orders which may be cancelled at any time and which provide for automatic billing and shipping of each title in the series upon publication. Please write for written details.

Advances in Stability Theory at the End of the 20th Century

Edited by
A.A. Martynyuk
Institute of Mechanics,
Kiev, Ukraine

Taylor & Francis Group

LONDON AND NEW YORK

First published 2003
by Taylor & Francis
11 New Fetter Lane, London EC4P 4EE

Simultaneously published in the USA and Canada
by Taylor & Francis Inc,
29 West 35th Street, New York, NY 10001

Taylor & Francis is an imprint of the Taylor & Francis Group

© 2003 Taylor & Francis

This book has been produced from camera ready copy
supplied by the authors

Printed and bound in Great Britain by
TJ International Ltd, Padstow, Cornwall

All rights reserved. No part of this book may be reprinted or reproduced or utilised in
any form or by any electronic, mechanical, or other means, now known or hereafter invented,
including photocopying and recording, or in any information storage or retrieval system,
without permission in writing from the publishers.

Every effort has been made to ensure that the advice and information in this book is
true and accurate at the time of going to press. However, neither the publisher nor
the authors can accept any legal responsibility or liability for any errors or omissions
that may be made. In the case of drug administration, any medical procedure or
the use of technical equipment mentioned within this book, you are strongly advised
to consult the manufacturer's guidelines.

British Library Cataloguing in Publication Data
A catalogue record for this book is available from the British Library

Library of Congress Cataloging in Publication Data
A catalog record for this book has been requested

ISBN 0–415–26962–8

Contents

Introduction to the Series ... ix

Preface ... xi

An Overview ... xiii

Part 1 Progress in Stability Theory by the First Approximation

1.1 Invariant Foliations for Carathéodory Type Differential Equations in Banach Spaces ... 1
B. Aulbach and T. Wanner

1.2 On Exponential Asymptotic Stability for Functional Differential Equations with Causal Operators ... 15
C. Corduneanu and Yizeng Li

1.3 Lyapunov Problems on Stability by Linear Approximation ... 25
N.A. Izobov

Part 2 Contemporary Development of Lyapunov's Ideas of Direct Method

2.1 Vector Lyapunov Functions Nonlinear, Time-Varying, Ordinary and Functional Differential Equations ... 49
P. Borne, M. Dambrine, W. Perruquetti and J.P. Richard

2.2 Some Results on Total Stability Properties for Singular Systems ... 75
A. D'Anna

2.3 Stability Theory of Volterra Difference Equations ... 89
F. Dannan, S. Elaydi and P. Li

2.4 Consistent Lyapunov Methodology for Exponential Stability: PCUP Approach ... 107
Ly.T. Gruyitch

2.5 Advances in Stability Theory of Lyapunov: Old and New ... 121
V. Lakshmikantham and S. Leela

2.6 Matrix Liapunov Functions and Stability Analysis of
Dynamical Systems 135
A.A. Martynyuk

2.7 Stability Theorems in Impulsive Functional Differential Equations
with Infinite Delay 153
A.A. Martynyuk, J.H. Shen and I.P. Stavroulakis

2.8 The Asymptotic Behaviour of Solutions of Stochastic Functional
Differential Equations with Finite Delays by Liapunov-Razumikhin
Method 175
T. Taniguchi

2.9 A Non-Standard Approach to the Study of the Dynamic System
Stability 189
V.A. Vujičić

Part 3 Stability of Solutions to Periodic Differential Systems

3.1 A Survey of Starzhinskii's Works on Stability of Periodic
Motions and Nonlinear Oscillations 201
Yu.A. Mitropol'skii, A.A. Martynyuk and V.I. Zhukovskii

3.2 Implications of the Stability of an Orbit for Its Omega Limit Set 217
J.S. Muldowney

3.3 Some Concepts of Periodic Motions and Stability Originated by
Analysis of Homogeneous Systems 231
V.N. Pilipchuk

3.4 Stability Criteria for Periodic Solutions of Autonomous Hamiltonian
Systems 243
A.A. Zevin

Part 4 Selected Applications

4.1 Stability in Models of Agriculture–Industry–Environment
Interactions 255
H.I. Freedman, M. Solomonovich, L.P. Apedaile and A. Hailu

4.2 Bifurcations of Periodic Solutions of the Three Body Problem 267
V.I. Gouliaev

4.3 Complex Mechanical Systems: Steady-State Motions, Oscillations,
Stability 289
A.Yu. Ishlinsky, V.A. Storozhenko and M.E. Temchenko

4.4 Progress in Stability of Impulsive Systems with Applications to Population Growth Models 321
Xinzhi Liu

Index **339**

Introduction to the series

The problems of modern society are both complex and interdisciplinary. Despite the apparent diversity of problems, tools developed in one context are often adaptable to an entirely different situation. For example, consider the Lyapunov's well known second method. This interesting and fruitful technique has gained increasing significance and has given a decisive impetus for modern development of the stability theory of differential equations. A manifest advantage of this method is that it does not demand the knowledge of solutions and therefore has great power in application. It is now well recognized that the concept of Lyapunov-like functions and the theory of differential and integral inequalities can be utilized to investigate qualitative and quantitative properties of nonlinear dynamic systems. Lyapunov-like functions serve as vehicles to transform the given complicated dynamic systems into a relatively simpler system and therefore it is sufficient to study the properties of this simpler dynamic system. It is also being realized that the same versatile tools can be adapted to discuss entirely different nonlinear systems, and that other tools, such as the variation of parameters and the method of upper and lower solutions provide equally effective methods to deal with problems of a similar nature. Moreover, interesting new ideas have been introduced which would seem to hold great potential.

Control theory, on the other hand, is that branch of application-oriented mathematics that deals with the basic principles underlying the analysis and design of control systems. To control an object implies the influence of its behavior so as to accomplish a desired goal. In order to implement this influence, practitioners build devices that incorporate various mathematical techniques. The study of these devices and their interaction with the object being controlled is the subject of control theory. There have been, roughly speaking, two main lines of work in control theory which are complementary. One is based on the idea that a good model of the object to be controlled is available and that we wish to optimize its behavior, and the other is based on the constraints imposed by uncertainty about the model in which the object operates. The control tool in the latter is the use of feedback in order to correct for deviations from the desired behavior. Mathematically, stability theory, dynamic systems and functional analysis have had a strong influence on this approach.

Volume 1, *Theory of Integro-Differential Equations*, is a joint contribution by V. Lakshmikantham (USA) and M. Rama Mohana Rao (India).

Volume 2, *Stability Analysis: Nonlinear Mechanics Equations*, is by A.A. Martynyuk (Ukraine).

Volume 3, *Stability of Motion of Nonautonomous Systems: The Method of Limiting Equations*, is a collaborative work by J. Kato (Japan), A.A. Martynyuk (Ukraine) and A.A. Shestakov (Russia).

Volume 4, *Control Theory and its Applications*, is by E.O. Roxin (USA).

Volume 5, *Advances in Nonlinear Dynamics*, is edited by S. Sivasundaram (USA) and A.A. Martynyuk (Ukraine) and is a multiauthor volume dedicated to Professor S. Leela.

Volume 6, *Solving Differential Problems by Multistep Initial and Boundary Value Methods*, is a joint contribution by L. Brugnano (Italy) and D. Trigiante (Italy).

Volume 7, *Dynamics of Machines with Variable Mass*, is by L. Cveticanin (Yugoslavia).

Volume 8, *Optimization of Linear Control Systems: Analytical Methods and Computational Algorithms*, is a joint work by F.A. Aliev (Azerbaijan) and V.B. Larin (Ukraine).

Volume 9, *Dynamics and Control*, is edited by G. Leitmann (USA), F.E. Udwadia (USA) and A.V. Kryazhimskii (Russia) and is a multiauthor volume.

Volume 10, *Volterra Equations and Applications*, is edited by C. Corduneanu (USA) and I.W. Sandberg (USA) and is a multiauthor volume.

Volume 11, *Nonlinear Problems in Aviation and Aerospace*, is edited by S. Sivasundaram (USA) and is a multiauthor volume.

Volume 12, *Stabilization of Programmed Motion*, is by E.Ya. Smirnov (Russia).

Volume 13, *Advances in Stability Theory at the end of the 20th Century*, is edited by A.A. Martynyuk (Ukraine) and is a multiauthor volume.

Due to the increased interdependency and cooperation among the mathematical sciences across the traditional boundaries, and the accomplishments thus far achieved in the areas of stability and control, there is every reason to believe that many breakthroughs await us, offering existing prospects for these versatile techniques to advance further. It is in this spirit that we see the importance of the 'Stability and Control' series, and we are immensely thankful to Taylor & Francis publishers for their interest and cooperation in publishing this series.

Preface

The development of stability theory in the twentieth century has been closely connected with the solution of major problems of science and engineering and also with the modelling and investigation of more complex phenomena of the real world. The peculiar features of progress in this field of the natural sciences are:

* the variety of engineering and scientific problems whose solution by the methods of motion stability theory has allowed numerous projects to be carried out in aviation, rocket engineering, submarine dynamics, economics, traffic, construction, etc.
* the intensive development of the ideas and methods proposed by the creators of stability theory such that Euler, Poincaré, and Lyapunov within the framework of modern achievements of the analytical and qualitative theory of equations;
* the integration of the efforts of scientists world-wide in solving the concrete scientific, engineering and general problems of stability theory.

The main idea of the present volume of the International Series of Scientific monographs is to present surveys and research papers from the various branches of the modern theory of stability written by scientists from around the world. Meanwhile, the application areas of stability theory are very diverse and an attempt to embrace as many of them as possible would inevitably result in the creation of several volumes. So, this volume presents only some of the applications of motion stability theory.

The papers collected in this volume are written by the scientists who are deeply involved in current research and they provide a general insight into the present-day state of stability theory.

This volume consists of four sections presenting the following areas of the development of stability theory.

Part 1. Progress in Stability Theory by the First Approximation.
Part 2. Contemporary Development of Lyapunov's Idea of the Direct Method.
Part 3. Stability of Solutions to Periodic Differential Systems.
Part 4. Selected Applications.

This volume is of considerable importance to young investigators in the field and should give an impetus to the statement of new problems in stability theory.

Acknowledgements

First of all I would like to express my sincere gratitude to contributors to this volume who have supported the idea of its creation and have kindly sent their original papers for this volume.

Special thanks to Professor Dr. Bernd Wegner and Mrs. Barbara Strazzabosco who provided us with CompactMATH from Zentralblatt MATH containing 1.700.000 titles and reviews of scientific works. This allowed us to make many references in this volume more accurate.

Collaborators from the Stability of Processes Department at the Institute of Mechanics, L.N.Chernetskaja, A.N.Chernienko and S.N.Rasshivalova have done great work to prepare this volume for publication.

Finally, I am grateful to the editors at Taylor & Francis for their enthusiastic support and many ideas which have promoted the successful work of this volume.

A.A.Martynyuk

An Overview

The results of the development of science at the end of the twentieth century are in no way a gathering of memorial tablets over the buried-for-ever ideas of prominent scholars. Most likely they are a collection of incomplete architectural ensembles many of which remain unfinished not because of the imperfection of the concepts but due to their engineering or economic prematurity.

The present volume introduces the reader to the further developments of outstanding ideas of the creators of the qualitative theories of equations and nonlinear dynamics of systems in a modern interpretation and provides a unique picture of the world-wide development of stability theory at the end of the twentieth century.

Thirty-five scientists from 12 countries have contributed to this volume. In result the volume contains 6 surveys and 14 research papers arranged in 4 sections.

In their surveys and papers the authors have formulated many open problems of different complexities which would seem to be inspiring starting points for new investigations by beginners and experts in the field. The lists of references to the papers comprise titles of papers and monographs by authors who have contributed to the development of stability theory and related directions.

Below we outline the contents of the volume.

Survey of Part 1

In *Section 1.1*, *B. Aulbach* and *T. Wanner* consider differential equations which explicitly but discontinuously depend on time and are rarely studied objects. Even so, they promise important applications, e.g. in control theory or in the theory of random dynamical systems. The authors continue a previous study of qualitative properties of so-called Carathéodory type differential equations whose feature is a measurable dependence of the right-hand side on time. In fact, they show that the fundamental theorem on the existence of integral manifolds can be generalized to a result providing two complete foliations of the extended state space by integral manifolds. This detailed information about the dynamical structure of the extended state space, on the other hand, can be used to construct transformations establishing the topological equivalence between certain weakly coupled and completely decoupled systems.

In *Section 1.2*, *C. Corduneanu* and *Yizeng Li* considered stability of perturbed functional differential equations which involve abstract Volterra (causal) operators.

The unperturbed system is assumed to be linear, while the perturbations are generally-nonlinear. The exponential asymptotic stability is mainly discussed.

In *Section 1.3*, *N.A. Izobov* presents a survey which contains a rather complete description of the results related to the Lyapunov problems on the exponential stability of the zero solution of differential systems under higher-order perturbations. He considers both the special and the general Lyapunov problems as well as their linear analogs.

Survey of Part 2

In *Section 2.1* by *P. Borne*, *M. Dambrine*, *W. Perruquetti* and *J.P. Richard* the considered comparison approach constitutes an alternative or, at least, an interesting complement to the standard Lyapunov method. It involves the use of a simpler model with the aim of obtaining conclusions available for a more complex one. This investigation sums up several recent results concerning comparison systems, and puts them into a more general framework: both ordinary and functional differential equations are dealt with, using a unified writing. The work is divided into six parts. The first three parts introduce the context and some notations. Part IV defines the comparison systems for general functional differential equations. Part V applies this concept to the analysis of some qualitative properties of sets (such as, for instance, stability or positive invariance). The last part concerns exclusively the case of ordinary differential systems, and provides some stability criteria that are easy to check.

In *Section 2.2*, *A. D'Anna* discusses the problem of stability properties when the vector function f in the system $\dot{x} = f(t,x)$ is not defined on a closed set M of R^n. Also the case of perturbations acting along the motion is considered and new conditions of stability and asymptotic behavior for M are obtained. The results attained are illustrated by an example concerning the two-body problem, when the motion takes place in an atmosphere and the reference frame is non-inertial.

Section 2.3 by *F. Dannan*, *S. Elaydi* and *P. Li* provides a survey of recent results and trends in the stability of Volterra difference equations of both convolution and nonconvolution type. The resolvent matrix and a variation of constants formula will be developed.

In *Section 2.4* by *Ly.T. Gruyitch* a new physical principle – the Physical Continuity and Uniqueness Principle is combined with the nature of time by expressing the continuous nature of physical variables relative to time. Their features are reflected in the mathematical model used in the Cauchy form. The new Lyapunov methodology for nonlinear systems called the consistent Lyapunov methodology is broadened in this paper to the exponential stability of compact connected invariant sets of such time-invariant nonlinear mathematical models. It provides the new necessary and sufficient conditions for the exponential stability of a compact connected invariant set A. The conditions are also necessary and sufficient for the accurate single-step construction of a system of Lyapunov functions proving the set of exponential stability.

In *Section 2.5* by *V. Lakshmikantham* and *S. Leela* some new concepts are discussed that have appeared in stability theory in recent years. It is recognized that the concept of Lyapunov-like functions together with the general comparison principles offer versatile tools to investigate the qualitative properties of solutions of differential equations. Furthermore, some new ideas and approaches which might provide an exciting prospect of further advancement are still in the initial stages of investigation.

In *Section 2.6* by *A.A. Martynyuk* a version of the matrix Lyapunov function method to analyse motion stability of dynamical systems is proposed. New results for stability, asymptotic stability, exponential stability and instability are established. In the case large-scale system the matrix-valued Lyapunov function is applied and a new stability conditions are obtained.

In *Section 2.7* by *A.A. Martynyuk, J.H. Shen* and *I.P. Stavroulakis* considered the system of impulsive functional differential equations with infinite delay and nonlinear impulsive perturbations. The authors extended a uniform asymptotic stability result by Burton and Zhang by employing the Lyapunov functional and examine the persistence of uniform asymptotic stability under the impulsive perturbations. Also obtained an impulsive stabilization result by employing the Lyapunov function and the Razumikhin technique.

Section 2.8 by *T. Taniguchi* discusses the sufficient conditions for $EW(t, X(t))$ to approach zero as $t \to \infty$, where $X(t)$ denotes a solution of a stochastic functional differential equation with finite delay $r > 0$ and W denotes a so-called Lyapunov function. As the application the author considers the almost sure Lyapunov exponent $\limsup_{t \to \infty} \frac{1}{t} \log |X(t)|$ of the solutions under more general assumptions than in Taniguchi (see Report, 58 (1996), 191–208). In this paper he uses the Lyapunov-Razumikhin method. Then, the difficulty of the constructions of the Lyapunov functionals is avoided.

In *Section 2.9*, *V.A. Vujičić* proves the general invariant criterion of the balanced state stability and mechanical system motion. Lagrange's theorem about the system's balanced state stability is generalized while its application to one characteristic instance of the rheonomic system is shown.

Survey of Part 3

In *Section 3.1*, *Y.A. Mitropol'skii, A.A. Martynyuk* and *V.I. Zhukovskii* survey Starzhinskii's works on the stability of periodic motions and nonlinear oscillations.

The paper contains the following sections: stability of periodic motions; the mathematical theory of parametric resonance and its applications; oscillations in substantially nonlinear systems. Six open problems for investigation in this area are formulated.

Section 3.2 by *J.S. Muldowney* provides a survey of results which use the attraction of an orbit for its neighbors to detect an omega limit which is a stable equilibrium or a periodic orbit.

Section 3.3 by *V.N. Pilipchuk* deals with so-called homogeneous mechanical systems and some related ideas in nonlinear mechanics. The simplest homogeneous system such as a nonlinear oscillator of power form characteristic was considered by Lyapunov in the theory of stability of motion to be a special degenerated case. The ideas and methods growing around the homogeneous systems appeared to be of more general sense then the systems themselves.

In *Section 3.4* by *A.A. Zevin* new stability criteria for a periodic solution lying on a convex energy surface of a Hamiltonian and associated with the minimum of a dual functional are found. Other conditions (feasible also for nonconvex Hamiltonians) guarantee the unique continuation and stability of Lyapunov families of periodic solutions within a given region. All conditions are checked through the Hessian of the Hamiltonian.

Survey of Part 4

In *Section 4.1*, *H.I.Freedman, M. Solomonovich, L.P. Apedaile* and *A. Hailu* considered a system of three ordinary differential equations as a model of agricultural and industrial wealth interacting with the environment. Two subsystems representing agriculture-industry and agriculture-environment are discussed. Criteria for the existence and stability of equilibria are given for each model.

In *Section 4.2*, *V.I. Gouliaev* scrutinizes some problems of bifurcation state modelling and modification of periodic orbits of bounded three body problems. Investigated are the stability of periodic solutions and the stability of triangular libration points in an elliptic bounded problem of three bodies. The appearance of chaotic solutions of the plane bounded problem of three bodies is discussed.

In *Section 4.3* by *A.Yu. Ishlinsky, V.A. Storozhenko* and *M.E. Temchenko* a survey is made of research carried out mainly at the Institute of Mathematics of the National Academy of Sciences of the Ukraine. The works deal with the motion investigation of mechanical systems. Major attention is paid to the new direction of rational mechanics which has recently been developed. It presents the motion investigation of an absolute solid suspended on an inextensible inertialess thread being a string and a "string" suspension composed of freely joint weighted rods. Alongside the theoretical problems such as the investigation of stationary motions, their stability and bifurcation, problems associated with the practical application of the obtained results are discussed. In particular, the application areas are the creation of large-size centrifuges, balancing techniques, the experimental determination of the dynamical characteristics of a solid of arbitrary form.

In *Section 4.4, Xinzhi Liu* establishes some stability criteria for impulsive differential systems. It is shown that impulses do contribute to yield stability properties even when the corresponding differential system without impulses does not enjoy any stability behavior.

Part 1
PROGRESS IN STABILITY THEORY BY THE FIRST APPROXIMATION

1.1 INVARIANT FOLIATIONS FOR CARATHÉODORY TYPE DIFFERENTIAL EQUATIONS IN BANACH SPACES*

B. AULBACH[1] and T. WANNER[2]

[1] Department of Mathematics, University of Augsburg, Augsburg, Germany
[2] Department of Mathematics and Statistics, University of Maryland, Baltimore, USA

1 Introduction

Throughout this paper we consider systems of differential equations of the form

$$\begin{aligned} \dot{x}_1 &= A_1(t)x_1 + F_1(t, x_1, x_2), \\ \dot{x}_2 &= A_2(t)x_2 + F_2(t, x_1, x_2), \end{aligned} \quad (1)$$

where x_1 and x_2 are elements of some Banach spaces $\mathcal{X}_1$ and $\mathcal{X}_2$, respectively, and $A_i \colon \mathbb{R} \to L(\mathcal{X}_i)$ and $F_i \colon \mathbb{R} \times \mathcal{X}_1 \times \mathcal{X}_2 \to \mathcal{X}_i$, $i = 1, 2$, are mappings satisfying the following two sets of hypotheses:

(H1) **Hypothesis on linear part:** The mappings $A_i \colon \mathbb{R} \to L(\mathcal{X}_i)$, $i = 1, 2$, are locally integrable and there exist real constants $K \geq 1$ and $\alpha < \beta$ such that the evolution operators $\Phi_i(t, s)$ of the homogeneous linear equations $\dot{x}_i = A_i(t)x_i$, respectively, satisfy the estimates

$$\|\Phi_1(t, s)\| \leq K e^{\alpha(t-s)} \quad \text{for all} \quad t \geq s,$$

$$\|\Phi_2(t, s)\| \leq K e^{\beta(t-s)} \quad \text{for all} \quad t \leq s.$$

(H2) **Hypothesis on nonlinear part:** The mappings $F_i \colon \mathbb{R} \times \mathcal{X}_1 \times \mathcal{X}_2 \to \mathcal{X}_i$,

---

*Advances in Stability Theory (Ed.: A.A. Martynyuk). Stability and Control: Theory, Methods and Applications, Taylor & Francis, London, **13** (2003) 1–14.

$i = 1, 2$, have the Carathéodory property[1] and they satisfy the identities $F_i(t, 0, 0) = 0$ for all $t \in \mathbb{R}$. Furthermore, there exists a constant $L \geq 0$ such that the estimates

$$\|F_i(t, x_1, x_2) - F_i(t, \bar{x}_1, \bar{x}_2)\| \leq L \|x_1 - \bar{x}_1\| + L \|x_2 - \bar{x}_2\|$$

hold for all $t \in \mathbb{R}$ and $x_i, \bar{x}_i \in \mathcal{X}_i$, $i = 1, 2$.

For the concept of solutions as well as other properties of this type of equations we refer to Aulbach and Wanner [1]. We also adopt the notation introduced in this paper such as

$$\lambda(t; \tau, \xi, \eta) = (\lambda_1(t; \tau, \xi, \eta), \lambda_2(t; \tau, \xi, \eta)) \in \mathcal{X} := \mathcal{X}_1 \times \mathcal{X}_2$$

for the so-called *cocycle* of (1), the solution satisfying the initial condition $x_1(\tau) = \xi$, $x_2(\tau) = \eta$. As shown in [1] this cocycle exists for all $t \in \mathbb{R}$ if system (1) satisfies the standing hypotheses (H1) and (H2).

The analysis of the present paper is based on the so-called Fundamental Existence Theorem for Integral Manifolds (see [1, Theorem 4.1]) which – roughly speaking – says the following: System (1) has two global integral manifolds of the form $x_2 = s_0(t, x_1)$ and $x_1 = r_0(t, x_2)$ with $s_0(t, 0) \equiv 0$ and $r_0(t, 0) \equiv 0$ if the coupling of the two equations in (1) is small in the sense that the Lipschitz constant L satisfies the estimate $L < \frac{\beta - \alpha}{4K}$. Moreover, those integral manifolds may be characterized by the asymptotic behaviour of the solutions they are made of. This asymptotic behaviour can be described in terms of the so-called quasiboundedness which also plays a crucial role in the present paper. We therefore recall that a function g from the reals to some Banach space is called γ^+-*quasibounded* if $\sup_{t \geq \tau} \|g(t)\| e^{-\gamma t} < \infty$ for some $\tau \in \mathbb{R}$. Accordingly, the function g is called γ^--*quasibounded* or $\gamma^{\pm}$-*quasibounded* if $\sup_{t \leq \tau} \|g(t)\| e^{-\gamma t} < \infty$ for some $\tau \in \mathbb{R}$ or $\sup_{t \in \mathbb{R}} \|g(t)\| e^{-\gamma t} < \infty$, respectively.

It is the purpose of this paper to extend the Fundamental Existence Theorem for Integral Manifolds to a result establishing invariant foliations of the extended state space $\mathbb{R} \times \mathcal{X}_1 \times \mathcal{X}_2$ of system (1) by integral manifolds. This result, on the other hand, allows to prove that (under a suitably strengthened coupling condition) system (1) is topologically equivalent to a decoupled system whose two equations are completely independent of each other.

---

[1]That means that for each $t \in \mathbb{R}$ the mappings $F_i(t, \cdot, \cdot)$ are continuous and that for each $(x_1, x_2) \in \mathcal{X}_1 \times \mathcal{X}_2$ the mappings $F_i(\cdot, x_1, x_2)$ are strongly measurable, $i = 1, 2$ (see [1, Definition 2.1]).

2 Integral Manifolds through Solutions

As mentioned in the previous section there exist two integral manifolds for system (1) having the zero solution in common. In the present section we extend this result by showing that *any* solution of system (1) is associated with a pair of integral manifolds which intersect along this (and no other) solution. The proof of this extension is very simple thanks to the fact that we are working in a *non*autonomous setting. Indeed, we simply transform the given solution in a canonical way to the zero solution of an equation which traditionally is called the *differential equation of perturbed motion*. We then apply the previous result.

It is worth mentioning here that this simple trick does not work in a purely autonomous context because the differential equation of perturbed motion is not autonomous if the perturbed motion is not constant.

Theorem 2.1 *Consider a differential equation of the form (1) on a Banach space $\mathcal{X} = \mathcal{X}_1 \times \mathcal{X}_2$ satisfying the assumptions (H1) and (H2) involving the growth constants $K \geq 1$, $\alpha < \beta$ and the Lipschitz constant $L \geq 0$. We assume that for a fixed δ with $0 < \delta < \frac{\beta - \alpha}{2}$ the Lipschitz constant L satisfies the estimate*

$$0 \leq L < \frac{\delta}{2K}.$$

Then for every choice of $\gamma \in [\alpha + \delta, \beta - \delta]$ we get the following:

(a) *There exists a uniquely determined mapping $s \colon \mathbb{R} \times \mathcal{X}_1 \times \mathbb{R} \times \mathcal{X} \to \mathcal{X}_2$ such that for every point $(\tau_*, \xi_*) \in \mathbb{R} \times \mathcal{X}$ the graph*

$$S_{\tau_*, \xi_*} := \big\{ (\tau, \xi_1, s(\tau, \xi_1, \tau_*, \xi_*)) \in \mathbb{R} \times \mathcal{X}_1 \times \mathcal{X}_2 : \tau \in \mathbb{R},\ \xi_1 \in \mathcal{X}_1 \big\}$$

of the mapping $s(\cdot, \cdot, \tau_*, \xi_*) \colon \mathbb{R} \times \mathcal{X}_1 \to \mathcal{X}_2$ allows the representation

$$S_{\tau_*, \xi_*} = \big\{ (\tau, \xi) \in \mathbb{R} \times \mathcal{X} : \lambda(\cdot; \tau, \xi) - \lambda(\cdot; \tau_*, \xi_*)\ \text{is}\ \gamma^+\text{-quasibounded} \big\}.$$

Moreover, the estimate

$$\|s(\tau, \xi_1, \tau_*, \xi_*) - s(\tau, \bar{\xi}_1, \tau_*, \xi_*)\| \leq \frac{K^2 L(\delta - KL)}{\delta(\delta - 2KL)} \|\xi_1 - \bar{\xi}_1\|$$

holds for all $\tau, \tau_* \in \mathbb{R}$, $\xi_1, \bar{\xi}_1 \in \mathcal{X}_1$ and $\xi_* \in \mathcal{X}$, and the mapping s is continuous. Finally, S_{τ_*, ξ_*} is an integral manifold for system (1), the so-called S-manifold through the point (τ_*, ξ_*) or through the solution $\lambda(\cdot; \tau_*, \xi_*)$.

(b) *There is a uniquely determined mapping $r \colon \mathbb{R} \times \mathcal{X}_2 \times \mathbb{R} \times \mathcal{X} \to \mathcal{X}_1$ such that for every $(\tau_*, \xi_*) \in \mathbb{R} \times \mathcal{X}$ the graph*

$$R_{\tau_*, \xi_*} := \big\{ (\tau, r(\tau, \xi_2, \tau_*, \xi_*), \xi_2) \in \mathbb{R} \times \mathcal{X}_1 \times \mathcal{X}_2 : \tau \in \mathbb{R},\ \xi_2 \in \mathcal{X}_2 \big\}$$

of the mapping $r(\cdot,\cdot,\tau_*,\xi_*)$ may be represented in the form

$$R_{\tau_*,\xi_*} = \{(\tau,\xi) \in \mathbb{R} \times \mathcal{X} : \lambda(\cdot;\tau,\xi) - \lambda(\cdot;\tau_*,\xi_*) \text{ is } \gamma^- \text{-quasibounded}\}.$$

Moreover, the estimate

$$\|r(\tau,\xi_2,\tau_*,\xi_*) - r(\tau,\bar\xi_2,\tau_*,\xi_*)\| \le \frac{K^2 L(\delta - KL)}{\delta(\delta - 2KL)} \|\xi_2 - \bar\xi_2\|$$

holds for all $\tau, \tau_* \in \mathbb{R}$, $\xi_2, \bar\xi_2 \in \mathcal{X}_2$ and $\xi_* \in \mathcal{X}$, and r is continuous. Finally, R_{τ_*,ξ_*} is an integral manifold for system (1), the so-called R-manifold through the point (τ_*,ξ_*) or through the solution $\lambda(\cdot;\tau_*,\xi_*)$.

Proof For every $(\tau_*,\xi_*) \in \mathbb{R} \times \mathcal{X}$ we consider the system

$$\begin{aligned}\dot x_1 &= A_1(t)x_1 + \tilde F_1(t,x_1,x_2,\tau_*,\xi_*), \\ \dot x_2 &= A_2(t)x_2 + \tilde F_2(t,x_1,x_2,\tau_*,\xi_*)\end{aligned} \qquad (2)$$

where the functions on the right-hand side are defined by

$$\tilde F_i(t,x_1,x_2,\tau_*,\xi_*) := F_i(t, x_1 + \lambda_1(t;\tau_*,\xi_*), x_2 + \lambda_2(t;\tau_*,\xi_*))$$
$$- F_i(t, \lambda_1(t;\tau_*,\xi_*), \lambda_2(t;\tau_*,\xi_*)), \quad i=1,2,$$

$\lambda = (\lambda_1,\lambda_2)$ denoting the cocycle of system (1). That system (2) indeed satisfies the assumptions of the Fundamental Existence Theorem for Integral Manifolds ([1, Theorem 4.1]) can be seen as follows:

- Due to the continuity of the cocycle the mapping $(\tilde F_1, \tilde F_2) : \mathbb{R} \times \mathcal{X}_1 \times \mathcal{X}_2 \times \mathbb{R} \times \mathcal{X} \to \mathcal{X}_1 \times \mathcal{X}_2$ is continuous. The parameter space is given by $\mathcal{P} := \mathbb{R} \times \mathcal{X}$.
- We have $\tilde F_1(t,0,0,\tau_*,\xi_*) \equiv 0$ and $\tilde F_2(t,0,0,\tau_*,\xi_*) \equiv 0$ on $\mathbb{R} \times \mathbb{R} \times \mathcal{X}$.
- The remaining assumptions of [1, Theorem 4.1] follow immediately from the above assumptions (H1) and (H2).

Applying Part (a) of [1, Theorem 4.1] to system (2) we get a uniquely determined mapping $s_0 : \mathbb{R} \times \mathcal{X}_1 \times \mathbb{R} \times \mathcal{X} \to \mathcal{X}_2$ describing an integral manifold for this system. Now we define the function

$$s(\tau,\xi_1,\tau_*,\xi_*) := s_0(\tau, \xi_1 - \lambda_1(\tau;\tau_*,\xi_*), \tau_*, \xi_*) + \lambda_2(\tau;\tau_*,\xi_*)$$

which is obviously continuous. Moreover, the estimate claimed in (a) follows immediately from the corresponding estimate in [1, Theorem 4.1].

In order to prove the claimed representation of the graph of s let us fix an arbitrary point $(\tau_*,\xi_*) \in \mathbb{R} \times \mathcal{X}$. Moreover, let $(\tau,\xi) = (\tau,\xi_1,\xi_2) \in \mathbb{R} \times \mathcal{X}$ be a point such that $\mu := \lambda(\cdot;\tau,\xi) - \lambda(\cdot;\tau_*,\xi_*)$ is γ^+-quasibounded. Then μ is a solution of (2) which is γ^+-quasibounded. Therefore we have $\mu_2(\tau) = s_0(\tau, \mu_1(\tau), \tau_*, \xi_*)$. Together with

$$\mu_1(\tau) = \xi_1 - \lambda_1(\tau;\tau_*,\xi_*) \quad \text{and} \quad \mu_2(\tau) = \xi_2 - \lambda_2(\tau;\tau_*,\xi_*)$$

this yields the relation

$$\xi_2 = \mu_2(\tau) + \lambda_2(\tau;\tau_*,\xi_*) = s_0(\tau, \mu_1(\tau), \tau_*, \xi_*) + \lambda_2(\tau;\tau_*,\xi_*)$$
$$= s_0(\tau, \xi_1 - \lambda_1(\tau;\tau_*,\xi_*), \tau_*, \xi_*) + \lambda_2(\tau;\tau_*,\xi_*) = s(\tau, \xi_1, \tau_*, \xi_*).$$

This means that $(\tau, \xi) \in S_{\tau_*,\xi_*}$. Conversely, let (τ, ξ) be an arbitrary point belonging to S_{τ_*,ξ_*}. Then for $\mu := \lambda(\cdot;\tau,\xi) - \lambda(\cdot;\tau_*,\xi_*)$ we deduce that μ is a solution of (2) with

$$\mu_1(\tau) = \xi_1 - \lambda_1(\tau;\tau_*,\xi_*) \quad \text{and} \quad \mu_2(\tau) = \xi_2 - \lambda_2(\tau;\tau_*,\xi_*).$$

The identity $\xi_2 = s(\tau, \xi_1, \tau_*, \xi_*)$ now implies

$$\mu_2(\tau) = \xi_2 - \lambda_2(\tau;\tau_*,\xi_*) = s(\tau, \xi_1, \tau_*, \xi_*) - \lambda_2(\tau;\tau_*,\xi_*)$$
$$= s_0(\tau, \xi_1 - \lambda_1(\tau;\tau_*,\xi_*), \tau_*, \xi_*) = s_0(\tau, \mu_1(\tau), \tau_*, \xi_*),$$

i.e. $\mu = \lambda(\cdot;\tau,\xi) - \lambda(\cdot;\tau_*,\xi_*)$ is γ^+-quasibounded. Since the invariance of S_{τ_*,ξ_*} can easily be deduced as in the proof of [1, Theorem 4.1], this concludes the proof of (a). Part (b) can be proved analogously.

Remark 2.1 If in hypothesis (H1) we additionally assume that the mappings $\|A_1(\cdot)\|$ and $\|A_2(\cdot)\|$ are locally bounded, then the mappings $s(\cdot,\cdot,\tau_*,\xi_*)$ and $r(\cdot,\cdot,\tau_*,\xi_*)$ are even Lipschitz continuous. This follows from [1, Remark 4.2] and the fact that in this case the solution $\lambda(\cdot;\tau_*,\xi_*)$ of system (1) is Lipschitz continuous. In order to see the latter we abbreviate system (1) in the form $\dot{x} = A(t)x + F(t,x)$ and notice that for arbitrary $t, s \in U$, where $U \subset \mathbb{R}$ is bounded, we have

$$\|\lambda(t) - \lambda(s)\| \leq \left| \int_s^t (\|A(\tau)\| \cdot \|\lambda(\tau)\| + \underbrace{\|F(\tau,\lambda(\tau))\|}_{\leq 2L\|\lambda(\tau)\|})\, d\tau \right|$$
$$\leq \left(\sup_{\tau \in U} \|A(\tau)\| + 2L \right) \cdot \sup_{\tau \in U} \|\lambda(\tau)\| \cdot |t - s|,$$

proving the desired Lipschitz continuity.

Theorem 2.1 provides two partitions of the extended phase space $\mathbb{R} \times \mathcal{X}$ into disjoint unions of S- and R-manifolds, respectively. In order to see this we consider the following two equivalence relations on the set of all solutions of (1):

$$\mu \sim_S \nu \quad :\Longleftrightarrow \quad \mu - \nu \text{ is } \gamma^+\text{-quasibounded},$$

$$\mu \sim_R \nu \quad :\Longleftrightarrow \quad \mu - \nu \text{ is } \gamma^-\text{-quasibounded}.$$

The characterization of the S- and R-manifolds in terms of the asymptotic behaviour of the solutions of which they are made implies the following:

- For any $(\tau_*, \xi_*) \in \mathbb{R} \times \mathcal{X}$ the equivalence class of $\lambda(\cdot; \tau_*, \xi_*)$ with respect to $\sim_S$ is exactly the S-manifold through (τ_*, ξ_*).
- For any $(\tau_*, \xi_*) \in \mathbb{R} \times \mathcal{X}$ the equivalence class of $\lambda(\cdot; \tau_*, \xi_*)$ with respect to $\sim_R$ is exactly the R-manifold through (τ_*, ξ_*).

Thus, two arbitrary S-manifolds are either disjoint or they coincide, i.e. the whole extended phase space can be partitioned into S-manifolds. An analogous statement is true for the set of R-manifolds.

3 Invariant Foliations of the Extended Phase Space

In the previous section we have seen that Theorem 2.1 enables us to partition the extended phase space of (1) into S- and R-manifolds and that the integral manifolds occurring in those two partitions are exactly the equivalence classes of the two equivalence relations $\sim_S$ and $\sim_R$. It is our next aim to find two suitable systems of representatives for those partitions. To this end we exploit our knowledge on the geometric structure of the extended phase space of (1) and show that the R_0-manifold is a system of representatives for the partition into S-manifolds, and that the S_0-manifold is a system of representatives for the partition into R-manifolds.

Definition 3.1 Employing the above notation we define:

(a) The mapping $f_H \colon \mathbb{R} \times \mathcal{X}_1 \times \mathbb{R} \times \mathcal{X}_2 \to \mathcal{X}_2$ defined by

$$f_H(\tau, \xi_1, \tau^*, \xi_2^*) := s(\tau, \xi_1, \tau^*, r_0(\tau^*, \xi_2^*), \xi_2^*),$$

is called the *horizontal foliation of* (1), and for each $(\tau^*, \xi_2^*) \in \mathbb{R} \times \mathcal{X}_2$ the set

$$F_H(\tau^*, \xi_2^*) := \{(\tau, \xi_1, f_H(\tau, \xi_1, \tau^*, \xi_2^*)) \colon \tau \in \mathbb{R}, \ \xi_1 \in \mathcal{X}_1\}$$

is called the *horizontal fiber through* (τ^*, ξ_2^*).

(b) The mapping $f_V \colon \mathbb{R} \times \mathcal{X}_2 \times \mathbb{R} \times \mathcal{X}_1 \to \mathcal{X}_1$ defined by

$$f_V(\tau, \xi_2, \tau^*, \xi_1^*) := r(\tau, \xi_2, \tau^*, \xi_1^*, s_0(\tau^*, \xi_1^*)),$$

is called the *vertical foliation of* (1), and for each $(\tau^*, \xi_1^*) \in \mathbb{R} \times \mathcal{X}_1$ the set

$$F_V(\tau^*, \xi_1^*) := \{(\tau, f_V(\tau, \xi_2, \tau^*, \xi_1^*), \xi_2) \colon \tau \in \mathbb{R}, \ \xi_2 \in \mathcal{X}_2\}$$

is called the *vertical fiber through* (τ^*, ξ_1^*).

In order to make this definition geometrically transparent we consider an arbitrary point $(\tau^*, \xi_1^*) \in \mathbb{R} \times \mathcal{X}_1$. Then the point $(\tau^*, \xi^*) := (\tau^*, \xi_1^*, s_0(\tau^*, \xi_1^*))$ is contained in the S_0-manifold of equation (1) and we have the following:

- According to Theorem 2.1 the S-manifold through the point (τ^*, ξ^*) consists exactly of those solutions μ of (1) for which the difference $\mu - \lambda(\cdot; \tau^*, \xi^*)$ is γ^+-quasibounded for some $\gamma \in [\alpha + \delta, \beta - \delta]$. Since $\lambda(\cdot; \tau^*, \xi^*)$ is γ^+-quasibounded as well, the S-manifold through (τ^*, ξ^*) consists of all γ^+-quasibounded solutions of (1), i.e. it actually is the S_0-manifold of (1).
- Due to Definition 3.1 the R-manifold through the above point (τ^*, ξ^*) is exactly the vertical fiber $F_V(\tau^*, \xi_1^*)$.

Analogous statements are true for points on the R_0-manifold of equation (1).

For autonomous, finite-dimensional differential equations the above-defined vertical foliation generalizes the corresponding notion used in Kirchgraber and Palmer [2], where the real parts of the eigenvalues of the matrix A_1 are supposed to be less than or equal to zero, and the real parts of the eigenvalues of A_2 are assumed strictly positive. Our horizontal foliation, however, differs from the notion of horizontal foliation used in [2].

In order to prove that the horizontal and vertical fibers introduced in Definition 3.1 in fact foliate the extended phase space $\mathbb{R} \times \mathcal{X}$ of (1) we first examine whether two different fibers are indeed disjoint.

Lemma 3.1 *Suppose system (1) satisfies the assumptions (H1) and (H2) as well as the condition $0 \leq L < \frac{\delta}{2K}$. Then for any $\tau \in \mathbb{R}$ we have:*

(a) *For all $\xi_2, \eta_2 \in \mathcal{X}_2$ with $\xi_2 \neq \eta_2$ the horizontal fibers $F_H(\tau, \xi_2)$ and $F_H(\tau, \eta_2)$ are disjoint.*

(b) *For all $\xi_1, \eta_1 \in \mathcal{X}_1$ with $\xi_1 \neq \eta_1$ the vertical fibers $F_V(\tau, \xi_1)$ and $F_V(\tau, \eta_1)$ are disjoint.*

Proof In order to prove (a) we choose an arbitrary $\gamma \in [\alpha + \delta, \beta - \delta]$ and suppose there is a $\zeta_1 \in \mathcal{X}_1$ such that

$$(\tau, \zeta_1, f_H(\tau, \zeta_1, \tau, \xi_2)) = (\tau, \zeta_1, f_H(\tau, \zeta_1, \tau, \eta_2)).$$

Let $\zeta_2 := f_H(\tau, \zeta_1, \tau, \xi_2) = f_H(\tau, \zeta_1, \tau, \eta_2)$, $\xi_1 := r_0(\tau, \xi_2)$ and $\eta_1 := r_0(\tau, \eta_2)$. Then the mappings $\lambda(\cdot; \tau, \zeta) - \lambda(\cdot; \tau, \xi)$ and $\lambda(\cdot; \tau, \zeta) - \lambda(\cdot; \tau, \eta)$ are γ^+-quasibounded according to Theorem 2.1, and thus $\lambda(\cdot; \tau, \eta) - \lambda(\cdot; \tau, \xi)$ is γ^+-quasibounded as well. On the other hand, $\lambda(\cdot; \tau, \eta)$ and $\lambda(\cdot; \tau, \xi)$ are γ^--quasibounded according to our construction, and so $\lambda(\cdot; \tau, \eta) - \lambda(\cdot; \tau, \xi)$ is γ^--quasibounded.

Altogether we see that $\lambda(\cdot; \tau, \eta) - \lambda(\cdot; \tau, \xi)$ is a $\gamma^\pm$-quasibounded solution of the differential equation of perturbed motion of (1) with respect to the solution $\lambda(\cdot; \tau, \xi)$. Then [1, Theorem 4.1(c)] yields the identity $\lambda(t; \tau, \eta) - \lambda(t; \tau, \xi) \equiv 0$ on $\mathbb{R}$ and consequently we get $\xi_2 = \eta_2$. This contradicts the above assumption.

Part (b) of the Lemma can be proved analogously.

Lemma 3.1 allows to draw the conclusion that every point of the extended phase space of (1) lies on at most one horizontal fiber and on at most one vertical fiber. It is our next aim to show that each point $(\tau^*, \xi^*) \in \mathbb{R} \times \mathcal{X}$ in fact lies on *exactly* one horizontal and on *exactly* one vertical fiber. Obviously, this is the case if and only if the S-manifold through (τ^*, ξ^*) intersects the R_0-manifold of (1) and the R-manifold through (τ^*, ξ^*) intersects the S_0-manifold of (1). That this is indeed the case is not obvious because even though the integral manifolds under consideration satisfy global Lipschitz conditions they might not have points in common if the Lipschitz constants are greater than 1.

This problem, however, can be overcome by noticing that the common Lipschitz constant $\frac{K^2 L(\delta - KL)}{\delta(\delta - 2KL)}$ for the mappings s_0, r_0, s and r (cf. [1, Theorem 4.1] and Theorem 2.1 of the present paper) converge to 0 as L tends to 0. As the following theorem and its proof show, the desired result can be derived by means of an application of the uniform contraction principle.

Theorem 3.1 *Consider a differential equation of the form (1) on a Banach space $\mathcal{X} = \mathcal{X}_1 \times \mathcal{X}_2$ satisfying the assumptions (H1) and (H2) involving the growth constants $K \geq 1$, $\alpha < \beta$ and the Lipschitz constant $L \geq 0$. We assume that for some fixed δ with $0 < \delta < \frac{\beta - \alpha}{2}$ the Lipschitz constant L satisfies the estimate*

$$0 \leq L < C(K, \delta) := \frac{\delta}{2K^2} \left(K + 2 - \sqrt{K^2 + 4} \right). \tag{3}$$

Then we get the following:

(a) *There exists a uniquely determined mapping*

$$\mathcal{F}_1 = (\mathcal{F}_{11}, \mathcal{F}_{12}) : \mathbb{R} \times \mathcal{X} \to \mathcal{X} = \mathcal{X}_1 \times \mathcal{X}_2$$

such that for any $(\tau^*, \xi^*) \in \mathbb{R} \times \mathcal{X}$ the inclusion

$$(\tau^*, \xi^*) \in F_V(\tau^*, \mathcal{F}_{11}(\tau^*, \xi^*)) \cap F_H(\tau^*, \mathcal{F}_{12}(\tau^*, \xi^*))$$

holds. Hence, every point of the extended phase space of (1) lies on exactly one horizontal and on exactly one vertical fiber. Furthermore, the mapping $\mathcal{F}_1$ is continuous, for all $\tau \in \mathbb{R}$ we have $\mathcal{F}_1(\tau, 0) = 0$, and for all $\tau \in \mathbb{R}$ and $\xi \in \mathcal{X}$ the estimate

$$\|\mathcal{F}_1(\tau, \xi)\| \leq \frac{2}{1 - C(L)} \|\xi\| \tag{4}$$

holds with $C(L) := \frac{K^2 L(\delta - KL)}{\delta(\delta - 2KL)}$. Finally, suppose that μ is an arbitrary solution of (1). Then $\mathcal{F}_1(\cdot, \mu(\cdot))$ is a solution of the decoupled system

$$\begin{aligned} \dot{x}_1 &= A_1(t)x_1 + F_1(t, x_1, s_0(t, x_1)), \\ \dot{x}_2 &= A_2(t)x_2 + F_2(t, r_0(t, x_2), x_2). \end{aligned} \tag{5}$$

(b) *There is a continuous mapping $\mathcal{F}_2 \colon \mathbb{R} \times \mathcal{X} \to \mathcal{X}$ such that for every $(\tau^*, \xi^*) \in \mathbb{R} \times \mathcal{X}$ the point $(\tau^*, \mathcal{F}_2(\tau^*, \xi^*))$ is the uniquely determined intersection point of the horizontal fiber $F_H(\tau^*, \xi_2^*)$ with the vertical fiber $F_V(\tau^*, \xi_1^*)$ in the hyperplane $t = \tau^*$. This in particular yields $\mathcal{F}_2(\tau, 0) = 0$ for all $\tau \in \mathbb{R}$. Moreover, for every solution ν of the decoupled system (5) the mapping $\mathcal{F}_2(\cdot, \nu(\cdot))$ is a solution of (1), and the estimate*

$$\|\mathcal{F}_2(\tau, \xi)\| \leq \frac{2}{1 - C(L)} \|\xi\| \qquad (6)$$

holds for all $\tau \in \mathbb{R}$ and $\xi \in \mathcal{X}$ with $C(L)$ as above.

(c) *For every $\tau \in \mathbb{R}$ the mappings $\mathcal{F}_1(\tau, \cdot)$ and $\mathcal{F}_2(\tau, \cdot)$ are inverse to each other, hence homeomorphisms on $\mathcal{X}$. Thus we call the two systems (1) and (5) topologically equivalent.*

(d) *If the right-hand side of (1) is periodic in t with period $\Theta > 0$, then so are the mappings $\mathcal{F}_1$ and $\mathcal{F}_2$. In particular, if the right-hand side of equation (1) is autonomous, the mappings $\mathcal{F}_1$ and $\mathcal{F}_2$ are independent of t, hence homeomorphisms on $\mathcal{X}$.*

Remark 3.1 Condition (3) is stronger than the condition on L required in Theorem 2.1, since for every $K \geq 1$ we have

$$\mathcal{C}(K, \delta) < \frac{\delta}{2K}.$$

Moreover, condition (3) has been chosen in such a way that the estimate

$$C(L) = \frac{K^2 L(\delta - KL)}{\delta(\delta - 2KL)} < 1 \qquad (7)$$

is satisfied. This latter estimate is crucial for the following proof of Theorem 3.1.

Proof The proof of Theorem 3.1 extensively uses the uniform contraction principle (cf. [1, Theorem B.2]).

(a) Let $(\tau^*, \xi^*) \in \mathbb{R} \times \mathcal{X}$ be arbitrary. We begin by showing that (τ^*, ξ^*) lies on exactly one vertical fiber $F_V(\tau^*, \xi_1)$. To this end we consider the mapping

$$T_1 \colon \begin{cases} \mathcal{X}_1 \times \mathbb{R} \times \mathcal{X} & \to \quad \mathcal{X}_1 \\ (\xi_1, \tau^*, \xi^*) & \mapsto \quad r(\tau^*, s_0(\tau^*, \xi_1), \tau^*, \xi^*). \end{cases}$$

Due to the continuity of r and s_0 also the mapping T_1 is continuous. Furthermore, for any $\tau^* \in \mathbb{R}$, $\xi^* \in \mathcal{X}$ and $\xi_1, \eta_1 \in \mathcal{X}_1$ [1, Theorem 4.1] and Theorem 2.1 provide the estimate

$$\|T_1(\xi_1, \tau^*, \xi^*) - T_1(\eta_1, \tau^*, \xi^*)\| \leq C(L) \|s_0(\tau^*, \xi_1) - s_0(\tau^*, \eta_1)\|$$
$$\leq C(L)^2 \|\xi_1 - \eta_1\|. \qquad (8)$$

According to (7), T_1 is a uniform contraction and [1, Theorem B.2] implies that for any (τ^*, ξ^*) there is a uniquely determined fixed point $\mathcal{F}_{11}(\tau^*, \xi^*)$ in $\mathcal{X}_1$, and that the mapping $\mathcal{F}_{11}$ is continuous. Moreover, $\xi_1 \in \mathcal{X}_1$ is a fixed point of $T_1(\cdot, \tau^*, \xi^*)$ if and only if

$$\xi_1 = r(\tau^*, s_0(\tau^*, \xi_1), \tau^*, \xi^*) \iff (\tau^*, \xi_1, s_0(\tau^*, \xi_1)) \in R_{\tau^*, \xi^*}$$
$$\iff (\tau^*, \xi^*) \in F_V(\tau^*, \xi_1).$$

In other words, ξ_1 is a fixed point of $T_1(\cdot, \tau^*, \xi^*)$ if and only if (τ^*, ξ^*) lies on the vertical fiber $F_V(\tau^*, \xi_1)$. Obviously, for $\xi^* = 0$ this fixed point is $\xi_1 = 0$, and this implies $\mathcal{F}_{11}(\tau^*, 0) = 0$.

Now let us fix any point $(\tau, \xi) \in \mathbb{R} \times \mathcal{X}$. Then [1, Theorems 4.1 and B.2], (8) and Theorem 2.1 imply with $C(L)$ as in (7)

$$\|\mathcal{F}_{11}(\tau, \xi)\| = \|\mathcal{F}_{11}(\tau, \xi) - \mathcal{F}_{11}(\tau, 0)\|$$
$$\leq \frac{1}{1 - C(L)^2} \|T_1(\mathcal{F}_{11}(\tau, 0), \tau, \xi) - T_1(\mathcal{F}_{11}(\tau, 0), \tau, 0)\|$$
$$= \frac{1}{1 - C(L)^2} \|T_1(0, \tau, \xi) - T_1(0, \tau, 0)\| = \frac{1}{1 - C(L)^2} \|r(\tau, 0, \tau, \xi)\| \quad (9)$$
$$\leq \frac{1}{1 - C(L)^2} (\|r(\tau, 0, \tau, \xi) - \underbrace{r(\tau, \xi_2, \tau, \xi)}_{=\xi_1}\| + \|\xi_1\|)$$
$$\leq \frac{1}{1 - C(L)^2} (C(L)\|\xi_2\| + \|\xi_1\|) \leq \frac{1}{1 - C(L)^2} \|\xi\| \leq \frac{1}{1 - C(L)} \|\xi\|.$$

Similarly it can be shown that every point $(\tau^*, \xi^*) \in \mathbb{R} \times \mathcal{X}$ lies on exactly one horizontal fiber $F_H(\tau^*, \mathcal{F}_{12}(\tau^*, \xi^*))$, where $\mathcal{F}_{12}$ is a continuous mapping with $\mathcal{F}_{12}(\tau^*, 0) = 0$ and

$$\|\mathcal{F}_{12}(\tau, \xi)\| \leq \frac{1}{1 - C(L)} \|\xi\|.$$

Altogether we obtain the desired estimate

$$\|\mathcal{F}_1(\tau, \xi)\| = \|\mathcal{F}_{11}(\tau, \xi)\| + \|\mathcal{F}_{12}(\tau, \xi)\| \leq \frac{2}{1 - C(L)} \|\xi\|.$$

It remains to verify that $\mathcal{F}_1$ maps solutions of equation (1) onto solutions of the decoupled equation (5). To this end let μ be an arbitrary solution of (1) with $\xi^* = \mu(\tau^*)$, and let

$$\xi_1 := \mathcal{F}_{11}(\tau^*, \xi^*) \quad \text{and} \quad \nu(t) := \lambda(t; \tau^*, \xi_1, s_0(\tau^*, \xi_1)).$$

According to the construction of $\mathcal{F}_{11}$ the mapping ν is a γ^+-quasibounded solution of (1) and $\mu - \nu$ is γ^--quasibounded, for any $\gamma \in [\alpha + \delta, \beta - \delta]$. Now let $\tau \in \mathbb{R}$

be arbitrary. Then $\eta_1 := \mathcal{F}_{11}(\tau, \mu(\tau))$ is the uniquely determined point in $\mathcal{X}_1$ for which the difference

$$\mu - \lambda(\cdot; \tau, \eta_1, s_0(\tau, \eta_1))$$

is γ^--quasibounded. Due to the above definition of ν as a solution on the S_0-manifold of (1) and the γ^--quasiboundedness of $\mu - \nu$ we therefore get $\nu = \lambda(\cdot; \tau, \eta_1, s_0(\tau, \eta_1))$, i.e. we have

$$\nu_1(\tau) = \mathcal{F}_{11}(\tau, \mu(\tau)) \quad \text{for every} \quad \tau \in \mathbb{R}.$$

Since ν is contained in the S_0-manifold of (1), we additionally have

$$\nu_2(\tau) = s_0(\tau, \nu_1(\tau)) \quad \text{for every} \quad \tau \in \mathbb{R}.$$

This implies that $\nu_1 = \mathcal{F}_{11}(\cdot, \mu(\cdot))$ is a solution of equation

$$\dot{x}_1 = A_1(t)x_1 + F_1(t, x_1, s_0(t, x_1)).$$

Analogously it can be shown that $\mathcal{F}_{12}(\cdot, \mu(\cdot))$ solves

$$\dot{x}_2 = A_2(t)x_2 + F_2(t, r_0(t, x_2), x_2).$$

This concludes the proof of (a).

(b) Let $(\tau^*, \xi^*) \in \mathbb{R} \times \mathcal{X}$ be arbitrary. In order to prove that the fibers $F_V(\tau^*, \xi_1^*)$ and $F_H(\tau^*, \xi_2^*)$ have exactly one common point within the hyperplane $t = \tau^*$ we define the mapping

$$T_2 : \begin{cases} \mathcal{X} \times \mathbb{R} \times \mathcal{X} & \to \quad \mathcal{X}_1 \times \mathcal{X}_2 = \mathcal{X} \\ (\xi, \tau^*, \xi^*) & \mapsto \quad (f_V(\tau^*, \xi_2, \tau^*, \xi_1^*), f_H(\tau^*, \xi_1, \tau^*, \xi_2^*)). \end{cases}$$

Due to the continuity of f_V and f_H the mapping T_2 is continuous, too, and for any $\tau^* \in \mathbb{R}$ and $\xi^*, \xi, \eta \in \mathcal{X}$ Theorem 2.1 implies

$$\|T_2(\xi, \tau^*, \xi^*) - T_2(\eta, \tau^*, \xi^*)\| = \|f_V(\tau^*, \xi_2, \tau^*, \xi_1^*) - f_V(\tau^*, \eta_2, \tau^*, \xi_1^*)\|$$
$$+ \|f_H(\tau^*, \xi_1, \tau^*, \xi_2^*) - f_H(\tau^*, \eta_1, \tau^*, \xi_2^*)\| \leq C(L)(\|\xi_1 - \eta_1\| + \|\xi_2 - \eta_2\|)$$
$$= C(L)\|\xi - \eta\|.$$

According to (7) the mapping T_2 is a uniform contraction. Hence [1, Theorem B.2] provides for every (τ^*, ξ^*) a uniquely determined fixed point $\mathcal{F}_2(\tau^*, \xi^*)$ in $\mathcal{X}$. Moreover, $\mathcal{F}_2$ is continuous and ξ is a fixed point of $T_2(\cdot, \tau^*, \xi^*)$ if and only if

$$\xi_1 = f_V(\tau^*, \xi_2, \tau^*, \xi_1^*) \quad \text{and} \quad \xi_2 = f_H(\tau^*, \xi_1, \tau^*, \xi_2^*)$$
$$\iff (\tau^*, \xi) \in F_V(\tau^*, \xi_1^*) \quad \text{and} \quad (\tau^*, \xi) \in F_H(\tau^*, \xi_2^*),$$

i.e. if and only if (τ^*, ξ) is an intersection point of the fibers $F_V(\tau^*, \xi_1^*)$ and $F_H(\tau^*, \xi_2^*)$ in the hyperplane $t = \tau^*$.

Now we choose any $\tau \in \mathbb{R}$ and $\xi \in \mathcal{X}$. Then as in the proof of (a) we get

$$\|\mathcal{F}_2(\tau, \xi)\| \leq \frac{1}{1 - C(L)} \|T_2(\mathcal{F}_2(\tau, 0), \tau, \xi) - T_2(\mathcal{F}_2(\tau, 0), \tau, 0)\|$$

$$= \frac{1}{1 - C(L)} \left(\|f_V(\tau, 0, \tau, \xi_1)\| + \|f_H(\tau, 0, \tau, \xi_2)\| \right)$$

$$\leq \frac{1}{1 - C(L)} \big(\|f_V(\tau, 0, \tau, \xi_1) - \underbrace{f_V(\tau, s_0(\tau, \xi_1), \tau, \xi_1)}_{=\xi_1}\| + \|\xi_1\|$$

$$+ \|f_H(\tau, 0, \tau, \xi_2) - \underbrace{f_H(\tau, r_0(\tau, \xi_2), \tau, \xi_2)}_{=\xi_2}\| + \|\xi_2\| \big)$$

$$\leq \frac{1}{1 - C(L)} \left(C(L)\|s_0(\tau, \xi_1)\| + \|\xi_1\| + C(L)\|r_0(\tau, \xi_2)\| + \|\xi_2\| \right)$$

$$\leq \frac{1}{1 - C(L)} \left(C(L)^2 \|\xi_1\| + \|\xi_1\| + C(L)^2 \|\xi_2\| + \|\xi_2\| \right)$$

$$\leq \frac{2}{1 - C(L)} \left(\|\xi_1\| + \|\xi_2\| \right) = \frac{2}{1 - C(L)} \|\xi\|.$$

The remaining claims of (b) can be proved similarly to (a).

(c) Due to our construction $\mathcal{F}_2(\tau^*, \xi^*)$ is the uniquely determined intersection point of the fibers $F_V(\tau^*, \xi_1^*)$ and $F_H(\tau^*, \xi_2^*)$ in the hyperplane $t = \tau^*$. On the other hand, for every (τ^*, ξ^*) in the extended phase space $F_V(\tau^*, \mathcal{F}_{11}(\tau^*, \xi^*))$ and $F_H(\tau^*, \mathcal{F}_{12}(\tau^*, \xi^*))$ are the uniquely determined vertical and horizontal fibers containing (τ^*, ξ^*). This immediately implies that $\mathcal{F}_1(\tau, \mathcal{F}_2(\tau, \xi)) = \xi$ and $\mathcal{F}_2(\tau, \mathcal{F}_1(\tau, \xi)) = \xi$ for all $\tau \in \mathbb{R}$ and $\xi \in \mathcal{X}$.

(d) This part of Theorem 3.1 can be deduced as in the proof of [1, Corollary 4.4].

Remark 3.2 It can readily be seen that the differential equation

$$\dot{x}_1 = A_1(t)x_1 + F_1(t, x_1, s_0(t, x_1)) \tag{10}$$

describes the behaviour of the solutions of system (1) on the S_0-manifold in the following sense:

- Suppose that $\mu: \mathbb{R} \to \mathcal{X}$ is a solution of (1) on the S_0-manifold. Then the mapping $\mu_1: \mathbb{R} \to \mathcal{X}_1$ is a solution of (10).
- Conversely, assume that $\nu: \mathbb{R} \to \mathcal{X}_1$ is a solution of (10). Then the mapping $(\nu(\cdot), s_0(\cdot, \nu(\cdot))): \mathbb{R} \to \mathcal{X}_1 \times \mathcal{X}_2 = \mathcal{X}$ is a solution of (1) which obviously lies on the S_0-manifold of this system.

Similarly, equation

$$\dot{x}_2 = A_2(t)x_2 + F_2(t, r_0(t, x_2), x_2)$$

describes the behaviour of (1) on the R_0-manifold. With these interpretations in mind, Theorem 3.1 states that – loosely speaking – the behaviour of system (1) is completely determined by its behaviour on the S_0- and the R_0-manifold.

In order to conclude this paper we want to mention that the invariant foliations we have constructed have some interesting applications. First of all, it is possible to obtain detailed information about the asymptotic behaviour of *all* solutions of system (1), in particular, one can derive the so-called reduction principle which for finite-dimensional autonomous systems has been established by Pliss [3] and Kelley [4], and for nonautonomous systems by Aulbach [5]. Furthermore, the precise information about the extended phase space of (1) makes it possible to construct topological mappings which – in addition to the decoupling achieved in this paper – allows to linearize at least one of the two equations of system (1). Finally, we can prove the generalizations of the theorems of Hartman-Grobman-type (cf. Grobman [6, 7], Hartman [8–10], Palmer [11, 12], and Šošitaĭsvili [13]) to our general nonautonomous setting in Banach spaces. All of this will be done in the forthcoming paper [14].

References

[1] Aulbach, B. and Wanner, T. (1996). Integral manifolds for Carathéodory type differential equations in Banach spaces. In:*Six Lectures on Dynamical Systems* (Eds.: B. Aulbach and F, Colonius). World Scientific, Singapore, 45–119.

[2] Kirchgraber, U. and Palmer, K. J. (1991). *Geometry in the Neighborhood of Invariant Manifolds of Maps and Flows and Linearization*. Pitman, London.

[3] Pliss, V. A. (1964). Principle reduction in the theory of the stability of motion. *Izv. Akad. Nauk S.S.S.R., Mat. Ser.* **28**, 1297–1324 (Russian).

[4] Kelley, A. (1967). Stability of the center-stable manifold. *J. of Mathematical Analysis and Applications*, **18**, 336–344.

[5] Aulbach, B. (1982). A reduction principle for nonautonomous differential equations. *Archiv der Mathematik*, **39**, 217–232.

[6] Grobman, D. M. (1959). Homeomorphisms of systems of differential equations. *Doklady Akademii Nauk SSSR*, **128**, 880 (Russian).

[7] Grobman, D. M. (1962). The topological classification of the vicinity of a singular point in n-dimensional space. *Math. SSSR, Sbornik*, **56**, 77–94 (Russian).

[8] Hartman, P. (1960). A lemma in the theory of structural stability of differential equations. *Proc. of the American Mathematical Society*, **11**, 610-620.

[9] Hartman, P. (1963). On the local linearization of differential equations. *Proc. of the American Mathematical Society*, **14**, 568–573.

[10] Hartman, P. (1982). *Ordinary Differential Equations*. Birkhäuser, Basel.

[11] Palmer, K. J. (1973). A generalization of Hartman's linearization theorem. *J. of Mathematical Analysis and Applications*, **41**, 753–758.

[12] Palmer, K. J. (1975). Linearization near an integral manifold. *J. of Mathematical Analysis and Applications*, **51**, 243–255.

[13] Šošitaĭšvili, A. N. (1975). Bifurcations of topological type of a vector field near a singular point. *Trudy Sem. Petrovsk.*, **1**, 279-309 (Russian).

[14] Aulbach, B. and Wanner, T. (2000). The Hartman–Grobman theorem for Carathéodory type differential equations in Banach spaces. *Nonlinear Analysis*, **40**, 91–104.

1.2 ON EXPONENTIAL ASYMPTOTIC STABILITY FOR FUNCTIONAL DIFFERENTIAL EQUATIONS WITH CAUSAL OPERATORS*

C. CORDUNEANU and YIZENG LI

Department of Mathematics, University of Texas at Arlington, Arlington, USA

1 Introduction

The notations used below are those found in our preceding papers [4, 5, 9].

The following linear (unperturbed) system is considered:

$$\dot{x}(t) = (Lx)(t), \quad t \in R_+, \tag{1}$$

with L a causal (abstract Volterra) operator acting on the space $L^2_{loc}(R_+, R^n)$. We assume L to be a continuous operator. For a general discussion of (1) and its nonhomogeneous counterpart, see the book [3].

The perturbed system has the form

$$\dot{x}(t) = (Lx)(t) + (Fx)(t), \quad t \in R_+, \tag{2}$$

with $F: L^2_{loc}(R_+, R^n) \to L^2_{loc}(R_+, R^n)$. Other choices for the underlying space are possible (see, for instance, [1, 10, 12]).

Assuming the zero solution to (1) (solution is understood in Carathéodory sense) has a particular kind of stability, it is desirable to find adequate conditions on the perturbing term $(Fx)(t)$, such that same kind of stability (or a weaker one) is valid for the zero solution of (2). One may regard this problem as a stability problem in the first approximation, or, as a problem of preservation of stability under perturbations.

---

*Advances in Stability Theory (Ed.: A.A. Martynyuk). Stability and Control: Theory, Methods and Applications, Taylor & Francis, London, **13** (2003) 15–23.

2 Definitions and Statement of Problems

In regard to the functional differential system

$$\dot{x}(t) = (Vx)(t), \quad t \in R_+, \tag{3}$$

in which V is a causal operator on the space $L^2_{\ell oc}(R_+, R^n)$, such that

$$(V\theta)(t) = \theta, \quad t \in R_+, \tag{4}$$

$\theta \in R^n$ being the zero vector, the initial value problem we shall consider is

$$x(t) = x_0(t), \quad t \in [0, t_0), \quad x(t_0) = x^0, \tag{5}$$

with $x_0 \in L^2([0, t_0], R^n)$, and $x^0 \in R^n$.

The definition of various types of Liapunov stability for the zero solution of (3) can be formulated as follows:

Stability. For each $\epsilon > 0$ and $t_0 > 0$, there exists $\delta = \delta(\epsilon, t_0)$ such that

$$|x^0| < \delta \quad \text{and} \quad |x_0|_2 < \delta \quad \text{imply} \quad |x(t; t_0, x^0, x_0)| < \epsilon \tag{6}$$

for $t \geq t_0$. Of course, $x(t; t_0, x^0, x_0)$ denotes the solution of (3), with initial data (5).

Uniform stability. The number $\delta(\epsilon, t_0)$ in the definition of stability must be independent of t_0: $\delta(\epsilon, t_0) \equiv \delta(\epsilon)$.

Asymptotic stability. The solution $x = \theta$ is stable, and there exists $\eta(t_0) > 0$ such that

$$\lim |x(t; t_0, x^0, x_0)| = 0 \quad \text{as} \quad t \to \infty, \tag{7}$$

as soon as

$$|x^0| + |x_0|_2 < \eta(t_0). \tag{8}$$

Uniform asymptotic stability. The solution $x = \theta$ is uniformly stable, and for each $\epsilon > 0$, there exists $T(\epsilon) > 0$, such that

$$|x(t; t_0, x^0, x_0)| < \epsilon \quad \text{for} \quad t > t_0 + T(\epsilon), \tag{9}$$

as soon as x^0 and x_0 satisfy (8), with some $\eta(t_0) = \text{const.}$

Exponential asymptotic stability. There exist two positive numbers K and λ, such that

$$|x(t; t_0, x^0, x_0)| \leq K(|x^0| + |x_0|_2)e^{-\lambda(t-t_0)}, \tag{10}$$

for $t \geq t_0$.

When V is a nonlinear operator, conditions (10) must be satisfied only for $|x^0|$ and $|x_0|_2$ sufficiently small.

It is obvious that each type of stability, defined above, implies the preceding types, excepting the case of asymptotic stability which does not imply, in general, the uniform stability (see, for instance, [2]).

The problems we shall investigate in this paper are concerned with perturbed systems of the form (2), and we shall be primarily dealing with uniform stability and exponential asymptotic stability. More precisely, we shall search under what conditions on the perturbing term $(Fx)(t)$, these two types of stability are preserved when we substitute (2) to (1).

The case of ordinary differential equations being a special case of the equations with causal operators, we cannot expect better results in the latter than those available in the first.

There are many open problems in the theory of stability of systems with causal operators. One such problem is to investigate the relationship between uniform asymptotic stability and the exponential asymptotic stability in the linear case. For ordinary differential systems, these concepts are equivalent under rather general assumptions. There are cases when the answer is positive for integro-differential systems, especially of the autonomous type (see, for instance [8], Theorems 2.2.1 and 2.2.2, and [13]).

Also, it is unknown how to characterize the various concepts of stability in terms of Liapunov functionals, excepting perhaps some particular cases (see again [8], Theorem 3.3.1). Even in case of linear systems, this problem is far from being solved.

Let us deal now with systems (1) and (2), and state conditions on $(Fx)(t)$ that will allow preservation of stability.

3 Two Results for Systems (1) and (2)

Let us begin with the case of uniform stability for the systems (1) and (2). The following result holds true.

Theorem 3.1 *Assume that L in (1) is a continuous causal operator on $L^2_{\ell oc}(R_+, R^n)$, and F is continuous on $L^2_{\ell oc}(R_+, R^n)$, taking bounded sets into bounded sets; moreover F is such that*

$$|(Fx)(t)| \leq \gamma(t)|x(t)|, \quad a.e. \text{ for } t \in R_+, \tag{11}$$

with $\gamma \in L(R_+, R)$.

Then, the solution $x = \theta$ of the system (2) is also uniformly stable.

Proof Since the solution of the initial value problem

$$\dot{x}(t) = (Lx)(t) + f(t), \quad t \in R_+, \tag{12}$$

with conditions (5), can be represented by the formula

$$x(t) = X(t,t_0)x^0 + \int_0^{t_0} \tilde{X}(t,s;t_0)x_0(s)\,ds + \int_{t_0}^t X(t,s)f(s)\,ds, \qquad (13)$$

assuming $f \in L^2_{loc}(R_+, R^n)$, where $X(t,s)$ is the Cauchy matrix attached to the operator L, we will consider now the integral equation

$$x(t) = X(t,t_0)x^0 + \int_0^{t_0} \tilde{X}(t,s;t_0)x_0(s)\,ds + \int_{t_0}^t X(t,s)(Fx)(s)\,ds. \qquad (14)$$

This equation is equivalent to the initial value problem (2), (5). Taking into account the properties of the matrices $X(t,s)$ and $\tilde{X}(t,s;t_0)$, as shown in [9], the following estimate is immediately obtained from (14):

$$|x(t)| \leq M(|x^0| + |x_0|_2) + M \int_{t_0}^t \gamma(s)|x(s)|\,ds, \qquad (15)$$

which holds true for $t \geq t_0$. The constant $M \geq 0$ in (15) exists on behalf of our assumption of uniform stability for the zero solution of (1), and is such that

$$|X(t,t_0)| \leq M \quad \text{for} \quad t \geq t_0 \geq 0, \qquad (16)$$

and

$$\int_0^{t_0} |\tilde{X}(t,s;t_0)|^2\,ds \leq M^2 \quad \text{for} \quad t \geq t_0 \geq 0. \qquad (17)$$

It is now easy to process the inequality (15) for $|x(t)|$, since it is of classical Gronwall type. One obtains the estimate

$$|x(t)| \leq M(|x^0| + |x_0|_2) \exp\left\{M \int_0^\infty \gamma(s)\,ds\right\}, \qquad (18)$$

which is valid for all $t \geq t_0$, provided $x(t)$ is defined on the whole semi-axis. But (18) shows that any solution of (2) is bounded on each compact interval of R_+. This implies the boundedness of the derivative $\dot{x}(t)$ on any compact interval. Therefore, if we assume $x(t)$ defined only on some interval (t_0, T), $(T < +\infty)$, we find out based on Cauchy's criterion for the existence of the limit that $\lim x(t)$ does exist as $t \uparrow T$. Then, the conclusion comes from the fact that such a solution can be continued to the right.

It is an elementary fact that the uniform stability of the zero solution of (2) follows from the estimate (18), valid on the whole semi-axis $t \geq t_0$.

Remark 3.1 The result stated in Theorem 3.1 is a straightforward extension of the classical result for ordinary differential equations (see, for instance, [2]). The condition (11) on F is, in the case of systems with causal operators, much more restrictive than it is in case of ordinary differential equations. A more reasonable assumption would be

$$|(Fx)(t)| \leq \gamma(t)|x|_t, \qquad (19)$$

with

$$|x|_t = \sup\{|x(s)|,\ 0 \leq s \leq t\}. \qquad (20)$$

We leave to the reader (as an open problem) the task of discussing the inequality similar to (15), when estimate (11) is replaced by (19), or pursue another approach to see if the theorem remains valid.

Remark 3.2 There are nontrivial examples of causal operators satisfying (11). For instance, one may choose

$$(Fx)(t) = \gamma(t)\left(m + \int_0^t |x(s)|\,ds\right)^{-1} |x(t)|, \qquad (21)$$

with $m > 0$ and $\gamma(t)$ as above.

We shall consider now the perturbed system (2), under the basic assumption that the zero solution of (1) is exponentially asymptotically stable.

Theorem 3.2 *Consider the system (2), and assume that the zero solution of (1) is exponentially asymptotically stable.*

Further, assume that the causal operator F is continuous on $L^2_{\ell oc}(R_+, R^n)$, and moreover, let F satisfy the estimate

$$|(Fx)(t)| \leq \mu |x(t)|, \qquad (22)$$

with $\mu > 0$ sufficiently small.

Then, the solution $x = \theta$ of (2) is also exponentially asymptotically stable.

Proof of Theorem 3.2 has been actually given in [9], under the assumption that F is a Nemytzky type operator: $(Fx)(t) = f(t, x(t))$. An estimate like (22) has been assumed. The proof is also a direct application of Theorem 1.5 in [11].

Of course, the result is valid without the assumption that F is a Nemytzky operator but the proof goes exactly on the same lines as in [9]. The formula for (21), with $\gamma(t) \equiv 1$ and large m provides an example of a causal operator satisfying (22).

4 Further Discussion

As mentioned above, in the section "Definitions and Statement of Problems", the question of whether or not uniform asymptotic stability of the zero solution of (1) is implying exponential asymptotic stability has been considered by various authors [8, 13] for special types of causal operators. The answer to this question is positive, under various sets of assumptions. Generally, speaking, the answer is negative.

We shall refer here to a result of Murkami [13]. We can illustrate the nature of the problem limiting our considerations to a particular case of that dealt with in [13].

Let us assume that

$$(Lx)(t) = Ax(t) + \int_0^t B(t-s)x(s)\,ds, \qquad (23)$$

with A a constant matrix of type $n \times n$, and $B(t)$, $t \in R_+$, a function matrix satisfying $|B| \in L(R_+, R)$. By $\hat{B}(s)$ one denotes the Laplace transform of B, i.e.,

$$\hat{B}(s) = \int_0^\infty e^{-ts} B(t)\,dt, \qquad (24)$$

which is surely defined for $\mathrm{Re}\,s \geq 0$.

The condition of asymptotic stability for the zero solution of (1), with L given by (2.2) is then (see [3], Theorem 6.3.1)

$$\det(sI - A - \hat{B}(s)) \neq 0 \quad \text{for} \quad \mathrm{Re}\,s \geq 0. \qquad (25)$$

As shown by Murakami [13], the exponential asymptotic stability occurs only in case

$$\int_0^\infty |B(t)|e^{\lambda t}\,dt < \infty, \qquad (26)$$

for some positive number λ.

Let us point out that the asymptotic stability for (1), with L given by (23), is actually uniform asymptotic stability, L being autonomous (time invariant).

Hence, even in rather special cases for L, such as (23), the uniform asymptotic stability and the exponential asymptotic stability are different concepts in the class of equations with causal operators.

There are many problems of stability theory of equations with causal operators that can be discussed based on the theory of integral inequalities. We shall illustrate this assertion considering again the system (2), but replacing the assumption (22) by the following ones:

$$|(Fx)(t) - (Fy)(t)| \leq \mu |x(t) - y(t)|, \tag{27}$$

$$(F\theta)(t) \in L^\infty(R_+, R^n), \tag{28}$$

where μ is a sufficiently small positive number.

We will rely on the Theorem 1.5 in the book [11], by Martynyuk, et al. Let us consider the inequality derived from the formula (14), taking into account (27) and (28):

$$|x(t)| \leq K(|x^0| + |x_0|_2)e^{-\lambda(t-t_0)} + L \int_{t_0}^{t} e^{-\lambda(t-s)}[\mu |x(s)| + A] \, ds. \tag{29}$$

We have assumed that the zero solution of (1) is exponentially asymptotically stable. We denoted $|F\theta|_\infty = A$.

The above mentioned theorem in [11] leads to the estimate

$$|x(t)| \leq K(|x^0| + |x_0|_2)e^{-(\lambda-\mu)(t-t_0)} + \frac{A}{\lambda - \mu}[1 - e^{-(\lambda-\mu)(t-t_0)}], \quad t \geq t_0. \tag{30}$$

The first conclusion we can draw from (30) is that *all solutions of (2) are bounded on R_+ if $\mu < \lambda$.*

Actually, one can easily prove, based on the same Theorem 1.5 in [11], that *each solution of (2) is exponentially asymptotically stable.*

Indeed, if $x(t)$ and $y(t)$ are two solutions of (2), corresponding to the initial data (x^0, x_0) resp. (y^0, y_0), then the following estimate holds true for $t \geq t_0$:

$$|x(t) - y(t)| \leq K(|x^0 - y^0| + |x_0 - y_0|_2)e^{-(\lambda-\mu)(t-t_0)}. \tag{31}$$

The conditions (27) and (28) are used to derive the estimate (31). It shows that each solution of (2), whom we know to be bounded on R_+ under condition (27) and (28), is exponentially asymptotically stable.

5 Some Open Problems

We want to call the reader's attention on some open problems related to the above discussed topics.

Presently, we are not aware of any stability result concerning the perturbation of (1), in case of uniform asymptotic stability (again, there are several results which regard only special cases of causal equations).

So far, to the best of our knowledge, no converse theorems on stability have been obtained in general case of causal equations.

A very important topic is stability theory is that of partial stability (or, stability with respect of part of the variables). The book [14] by Vorotnikov may serve as a source of inspiration. In the linear case, the system (1) should be rewritten as

$$\dot{y}(t) = (Ay)(t) + (Bz)(t), \quad \dot{z}(t) = (Cy)(t) + (Dz)(t),$$

in which A, B, C, D are linear causal operators on a given function space. The behaviour of y-variables is the main concern.

The case of neutral functional differential equations with causal operators is another important aspect of stability theory. The recent paper [7] by Kurbatov contains several ideas conducing towards stability results for systems that can be represented by

$$\frac{d}{dt}(Vx)(t) = (Wx)(t), \qquad (32)$$

with both V and W causal operators acting on convenient function spaces. See also [6].

Finally, a development of the Liapunov's function(al) method is still a matter of the future. As pointed out in [1], the study of stability in the nonlinear case is presenting some serious difficulties. See [4, 5] for a few considerations in this regard, related to comparison method.

References

[1] Azlebev, N. V. and Simonov, P. M. *Stability of Functional Differential Equations with Aftereffect*. Gordon and Breach Science Publishers, New York, (to appear).

[2] Corduneanu, C. (1988). *Principles of Differential and Integral Equations*. Chelsea, New York.

[3] Corduneanu, C. (1991). *Integral Equations and Applications*. Cambridge University Press, Cambridge.

[4] Corduneanu, C. (1995). Stability problems for Volterra functional differential equations. In: *Comparison Methods and Stability Theory* (Eds.: X. Liu and D. Siegel). Marcel Dekker, New York.

[5] Corduneanu, C. (1996). Some new trends in Liapunov's second method. In: *World Congress of Nonlinear Analysts* (Ed.: V. Lakshmikantham). Walter de Gruyter, Berlin, 1295–1302.

[6] Drakhlin, M. E. (1987). Questions in stability theory for functional differential equations of neutral type. *Differential Equations (Transl.)* **22**, 919–924.

[7] Kurbatov, V. (1995). Stability of neutral type equations in differential phase space. *Functional Differential Equations* (Israel) **3**, 99–133.

[8] Lakshmikantham, V. and Rao, M. (1995). *Theory of Integro-Differential Equations*, Gordon and Breach Science Publishers, New York.

[9] Li, Yizeng (1996). Stability problems of functional differential equations with abstract Volterra operators. *J. Integral Equations and Appl.* **8**, 47–63.

[10] Mahdavi, M. (1995). Linear functional differential equations with abstract Volterra operators. *Diff. and Integral Equations* **8**, 1517–1523.
[11] Martynyuk, A. A., Lakshmikantham, V. and Leela, S. (1989). *Stability of Motion: The Method of Integral Inequalities.* Naukova Dumka, Kiev. (Russian).
[12] Massera, J. L. and Schäffer, J. J. (1966). *Linear Differential Equations and Function Spaces.* Academic Press, New York.
[13] Murakami, S. (1991). Exponential stability for fundamental solution of some linear functional differential equations. In: *Functional Differential Equations* (Eds.: T. Yoshizawa and J. Kato). World Scientific, Singapore.
[14] Vorotnikov, V. I. (1998). *Partial Stability and Control.* Birkhäuser, Basel.

1.3 LYAPUNOV PROBLEMS ON STABILITY BY LINEAR APPROXIMATION*

N.A. IZOBOV

Institute of Mathematics, Academy of Sciences of Belorusia, Minsk, Belorusia

0 Introduction

Foundations of stability theory was laid by A.M.Lyapunov, the outstanding Russian mathematician. The first Lyapunov method for investigation of exponential stability of differential systems by linear approximation is based on the notion of the characteristic Lyapunov exponent [1, p.27] (see also [2, p.17]) $\lambda[f] \equiv \varlimsup_{t \to +\infty} t^{-1} \ln \|f(t)\|$, where a vector-valued or matrix-valued function f is piecewise continuous on $[0, +\infty)$.

Notation. Problems. We shall consider n-dimensional real differential systems: the linear approximation system

$$\dot{x} = A(t)x, \quad x \in R^n, \quad n \geq 2, \quad t \geq 0, \qquad (1_A)$$

with piecewise continuous (measurable) bounded ($\|A(t)\| \leq a < +\infty$ for $t \geq 0$) coefficients and the perturbed nonlinear systems

$$\dot{y} = A(t)y + f(t,y), \quad y \in R^n, \quad t \geq 0, \qquad (2)$$

with piecewise continuous in $t \geq 0$ and continuous in $y \in U_\rho \equiv \{y \in R^n : \|y\| < \rho = \rho(f)\}$ m-perturbations $f(t,y)$ of class

$$F_m \equiv \{f \colon \|f(t,y)\| \leq C_f \|y\|^m, \quad C_f = \text{const}, \quad (t,y) \in [0,+\infty) \times U_{\rho(f)}\},$$

---

*Advances in Stability Theory (Ed.: A.A. Martynyuk). Stability and Control: Theory, Methods and Applications, Taylor & Francis, London, **13** (2003) 25–48.

with fixed $m > 1$. This class of perturbations is most important in applications, it, evidently, contains the Lyapunov class [1, p.16] of the second order holomorphic perturbations

$$f(t,y) = \sum_{k_1+\cdots+k_n \geq 2} f_{k_1\ldots k_n}(t) y_1^{k_1} \ldots y_n^{k_n}, \quad k_i \geq 0, \qquad (3)$$

with continuous and bounded coefficients $f_{k_1\ldots k_n}: [0, +\infty) \to R^n$, $y = (y_1, \ldots, y_n) \in R^n$.

Along with non-linear systems (2) we also consider the perturbed linear systems (1_{A+Q}) with any piecewise continuous and exponentially decreasing (as $t \to +\infty$) $n \times n$ - matrices $Q(t)$ of class $E_\sigma \equiv \{Q: \lambda[Q] \leq -\sigma\}$, $\sigma = \text{const} > 0$. These matrices are called σ-matrices (similarly to m-perturbations).

Let $[1-5] : \lambda_1(A) \leq \cdots \leq \lambda_n(A)$ denote the characteristic Lyapunov exponents of (1_A) forming the characteristic aggregate of this system with the lower $\lambda_1(A)$ and the higher $\lambda_n(A)$ exponents; $\sigma_L(A)$, $\sigma_G(A)$ and $\sigma_P(A)$ Lyapunov, Grobman [6], and Perron [7] irregularity coefficients of system (1_A), respectively; σ_M the Millionshchikov [8] asymptotic number of (1_A); $X_A(t,\tau)$ the Cauchy matrix of (1_A); $\lambda(A, f) \equiv \varlimsup_{y_0 \to 0} \lambda[y(\cdot, y_0)]$ the higher exponent [9] of system (2), where $\lambda[y(\cdot, y_0)]$ is the Lyapunov exponent of a solution $y: [0, t_y) \to U_\rho \setminus \{0\}$ when $t_y = +\infty$ and it is equal to $+\infty$ when t_y is finite [10]. If $\lambda(A, f) < 0$, then this exponent defines the exact asymptotics for the norms of solutions of (2) as $t \to +\infty$.

To solve the stability problem by linear approximation is, as a rule, to investigate exponential stability of the zero solution of a differential system under consideration. Since the days of Lyapunov the following definition is universally recognized.

Definition 0.1 The solution $y = 0$ of system (2) is called *exponentially stable* if this solution is Lyapunov's stable and $\lambda(A, f) < 0$ (it is evident that to define the exponential stability of the solution $x = 0$ of linear system (1_A) it suffices to require $\lambda_n(A) < 0$).

Due to Lyapunov [1, p.52–55] the distinction is made [2, p.232–238; 5; 11] between the general and special Lyapunov problems on exponential stability of (2) by linear approximation (1_A). We also consider their linear analogs: the general and special Lyapunov problems on exponential stability of (1_{A+Q}) by linear approximation (1_A). So we formulate these problems simultaneously.

The *special Lyapunov problem (linear problem)* is to obtain a necessary and sufficient condition for exponential stability of the zero solution to perturbed system (2) with any perturbation $f \in E_{1+0} \equiv \bigcup_{m>1} F_m$ (perturbed system (1_{A+Q}) with any matrix $Q \in E_{+0} \equiv \bigcup_{\sigma>0} E_\sigma$) via linear approximation system (1_A) and then to calculate the exact in F_{1+0} asymptotics $\Omega_{1+0} \equiv \sup_{f \in F_{1+0}} \lambda(A, f)$ for the solution

$y(t, y_0)$ starting from some sufficiently small neighborhood of the origin at the initial moment $t = 0$ (the exact in E_{+0} asymptotics $\sup_{Q \in E_{+0}} \lambda_n(A+Q)$ of the solutions of (1_{A+Q})). $\Omega_{1+0}(A)$ is called [12] the higher order central exponent of system (1_A).

The *general Lyapunov problem (linear problem)* is to obtain a necessary and sufficient condition for exponential stability of the zero solution to perturbed system (2) with any m-perturbation $f \in F_m$ with fixed $m > 1$ (perturbed system (1_{A+Q}) with any σ-matrix $Q \in E_\sigma$, $\sigma =$ fix > 0) via linear approximation system (1_A) and then to calculate the exact in F_m asymptotics $\Omega_m(A) \equiv \sup_{f \in F_m} \lambda(A, f)$ for the solution $y(t, y_0)$ starting from some sufficiently small neighborhood of the origin at the initial moment $t = 0$ (the exact in E_σ asymptotics $\sup_{Q \in E_\sigma} \lambda_n(A+Q)$ for the solutions of (1_{A+Q})). The quantity $\Omega_m(A)$ was introduced in [9]. We call $\Omega_m(A)$ the a priori m-exponent of (1_A) according to [13, 14].

These four problems are connected by the methods of their solving. The methods for solving the special and general linear problems on exponential stability of (1_{A+Q}) with the matrix $Q \in E_{+0}$ and $Q \in E_\sigma$, $\sigma > 0$, are used, with necessary modifications, for solving the special and general Lyapunov problems on exponential stability of (2) with perturbations $f \in F_{1+0}$ and $f \in F_m$, $m > 1$.

The *simplified form of the general Lyapunov problem* (see [13, 14]) is to find, when the higher exponent $\lambda_n(A)$ of the system of linear approximation (1_A) is negative, the minimal value $m_0 = m_0(A) \geq 1$ for m-perturbation $f \in F_m$ such that the zero solution of (2) is exponential stable for any $f \in F_m$ and $m > m_0$. In the linear case the problem is to find the maximal value $-\sigma_0 = -\sigma_0(A) \leq 0$ for σ-matrix Q, $Q \in E_\sigma$ such that the zero solution of (1_{A+Q}) is exponential stable for any $Q \in E_\sigma$ and $\sigma > \sigma_0$.

History of the problems. The complete solution of the special Lyapunov problem on exponential stability of system (2) by linear approximation (1_A) when the latter is a *regular* [1, p.38] linear system is that $\lambda_n(A) < 0$ and $\lambda(A, f) = \lambda_n(A)$. This solution was given: 1) by Lyapunov [1, p.52] in class F_2 of holomorphic perturbations (3); 2) by Massera (essentially later) [15] in class F_{1+0} of the higher order perturbation.

In solving the general Lyapunov problem on exponential stability by linear approximation (1_A) the main classical achievements are the conditions of exponential stability of the zero solution of (2) with perturbations $f \in F_m$, $m > 1$, and the conditions of the realization of the equality $\lambda(A, f) = \lambda_n(A)$ (or a corresponding estimate). These conditions are that the value $L_m(\sigma) \equiv (m-1)\lambda_n(A) + \sigma$ (or its strict analog) is negative for $m > 1$ and various $\sigma \geq 0$. These conditions are:

1) *Lyaponov's condition* [1, p.54–55] $L_2(\sigma_L(A)) < 0$ in class F_2 of holomorphic perturbations (3);

2) *Massera's condition* [15] $L_m(\sigma_L(A)) < 0$;
3) *Grobman's condition* [6] $L_m(\sigma_G(A)) < 0$;
4) *Bolshakov–Prokhorova's condition* $L_m(\sigma_P(A)) < 0$ for $n = 2$ [17, 18] (the condition is not valid [18] for $n \geq 3$);
5) *Malkin's condition* [16, p.379] $(m - 1)R + r < 0$;
6) *Vinograd's condition* [2, p.233] $\overline{\lim_{t \to \infty}} t^{-1} \int_0^t [(m - 1)R(s) + r(s)]\, ds < 0$, where the piecewise continuous functions $r(s)$ and $R(s)$, $s \geq 0$, and their constant values $r > 0$ and $R < 0$ give the estimate $\ln \|X_A(t, \tau)\| \leq \text{const} + \int_0^\tau r(s)\, ds + \int_\tau^t R(s)\, ds$, $0 \leq \tau \leq t$, where $X_A(t, \tau)$ is Cauchy's matrix of (1_A) ($\lambda(A, f) \leq \overline{\lim_{t \to \infty}} t^{-1} \int_0^t R(s)\, ds$ for both the conditions).

For $m_0(A) \geq 1$ the following estimate $m_0(A) \leq +\sigma_G(A)/|\lambda_n(A)|$ (or the estimate $m_0(A) \leq 1 + r/|R|$ for Malkin's condition) is valid.

For linear systems (1_{A+Q}), the Grobman theorem [6] holds:

$$\lambda[Q] < -\sigma_G(A) \Rightarrow \lambda(A + Q) = \lambda(A). \tag{4}$$

This theorem gives the solution of the special Lyapunov problem on exponential stability of (1_{A+Q}) by regular linear approximation, i.e. the condition $\lambda_n(A) < 0$. The Grobman theorem gives the sufficient condition for the solution of the general linear Lyapunov problem. For the Perron perturbations the statement $n = 2$, $\lambda[Q] < -\sigma_P(A) \Rightarrow \lambda_2(A + Q) = \lambda_2(A)$ holds [19].

The asymptotic number σ_M of Millionshchikov [8] (see also [3]) gives the exact correspondence between the solutions $x(t)$ and $y(t)$ of systems (1_A) and (1_{A+Q}):

1) $\lambda[Q] < -\sigma_M(A) \Rightarrow \|y(t) - x(t)\| \leq q\|x(t)\|$, $q < 1$, $t \geq t_0$;
2) for any $x(t)$ there exist a system (1_{A+Q}) with $\lambda[Q] \leq -\sigma$, $\sigma < \sigma_M(A)$, such that for any solution $y(t)$ of (1_{A+Q}) the previous relation is not valid.

The essential advances in solving the special and general Lyapunov problems and their linear analogs are connected with the rotation method proposed by Millionshchikov ([20, 21], see also the survey [3]). For the first time this method was applied to systems (2) with m-perturbations f of the order $m > 1$ by to Vinograd [22]. The necessary conditions of the criteria stated below in Sections 1–5 and the attainability of estimates from many theorems of this survey were proved by Millionshchikov's rotation method. Moreover, the rotation method initiates the rapid development of the theory of Lyapunov's exponents over the last 30 years. Using this method Millionshchikov solved many problems of long standing: the problem on attainability of the central Vinograd exponents, Perron's problem on the stability of characteristic exponents, Erugin's problem on the existence of the linear quasi-periodic and almost periodic irregular systems and so on.

The following exponents calculated from the Cauchy matrix $X_A(t,\tau)$ of linear approximation system (1_A) are the basic tools for solving of these problem. These exponents are: 1) the estimating m-exponent $\Omega'_m(A)$ proposed by Vinograd [22]; 2) the lower $\Delta_0(A)$ and the higher $\nabla_0(A)$ exponential indices, the higher sigma-exponent $\nabla_\sigma(A)$, $\sigma > 0$, and the constructive m-exponent proposed by the author of this survey [3, 4, 9, 11–14, 23, 24]. In this connection we have the following structure of our review. The first three sections deal with definitions of these exponents and investigations of their properties. In the next two sections we consider the applications of the exponents for solving four Lyapunov's problems formulated above. The last section contains some unsolved problems.

Remark 0.1 Note that, in general, system (2) with m-perturbation $f \in F_m$, $m > 1$ (including an exponential stable system), has not the property of uniqueness of its solutions at any moment $t = t_0 \geq 0$ and on an arbitrary small neighborhood of the origin $y = 0$ (see [25]). But this phenomenon has not an essential effect on the investigation of the stability because the zero solution of (2) is unique.

Remark 0.2 This review does not contain the results by Bogdanov on stability by essentially non-linear approximation (see [3]) and the recent results of Millionshchikov on conditional stability announced in the journal "Differentsial'nye Uravneniya".

Remark 0.3 Lyapunov characteristic exponents theory for linear systems is surveyed in [3], some parts of this theory is treated in surveys [4] and [5]. However, the stability by linear approximation is touched very slightly in these our papers.

1 Exponential Indices of Linear Systems

Using Cauchy's matrix $X_A(t,\tau)$ of linear system (1_A) we define [23] the *lower and the higher exponential indices* of (1_A) as

$$\nabla_0(A) \equiv \lim_{\theta \to 1+0} \varlimsup_{k \to \infty} \theta^{-k} \sum_{i=1}^{k} \ln \|X_A(\theta^i, \theta^{i-1})\|,$$

$$\Delta_0(A) \equiv -\lim_{\theta \to 1+0} \varlimsup_{k \to +\infty} \theta^{-k} \sum_{i=1}^{k} \ln \|X_A^{-1}(\theta^i, \theta^{i-1})\|.$$

The existence of $\lim_{\theta \to 1+0}$ is proved in [23], so our exponential indices are well defined. Since the matrix $A(t)$ is bounded by a on the semiaxis $t \geq 0$ we have $-a \leq \Delta_0(A) \leq \nabla_0(A) \leq a$.

The following theorem gives the first basic property of the exponential indices.

Theorem 1.1 [23] *The relations* $\Delta_0(A) = \inf_{Q \in E_{+0}} \lambda_1(A+Q)$, $\nabla_0(A) = \sup_{Q \in E_{+0}} \lambda_n(A+Q)$ *are valid.*

The second basic property of the higher exponential index $\nabla_0(A)$ of (1_A) gives the relation between $\nabla_0(A)$ and the central exponent $\Omega_{1+0}(A)$ of the higher order.

Theorem 1.2 [12] $\nabla_0(A) < 0 \Rightarrow \Omega_{1+0}(A) = \nabla_0(A)$; $\nabla_0(A) > 0 \Rightarrow \Omega_{1+0}(A) = +\infty$.

For the relation between $\Delta_0(A)$ and the exponent $\omega_{1+0}(A) \equiv \inf_{f \in F_{1+0}} \varlimsup_{y_0 \to 0} \lambda[y(\cdot, y_0)]$ (this exponent is used to investigate conditional stability and instability of the zero solution of (2) with perturbations $f \in F_{1+0}$) see Problem 8. If system (1_A) is regular, then $\Delta_0(A) = \lambda_1(A)$ and $\nabla_0(A) = \lambda_n(A)$. In general, these relations are not valid for irregular system (1_A). The inequalities $\lambda_1(A) > \Delta_0(A)$ and $\lambda_n(A) < \nabla_0(A)$ are valid, e.g., for the following two-dimensional system (1_A): $A(t) = \text{diag}[a(t), -a(t)]$, where $a \colon [1, +\infty) \to \{-1, 1\}$ is a piecewise-constant function such that $a(t) = (-1)^i$ for $t \in [\theta^{2k+i}, \theta^{2k+i+1})$, $\theta = \text{const} > 1$, $i = 0, 1$, $k \geq 0$. For this system we have $\lambda_1(A) = \lambda_2(A) = (\theta-1)/(\theta+1)$ and $-\Delta_0(A) = \nabla_0(A) = 1$.

Note that as well as for Vinograd's central exponents the equalities $\Delta_0(P) = \Delta_0(P_d)$, $\nabla_0(P) = \nabla_0(P_d)$ hold [26] for triangle system (1_P) and its diagonal approximation (1_{P_d}). Millionshchikov [27] and Agafonov [28, 29] establish one more important property of exponential indices, namely that $\Delta_0(\cdot)$ and $\nabla_0(\cdot)$ belong to the second Baire class and do not belong to the first one.

In general case it is very difficult problem to give a complete description of the mutual position of the exponential indices $\Delta_0(A)$, $\nabla_0(A)$, the characteristic Lyapunov exponents $\lambda_1(A) \leq \cdots \leq \lambda_n(A)$, the lower and the higher central Vinograd exponents [2, p.116–117] $\omega(A) \leq \Omega(A)$, and the general Bohl exponents [30] $\omega_0(A) \leq \Omega_0(A)$. The solution of this problem for the two-dimensional case is given by the following theorem.

Theorem 1.3 [31, 32] *Arbitrary real numbers λ_1 and λ_2, Δ and ∇, ω and Ω, ω_0 and Ω_0 are, respectively, the lower and the higher characteristic exponents, exponential indices, central, regular exponents of some two-dimensional system (1_A) if and only if $\omega_0 \leq \omega \leq \Delta \leq \lambda_1 \leq \lambda_2 \leq \nabla \leq \Omega \leq \Omega_0$; $\omega_0 + \Omega \leq \min\{\lambda_1 + \lambda_2, \Delta + \nabla\}$; $2\Delta \leq \omega + \Omega$; $\lambda_1 + \lambda_2 = \omega_0 + \Omega = \omega_0 + \Omega_0 \Rightarrow \Delta = \lambda_1$.*

The following relations give the complete description of the mutual position of the exponential indices and two other exponents of two-dimensional system (1_A) [31]:

1) $\omega(A) \leq \Delta_0(A) \leq \lambda_1(A) \leq \lambda_2(A) \leq \nabla_0(A) \leq \Omega(A)$ and $2\Delta_0(A) \leq \omega(A) + \Omega(A)$;

2) $\omega_0(A) \leq \Delta_0(A) \leq \lambda_1(A) \leq \lambda_2(A) \leq \nabla_0(A) \leq \Omega_0(A)$; $\omega_0(A) + \nabla_0(A) \leq \lambda_1(A) + \lambda_2(A)$ and $\Delta_0(A) = \lambda_1(A)$ in the case $\lambda_1(A) + \lambda_2(A) = \omega_0(A) + \nabla_0(A) = \omega_0(A) + \Omega_0(A)$.

For the set of two-dimensional systems (1_A) the following inequalities and only they describe the mutual position of characteristic exponents and the exponential indices: $\Delta_0(A) \leq \lambda_1(A) \leq \lambda_2(A) \leq \nabla_0(A)$.

For the complete description of the mutual position of characteristic, central, general exponents and exponential indices of n-dimensional linear systems see [33].

The higher exponential index $\nabla_0(A)$ of system (1_A) is used in solving the special linear problem on exponential stability of system (1_{A+Q}) with any matrix $Q \in E_+$ and the Lyapunov problem on exponential stability of the zero solution of nonlinear systems (2) with the perturbations $f \in F_{1+0}$ by the same linear approximation (1_A).

2 The Sigma-Exponent of a Linear System

As in Section 1, using the Cauchy matrix $X_A(t,\tau)$ of system (1_A), we define for (1_A) so-called the higher sigma exponent $\nabla_\sigma(A)$. In Section 5 this exponent will be used in solving the general Lyapunov problem formulated in Introduction. In Section 3 the algorithm of calculating of the sigma-exponent will be used (with necessary modifications) for calculating of the constructive m-exponent $\Omega''_m(A)$, which is one of the basic tools to solve the general Lyapunov problem in the non-critical case (Section 5).

Definition 2.1 [24] We say that $\nabla_\sigma(A) \equiv \varlimsup_{k\to\infty} \xi_k(\sigma)/k$, $\sigma > 0$ is the *higher sigma-exponent of system* (1_A) if the recurrent sequence $\{\xi_k(\sigma)\}$ is defined as $\xi_k(\sigma) = \max_{0 \leq i < k} \{\ln \|X_A(k,i)\| + \xi_i(\sigma) - \sigma i\}$, $\xi_0(\sigma) = 0$, $k \in \mathbb{N}$.

The following theorem gives the basic property of the higher sigma-exponent.

Theorem 2.1 [24] *The relation* $\nabla_\sigma(A) = \sup_{Q \in E_\sigma} \lambda_n(A+Q)$, $\sigma > 0$ *is valid.*

If we consider the higher sigma-exponent $\nabla_\sigma(A)$ of system (1_A) as a function of the parameter σ, $\sigma \in (0, +\infty)$, then:

1) the image of $(0, +\infty)$ under the map $\nabla_\sigma(A)$ is $[\lambda_n(A), \nabla_0(A))$ and $\lim_{\sigma \to +0} \nabla_\sigma(A) = \nabla_0(A)$ [23, 24];
2) $\nabla_\sigma(A)$ is a non-increasing continuous function that is strictly decreasing on the interval $(0, \sigma_1)$, $\sigma_1 \leq \sigma_G(A)$, and $\lambda_n(A) = \nabla_{\sigma_1}(A) < \nabla_\sigma(A)$ for all $\sigma \in (0, \sigma_1)$, $\sigma_1 \leq \sigma_G(A)$ [24].

By the above it follows that $\nabla_\sigma(A) = \lambda_n(A)$ for all $\sigma \geq \sigma_1$.

The following theorem gives the complete description of the properties of the sigma-exponent $\nabla_\sigma(A)$ as the function of $\sigma \in (0, +\infty)$ on the set of linear systems (1_A).

Theorem 2.2 [34] *A function $\varphi \colon [0, +\infty) \to R$ is the sigma-exponent $\nabla_\sigma(A)$ of some system (1_A) if and only if*

(1) φ *is bounded;*
(2) φ *is convex;*
(3) *there exist a constant $\sigma_1 \geq 0$ such that $\varphi(\sigma) \equiv$ const for all $\sigma \geq \sigma_1 \geq 0$.*

Note that the necessity of condition (2) of Theorem 2.2 were established independently in [35] and [36]. Since the sigma-exponent is a continuous and non-increasing function [24], Theorem 2.1 and the Grobman theorem (4) yield $\nabla_\sigma(A) = \lambda_n(A)$ for $\sigma \leq \sigma_1$.

Barabanov proved [37] that the sets of functions represented by the higher order sigma-exponents of general linear systems (1_A) and of systems corresponding to n-order linear homogeneous differential equation coincide. If we replace the set of general linear systems by the set of Hamiltonian systems then the set $\{\varphi(\sigma)\}$ that can be presented by the higher sigma-exponents Hamiltonian systems is completely characterized [38] by the properties (1)–(3) of Theorem 2.2 and the following additional condition:

(4) $\sigma_1 \leq 2\beta$, where $\beta = \lim\limits_{\sigma \to +\infty} \varphi(\sigma)$, and $\sigma_1 = \inf\{\sigma > 0 \colon \varphi(\sigma) = \beta\}$.

The following theorem gives the complete description of the dependence between the higher sigma-exponents $\nabla_\sigma(P)$ of triangular system (1_P) and $\nabla_\sigma(P_d)$ of its diagonal approximation (1_{P_d}).

Theorem 2.3 [39, 40] *Functions $\varphi_1 \colon (0, +\infty) \to R$ and $\varphi_0 \colon (0, +\infty) \to R$ are the higher sigma-exponents of triangular system (1_P) and its diagonal approximation (1_{P_d}), respectively, if and only if*

(1) *the functions $\varphi_i(\sigma)$, $i = 0, 1$, are bounded, convex, and they are constant on some interval infinite to the right;*
(2) $\varphi_1(\sigma) \geq \varphi_0(\sigma)$ *for all $\sigma > 0$;*
(3) $\lim\limits_{\sigma \to +0} \varphi_0(\sigma) = \lim\limits_{\sigma \to +0} \varphi_1(\sigma)$;
(4) *there exists a constant $c_\varphi \in [0, n-1)$ such that the inequality $\varphi_0(\sigma) \geq a + (nk - c_\varphi)\sigma$ holds for all $\sigma > 0$ provided that $l(\sigma) = a + k\sigma$ is a line of support for the graph of $\varphi_1(\sigma)$, $\sigma > 0$.*

Later Makarov [41] established that condition (4) in Theorem 2.3 is equivalent to the simpler condition that $\varphi_0(\sigma) \geq \varphi_1(n\sigma) - c_\varphi \sigma$ for all $\sigma > 0$ and some constant $c_\varphi \in [0, n-1)$.

One more interesting and important property of the higher sigma-exponent $\nabla_\sigma(\cdot)$ as a function of the matrix of coefficients A: The higher sigma-exponent belong to the second Baire class (Bykov [42]) and does not belong to the first class (Vetokhin [43]).

Since the higher order sigma-exponent $\nabla_\sigma(A)$ is continuous in $\sigma > 0$ [24] it can be calculated by the estimating method:

Theorem 2.4 [24] *Let $R_\sigma(A) = \{r(t)\}$ be the set of piecewise continuous functions $r\colon [0, +\infty) \to R$ such that $\ln \|X_A(t,\tau)\| \leq \text{const} + r(t) - r(\tau) + \sigma\tau$, $0 \leq \tau \leq t$, $\sigma > 0$, where $X_A(t,\tau)$ is Cauchy's matrix of system (1_A). Then $\nabla_\sigma(A) = \inf\limits_{r \in R_\sigma(A)} \varlimsup\limits_{t\to\infty} r(t)/t$, $\sigma > 0$.*

To investigate conditional stability of linear differential system (1_{A+Q}), the quantity $\bar{\Delta}_\sigma(A) \equiv \sup\limits_{Q \in E_\sigma} \lambda_1(A+Q)$ is used. Prokhorova established [44] that this quantity essentially differs from the higher sigma-exponent $\nabla_\sigma(A)$: the latter is a continuous function of the parameter $\sigma > 0$, whereas the function $\bar{\Delta}_\sigma(A)$ is discontinuous in general. For example, it is discontinuous at the point $\sigma = \sigma_P$ for the system (1_A) constructed in [44] (here $\sigma_P(A)$ is the Perron irregularity coefficient).

See Section 5 for applications the higher sigma-exponent $\nabla_\sigma(A)$ of system (1_A) in solving the general linear problem on exponential stability of the zero solution of system (1_{A+Q}) with any matrix $Q \in E_\sigma$, $\sigma > 0$.

3 Estimating and Constructive m-exponents of a Linear System

As in the first two sections, using the Cauchy matrix of system (1_A), we define for this system two new exponents mentioned in the title of this section. These exponents are used for solving the special and general Lyapunov problems on exponential stability.

3.1 The estimating m-exponent of a linear system

Definition 3.1 [22, 14] We attribute any pair (w, r) of a number $w \in R$ and a *piecewise continuous function* $r\colon [0, +\infty) \to (-\infty, 0]$ to a *class* $G_m(A)$ if $\ln\|X_A(t,\tau)\| \leq w(t-\tau) + r(t) - mr(\tau)$, $0 \leq \tau \leq t$, $m > 1$, where $X_A(t,\tau)$ is Cauchy's matrix of system (1_A).

We define the estimating m-exponent of system (1_A) as

$$\Omega'_m(A) = \begin{cases} \Lambda_m(A) \equiv \inf\limits_{G_m(A), w<0} \varlimsup\limits_{t\to+\infty} r(t)/t, & \text{if } w_m(A) < 0, \\ +\infty, & \text{if } w_m(A) > 0, \end{cases}$$

where $w_m(A) \equiv \inf\limits_{G_m(A)} \{w\}$ is an index of stability.

If $r(t)$ and $w \in R$ make the pair $(w, r) \in G_m(A)$, $m > 1$, then $r(t)$ is nonpositive for all $t \geq 0$, so the inclusion $G_{m_1}(A) \subseteq G_{m_2}(A)$ is valid for all $m_2 > m_1 > 1$. It follows that the estimating m-exponent $\Omega'_m(A)$ is not less than the higher exponent $\lambda_n(A)$ of (1_A) and is a non-increasing function of the argument $m > 1$, where m is an order of the perturbations $f \in F_m$ in system (2).

If $w_m(A) < 0$ then the estimating m-exponent $\Omega'_m(A) = \Lambda_m(A)$ is [14] a finite negative number. The following two theorems give the properties of this exponent regarded as the function of the parameter $m > 1$.

Theorem 3.1 [14] *The estimating m-exponent $\Omega'_m(A)$ of (1_A) is a continuous function for all m such that $\Omega'_m(A) < 0$.*

Theorem 3.2 [14] *The estimating m-exponent $\Omega'_m(A)$ of system (1_A) is a strictly decreasing function on the set $\{m > 1 : \lambda_n(A) < \Omega'_m(A) < 0\}$.*

The basic property of the estimating m-exponent is given by

Theorem 3.3 [14] *The a priori m-exponent $\Omega_m(A)$ of system (1_A) is not more than the estimating m-exponent $\Omega'_m(A)$ on the domain of $\Omega'_m(A)$.*

The estimating m-exponent $\Omega'_m(A)$ of system (1_A) is used for solving the general Lyapunov problem on exponential stability of the zero solution of system (2) with perturbations $f \in F_m$, $m > 1$, by linear approximation (1_A).

3.2 The constructive m-exponent of a linear system

We modify the definition of $\nabla_\sigma(A)$ of system (1_A) to define the notion of the constructive m-exponent of (1_A).

Definition 3.2 [9, 14] We say that the quantity

$$\Omega''_m(A) = \varliminf_{\alpha \to -\infty} \varlimsup_{k \to \infty} \xi_k(m,\alpha)/k, \quad m > 1,$$

is the *constructive m-exponent* of system (1_A). Here the sequence $\{\xi_k(m,\alpha)\}$ is determined by the following equalities

$$\xi_k(m,\alpha) = \max_{0 \leq i < k} \{ln\|X_A(k,i)\| + m\xi_i(m,\alpha)\},$$

$\xi_0(m,\alpha) = \alpha$, $k \in \mathbb{N}$.

The existence of $\lim_{\alpha \to -\infty}$ follows [14] from the inequalities $\xi_k(m,\alpha_1) \leq \xi_k(m,\alpha_2)$, which are valid for all $\alpha_1 \leq \alpha_2$ and $k \in \mathbb{N}$, so the constructive m-exponent is well defined.

The following theorems give some properties of the constructive m-exponent $\Omega''_m(A)$.

Theorem 3.4 [9, 14] *Let $\Omega''_m(A)$ be the constructive m-exponent $\Omega''_m(A)$ of system (1_A) with the higher exponent $\lambda_n(A)$. The set of images of $\Omega''_m(A)$ regarded as a function of m ($m > 1$ is an order of perturbations) is $+\infty$ if $\lambda_n(A) > 0$, and belongs to the set $+\infty \bigcup [\lambda_n(A), 0]$ if $\lambda_n(A) \leq 0$.*

Theorem 3.5 [14] *The constructive m-exponent $\Omega''_m(A)$ of system (1_A) coincides with its estimating m-exponent $\Omega'_m(A)$ on the domain of definition of $\Omega'_m(A)$.*

In the certain sense the next theorem gives the converse result.

Theorem 3.6 [14] *The estimating m-exponent $\Omega'_m(A)$ of system (1_A) coincides with its constructive m-exponent $\Omega''_m(A)$, if the latter is either positive or essentially negative ($\Omega''_m(A)$ is essentially negative at point $m = m_2$ if $\exists m_1 \in (1, m_2)$ such that $\Omega''_{m_1}(A) < 0$).*

The basic property of $\Omega''_m(A)$ is given by the following theorem.

Theorem 3.7 [9, 14] *The a priori m-exponent $\Omega_m(A)$ of system (1_A) is not less than the constructive m-exponent $\Omega''_m(A)$ for all $m > 1$.*

The constructive m-exponent $\Omega''_m(A)$ of system (1_A) are used for solving the special (see Section 4) and the general (see Section 5) Lyapunov problems on exponential stability of systems (2) with perturbations $f \in F_{1+0}$ and $f \in F_m$, respectively.

3.3 The complete description of m-exponents regarded as functions of the parameter m, $m > 1$

The m-exponents of system (1_A) are essentially used for solving the special and the general Lyapunov problems on exponential stability of systems (2) with perturbations $f \in F_{1+0}$ and $f \in F_m$. We also need calculate the exact in class F_m asymptotics $\Omega_m(A)$. In this connection, the important problem arises to give the complete description of the m-exponents of linear systems (1_A) regarded as functions of the argument $m > 1$ on the domain $(m_1, +\infty)$, $m_1 > 1$, where all these exponents are negative. Theorems 3.3, 3.5–3.7 give the important corollary on the coincidence of all m-exponents.

Corollary 3.1 [14] *The a priori $\Omega_m(A)$, estimating $\Omega'_m(A)$ and constructive $\Omega''_m(A)$ m-exponents, $m > 1$, of system (1_A) are equal:*
  (1) *and they are equal to $+\infty$ if at least one of them is positive;*
  (2) *for all $m > 1$ such that the estimating m-exponent $\Omega'_m(A)$ is defined;*
  (3) *for all $m > 1$ such that the constructive m-exponent $\Omega''_m(A)$ is either positive or essentially negative.*

From Corollary 3.1 it follows that if at least one of the m-exponents is negative on the whole interval $(m_1, m_2] \subset (1, +\infty)$ then all the m-exponents are equal on this interval. We keep the symbol $\Omega_m(A)$ for this common value of the m-exponents as well as for the value $\Omega_m(A) = +\infty$, and say that $\Omega_m(A)$ is the m-exponents of system (1_A). Note that from Grobman's theorem (4) it follows that for some $m_0 \geq 1$ the equality $\Omega_m(A) = \lambda_n(A) < 0$ is valid for all $m > m_0$.

To give the complete description of the set of functions that can be represented by negative m-exponents $\Omega_m(A)$ of linear systems on some interval $(m_1, m_2]$, we introduce convergent power series $R(m,c) = \sum_{i=0}^{N(c)} r_i(c) m^i$ depending on the parameter $c \in (m_1, m_2]$, polynomials of degree $N(c) \geq 0$ being including. We say that these power series are uniformly convergent in a two-dimensional domain M of variables $m > 1$ and $c > 1$ if for any $\varepsilon > 0$ there exist a number $n = n(\varepsilon)$ such that the inequality $\left| \sum_{i=k}^{N(c)} r_i(c) m^i \right| < \varepsilon$ holds for all $(m, c) \in M$ and $k \geq n$.

Theorem 3.8 [41, 42] *A function $\varphi(m) \leq \text{const} < 0$ is the m-exponent $\Omega_m(A)$ of some linear system (1_A) on the half-opened interval $(m_1, m_2]$, $m_1 > 1$, if and only if there exists a power series $R(m, c)$ uniformly convergent in the any region $M_\rho \equiv \{(m, c) \in R^2 \colon m_1 < m \leq c \in [\rho, m_2]\}$, $\rho \in (m_1, m_2)$, such that*

(1) $R(c, c) = \varphi(c)$ *for all* $c \in (m_1, m_2]$;

(2) *there exist $\varepsilon_0 \in (0, 1)$ and $\theta_i(c) \in (0, 1 - \varepsilon_0)$ $(i = 1, \ldots, N(c))$, $\theta_0(c) \equiv 1$ and $\theta_{N(c)+1}(c) \equiv 0$ for finite $N(c)$, such that the inequalities*

$$\varepsilon_0 |r_i(c)| \leq \prod_{j=0}^{i} \theta_j(c) \equiv v_i(c) \leq \frac{1}{\varphi(m)} \left(\sum_{j=i}^{k} r_j(c) m^{j-i} - \frac{1}{\varepsilon_0} v_{k+1}(c) m^{k-i+1} \right)$$

hold for all m, $c \in (m_1, m_2]$ and all finite $0 \leq i \leq k \leq N(c)$.

The corollaries of this general criterion establish new properties of the m-exponent $\Omega_m(A) = \varphi(m) \leq \text{const} < 0$ as a function of the argument $m \in (m_1, m_2]$, $m_1 > 1$.

Corollary 3.2 [45, 46] *$D^-\varphi(m) \leq D_+\varphi(m)$ at any point $m \in (m_1, m_2)$, where $D^-\varphi(m)$ and $D_+\varphi(m)$ are upper left and lower right Dini derivatives of the m-exponent $\varphi(m)$.*

Corollary 3.3 [45, 46] *The m-exponent $\varphi(m)$ is differentiable at all points where it is concave.*

Corollary 3.4 [47] *The polynomial $\varphi(m)$ of degree $q \in N$ in the neighborhood of some point $m = m_0 > 1$ is the m-exponent $\Omega_m(A)$ of some linear system (1_{A_φ}) if and only if $\varphi(m_0) < 0$, $\varphi'(m_0) < 0$.*

One more interesting and important property of the series $R(m, c)$ is given by the following corollary.

Corollary 3.5 [46] *Let the series $R(m, c)$ satisfy the conditions of Theorem 3.8. Then $R(m, c)$ is convergent in m on some non-empty right neighborhood (c, c_1) of any point $c \in (m_1, m_2)$.*

Examples of the m-exponents of (1_A):

1) convex non-increasing negative on $[m_1, m_2]$, $m_1 > 1$, function (note that the set of the higher sigma-exponents $\nabla_\sigma(A)$ of linear systems (1_A) is exhausted by convex non-increasing functions of the argument $\sigma > 0$ (see

Theorem 2.2), but this fact is not valid for the set of the m-exponents $\Omega_m(A)$);

2) a convex negative on $[m_1, m_2]$, $m_1 > 1$, polynomial $P_q(m)$ of degree $q \geq 0$ with nonpositive coefficients.

4 Special Linear Problem and Lyapunov Problem on Exponential Stability

In this section we solve the problems indicated in the heading of the section by means of the higher exponential index $\nabla_0(A)$ and the higher constructive m-exponent $\Omega_m''(A)$ for linear approximation system (1_A) ($\nabla_0(A)$ is introduced in Section 1 and $\Omega_m''(A)$ in Section 3). We also establish the relation between exponential stability of the zero solutions for linear system (1_{A+Q}) with any exponentially decreasing matrix $Q(\cdot)$ and for nonlinear system (2) with any perturbations of higher order.

4.1 The solution of the special linear problem on exponential stability

of linear system (1_{A+Q}) by linear approximation (1_A) is given by

Theorem 4.1 [11] *Let $Q \in E_{+0}$ be any exponentially decreasing as $t \to +\infty$ matrix. The zero solution of linear system (1_{A+Q}) is exponentially stable iff the higher characteristic exponent $\lambda_n(A)$ and the higher exponential index $\nabla_0(A)$ of (1_A) satisfy the inequalities $\lambda_n(A) < 0$, $\nabla_0(A) \leq 0$. Moreover, the exact in the class E_{+0} asymptotics $\sup_{Q \in E_{+0}} \lambda_n(A+Q)$ for solutions of (1_{A+Q}) equals to the higher exponential index $\nabla_0(A)$ of (1_A) in this case.*

Theorem 4.2 [11] *Let $Q \in E_{+0}$ be any exponentially decreasing as $t \to +\infty$ matrix. The zero solution of linear system (1_{A+Q}) is exponentially stable and the exact in the class E_{+0} asymptotics $\sup_{Q \in E_{+0}} \lambda_n(A+Q)$ for solutions of (1_{A+Q}) is negative iff the higher exponential index $\nabla_0(A)$ of (1_A) is negative. Under this condition the mentioned asymptotics coincides with the index $\nabla_0(A)$.*

Remark 4.1 [11, 23] Let us say that the zero solution of linear system (1_{A+Q}) is exponentially unstable if the higher exponent $\lambda_n(A+Q)$ of (1_{A+Q}) is positive. The following statements are valid:

(1) $\nabla_0(A) > 0 \Rightarrow \exists Q \in E_{+0}$, such that the solution $y = 0$ of (1_{A+Q}) is exponentially unstable;

(2) $\Delta_0(A) > 0 \Rightarrow$ the solution $y = 0$ of (1_{A+Q}) with any $Q \in E_{+0}$ is exponentially unstable.

4.2 The solution of the special linear problem on exponential stability

of nonlinear system (2) with any perturbation $f \in F_{1+0}$ by linear approximation (1_A) is given by

Theorem 4.3 [11, 14] *Let $f \in F_{1+0}$ be any perturbation of higher order. The zero solution of (2) is exponentially stable iff the m-exponent $\Omega_m(A)$ of (1_A) is negative for all $m > 1$. In addition, the following statements are valid:*

(i) $\Omega_m(A) = \Omega'_m(A) = \Omega''_m(A)$, *where $\Omega'_m(A)$ is estimating m-exponent and $\Omega''_m(A)$ is constructive m-exponent of (1_A);*

(ii) *the exact in the class F_{1+0} asymptotics $\Omega_{1+0} \equiv \sup\limits_{f \in F_{1+0}} \lambda(A, f)$ for solutions of (2) equals to the higher exponential index $\nabla_0(A)$ of linear approximation system (1_A).*

Theorem 4.4 [11, 12] *Let $f \in F_{1+0}$ be any perturbation of higher order. The zero solution of (2) is exponentially stable and the exact in the class F_{1+0} asymptotics Ω_{1+0} is negative iff the higher exponential index $\nabla_0(A)$ of (1_A) is negative. In addition, the equality $\Omega_{1+0} = \nabla_0(A)$ holds.*

Remark 4.2 [12] *If $\nabla_0(A) > 0$, then $\exists f \in F_{1+0}$, such that the zero solution of (2) with this f is unstable.*

4.3 Relation between exponential stability of the zero solutions of linear system (1_{A+Q}) with any matrix $Q \in E_{+0}$ and nonlinear system (2) with any perturbation $f \in F_{1+0}$ is given by

Theorem 4.5 [11] *If the zero solution of nonlinear system (2) with any perturbation $f \in F_{1+0}$ is exponentially stable, then the zero solution of linear system (1_{A+Q}) with any matrix $Q \in E_{+0}$ is exponentially stable.*

The reverse statement is no more valid (see [11]), i.e. exponential stability of the zero solution of (1_{A+Q}) with all matrix $Q \in E_{+0}$ *does not imply*, in general, exponential stability of the zero solution of (2) with any perturbation $f \in F_{1+0}$. This fact can be established by means of the following example. Let $a_i(t) = 0$, $t \in [\theta^{2k-i}, \theta^{2k+1-i})$, $\theta = \text{const} > 1$, $k \in N$, $i = 1, 2$; $a_1(t) + a_2(t) \equiv -1$, $t \geq 1$. Then for the system $\dot{x} = A(t)x = \text{diag}\,[a_1(t), a_2(t)]x$, $x \in R^2$, $t \geq 1$, we have $\lambda_2(A) = -(1+\theta)^{-1}$ and $\nabla_0(A) = 0$. Hence by Theorem 4.1, the zero solution of (1_{A+Q}) with arbitrary $Q \in E_{+0}$ is exponentially stable. On the other hand, for the Cauchy matrix $X_A(t, \tau)$ of (1_A) and for any numbers $\alpha < 0$ and $m \in (1, \theta)$, the condition $\theta^{-k}[\alpha m^k + \sum\limits_{i=1}^{k} m^{k-i} \ln \|X_A(\theta^i, \theta^{i-1})\|] \to 0$ as $k \to \infty$ holds, i.e. $\Omega''_m(A) \geq 0$ for all $m \in (1, \theta)$. Thus, the conditions of Theorem 4.3 are not satisfied and, therefore, there exists a perturbation $f \in F_{1+0}$ such that the zero solution of (2) with the perturbation f is not exponentially stable.

5 General Lyapunov Problem on Exponential Stability and its Linear Analog

In this section we state the solution of the general linear problem on exponential stability of linear systems (1_{A+Q}) with arbitrary matrices $Q \in E_\sigma$ by linear approximation (1_A). This solution use the notion of higher σ-exponent $\nabla_\sigma(A)$ introduced in Section 2. In addition, we give the complete solution of the general Lyapunov problem on exponential stability in the following cases:

1) when the linear approximation is diagonal;
2) for the simplified form of the problem (see Introduction);
3) in non-critical case, i.e. for all $m > 1$ such that at least one of the following conditions holds:
   i) the estimating m-exponent $\Omega'_m(A)$ of (1_A) is defined and negative;
   ii) 2) the constructive m-exponent $\Omega''_m(A)$ is essentially negative at $m > 1$ $(\exists m_1 \in (1, m): \Omega''_{m_1}(A) < 0)$.

We also investigate in detail the cases when the clasical conditions by Lyapunov, Grobman and Massera degenerate as well as the critical case of the general Lyapunov problem.

5.1 Solution of the general linear problem on exponential stability of systems (1_{A+Q}) with matrices $Q \in E_\sigma$ by arbitrary linear approximation (1_A) is given by

Theorem 5.1 [24] *The zero solution of perturbed linear system (1_{A+Q}) with any piecewise continuous matrix $Q \in E_\sigma$ is exponentially stable iff the higher σ-exponent $\nabla_\sigma(A)$ of (1_A) is negative. In this case the exact in E_σ asymptotics $\sup_{Q \in E_\sigma} \lambda_n(A + Q)$ for all solutions of systems (1_{A+Q}) coincides with $\nabla_\sigma(A)$.*

Remark 5.1 By means of the notion of exponential instability introduced in Subsection 4.1 the following criterion can be stated. The zero solution of perturbed system (1_{A+Q}) with some $Q \in E_\sigma$ is exponentially unstable iff the σ-exponent $\nabla_\sigma(A)$ of original system (1_A) is positive.

5.2 Solution of general Lyapunov problem on exponential stability in non-critical case

Vinograd [22] has obtained the following sufficient condition for exponential stability of system (2) under perturbations $f \in F_m$.

Theorem 5.2 [22] *If the estimating m-exponent $\Omega'_m(A) < 0$, then for each m-perturbation $f \in F_m$, the zero solution of (2) is exponentially stable and $\Omega'_m(A)$ is the common exponent of asymptotics, i.e. for any $\varepsilon > 0$ there exist $\delta = \delta(\varepsilon) > 0$ and $B = B(\varepsilon) > 0$ such that all solutions of (2) beginning at $t = 0$ in*

δ-neighborhood of the origin admits the common estimation $\|y(t)\| \le B\|y(0)\|$ $\exp[\Omega'_m(A) + \varepsilon]t$ for all $t \ge 0$. This exponent can not be diminished since for any $\Omega < \Omega'_m(A)$ one can find an m-perturbation f such that the previous estimation with small $\varepsilon > 0$ is not valid whatever $\delta > 0$ and $B > 0$ are given.

Remark 5.2 [22] If the estimating m-exponent $\Omega'_m(A)$ of (1_A) is positive, then there exist m-perturbations $f \in F_m$ such that the zero solution of (2) under this f is unstable.

Theorem 5.2 do not allow us to evaluate the exact in F_m asymptotics for solutions of (2) under m-perturbations, i.e. the a priori m-exponent $\Omega_m(A)$ of (1_A). So, we can not use this Theorem in order to solve the second part of the general Lyapunov problem on exponential stability. This defect can be removed by means of Theorem 3.7 stating that the a priori $\Omega_m(A)$ and constructive $\Omega''_m(A)$ m-exponents of (1_A) $m > 1$ satisfy the inequality $\Omega_m(A) \ge \Omega''_m(A)$ for all $m > 1$. Thus, Theorems 3.3 and 3.5 enable us to complete Theorem 5.2 by the necessary equality $\Omega_m(A) = \Omega'_m(A)$.

Some properties of constructive m-exponent enable us to obtain the following sufficient condition for exponential stability.

Theorem 5.3 [9, 14] *If the constructive m-exponent $\Omega''_m(A)$ of (1_A) is essentially negative, then the zero solution of (2) with any m-perturbation $f \in f_m$, $m > 1$, is exponentially stable and the a priori m-exponent $\Omega_m(A)$ of (1_A) coincides with the constructive m-exponent $\Omega''_m(A)$ of (1_A).*

Remark 5.3 [12] If $\Omega''_m(A) > 0$, then there exist m-perturbations $f \in F_m$, $m > 1$, such that the zero solution of (2) under these perturbations is unstable.

In the proof of the main theorem of [9] (see also Theorem 4 of joint paper [14]) when the constructive m-exponent $\Omega''_m(A)$ of (1_A) is nonpositive, we construct some fixed perturbation $f \in F_m$, $m > 1$ such that $\lambda(A, f) \ge \Omega''_m(A)$ for the higher exponent $\lambda(A, f)$ of (2). Hence we have the following necessary condition of exponential stability.

Theorem 5.4 [9, 14, 25] *If the zero solution of (2) with any perturbation $f \in F_m$, $m > 1$, is exponentially stable, then the constructive m-exponent $\Omega''_m(A)$ of (1_A) is negative.*

Theorems 5.2 and 5.4 are irreversible. This fact is established in the next Subsection 5.3 of this Section.

Theorems 5.3 and 5.4 yield

Theorem 5.5 (on complete solution of the general Lyapunov problem in simplified form) *Let $m_0 \equiv \inf\{m > 1 : \Omega''_m(A) < 0\}$. Then*

(1) *for $m > m_0$, the zero solution of (2) with any perturbation $f \in F_m$ is exponentially stable and the a priori m-exponent $\Omega_m(A) \equiv \sup_{f \in F_m} \lambda(A, f)$ of*

(1_A) coincides with the constructive m-exponent $\Omega''_m(A)$ (note that $\Omega''_m(A)$ is calculated on the base of the Cauchy matrix $X_A(t,\tau)$);

(2) for $m \in (1, m_0)$, there exists a perturbation $f \in F_m$ such that the solution $y = 0$ of (2) is not exponentially stable.

5.3 Critical cases

We start from the classical critical cases arising when the original inequalities are replaced by equalities in 1) Grobman Theorem (4) on coincidence of characteristic aggregates $\lambda(A) = (\lambda_1(A), \ldots, \lambda_n(A)) \in R^n$ of the original system and $\lambda(A+Q) \in R^n$ of perturbed system (1_{A+Q}); 2) Lyapunov, Massera, and Grobman conditions for exponential stability of the solution $y = 0$ of (2) with any m-perturbation $f \in F_m$.

5.3.1 Generalized Grobman perturbations critical case. These perturbations are of the class $E_{\sigma_1(A)} \equiv \{Q \colon \lambda[Q] \leq -\sigma_G(A)\}, \sigma_1(A) \equiv \sigma_G(A) > 0$. Here we have the following problems:

1) on determination of the whole class of systems (1_A) such that the characteristic aggregates λ_A of (1_A) and $\lambda(A+Q)$ of (1_{A+Q}) are the same for all $Q \in E_{\sigma_1(A)}$;
2) on existence of systems (1_A) and perturbations $Q \in E_{\sigma_1(A)}$, such that $\lambda(A+Q) \neq \lambda(A)$;
3) on stability of characteristic exponents of systems (1_A) under generalized Grobman perturbations;
4) on the structure of so-called Grobman spectral sets $\Gamma_n(A) = \{\lambda(A+Q) \in R^n \colon Q \in E_{\sigma_1(A)}\}$ of (1_A), including the problem on existing of system (1_A) of arbitrary order $n \geq 2$ such that the Lebesgue measure of $\Gamma_n(A)$ is positive.

Let $X_A(t) = [X_1(t), \ldots, X_n(t)]$ be some normal fundamental solutions system of (1_A). In order to solve Problem 1) we define angular irregularity coefficient [48, 49] $\sigma_0(A) \geq 0$ of system (1_A) by $\sigma_0(A) \equiv \max\limits_{1 \leq k \leq n} \{\lambda[1/\alpha_k]\}$, where $\alpha_k(t)$ is the angle (see [8]) between the solution $X_k(t)$ and the linear hull of the other $n-1$ solutions $X_i(t) \in X_A(t)$. Obviously, the angular irregularity coefficient $\sigma_0(A)$ equals to zero for a diagonal system and do not exceed the Grobman irregularity coefficient $\sigma_1(A)$ in the general case. Now we have the following statement.

Theorem 5.6 [48–50] *Let $Q \in E_{\sigma_1(A)}$. Characteristic aggregates $\lambda(A)$ of (1_A) and $\lambda(A+Q)$ of (1_{A+Q}) are the same if $\sigma_0(A) < \sigma_1(A)$.*

Thus, the inequality $\sigma_0(A) < \sigma_1(A)$ describes a class of systems (1_A) such that $\lambda(A) = \lambda(A+Q)$ for all $Q \in E_{\sigma_1(A)}$. This class is complete in the following sense.

Theorem 5.7 [49] *For any numbers $2 \leq n \in N$, $\lambda_1 \leq \cdots \leq \lambda_n$, $\alpha \in [\lambda_1, \lambda_n]$, $\sigma > 0$ there exist n-dimensional system (1_A) with characteristic exponents $\lambda_i(A) = \lambda_i$, $i = 1, \ldots, n$, and with coinciding angular irregularity coefficient and Grobman irregularity coefficient $\sigma_0(A) = \sigma_1(A) = \sigma$, and a matrix $Q \in E_{\sigma_1(A)}$ such that α is one of the characteristic exponents of perturbed system (1_{A+Q}).*

This theorem give also a solution for Problem 2) on existence of system (1_A) and a matrix $Q \in E_{\sigma_1(A)}$ such that $\lambda(A+Q) \neq \lambda(A)$. Other corollaries of Theorem 5.7 are the following: the lower exponent $\lambda_1(A)$ of (1_A) is upward unstable; each exponent $\lambda_k(A) \in (\lambda_1(A), \lambda_n(A))$ is unstable (upward as well as downward) if it is not multiple; the lower multiple exponent $\lambda_i(A) > \lambda_1(A)$ is downward unstable; The higher multiple exponent $\lambda_k(A) < \lambda_n(A)$ is upward unstable. These statements give a partial solution for Problem 3).

The next theorem establishes that for any Lyapunov irregular system (1_A) the lower exponent $\lambda_1(A)$ is stable downward and the higher exponent $\lambda_n(A)$ is stable upward under generalized Grobman perturbations.

Theorem 5.8 [51] *If $Q \in E_{\sigma_1(A)}$ and $\sigma_1(A) > 0$, then $\lambda_1(A) \leq \lambda_i(A+Q) \leq \lambda_n(A)$, $i = 1, \ldots, n$.*

In [51] we prove that there exists system (1_A) such that its higher and lower characteristic exponents are simultaneously unstable under perturbations $Q \in E_{\sigma_1(A)}$. This result completes the solution of Problem 3).

The next theorems give solution for Problem 4) on existence of linear systems (1_A) such that their Grobman spectral sets $\Gamma_m(A)$ has a positive Lebesgue m-measure:

Theorem 5.9.1 [51, 52] *For any parameters $\sigma > 0$, $\lambda_1 < \lambda_2$, $\theta \in (1, 1 + \sigma/(\lambda_2 - \lambda_1))$, there exists $2n$-dimensional, $n \in N$, system (1_A) such that $\lambda_1(A) = \lambda_1$, $\lambda_{2n}(A) = \lambda_2$, $\sigma_0(A) = \sigma_1(A) = \sigma$, and $\Gamma_{2n}(A) \subset R^{2n}$ consists of all points μ with coordinates $\mu_1 \leq \cdots \leq \mu_{2n}$, such that each point (μ_{2i-1}, μ_{2i}), $i = 1, \ldots, n$, lies in the triangle $\{(\mu_1, \mu_2): \mu_1 - \lambda_1 \geq \theta^2(\lambda_2 - \mu_2) \geq 0 \geq \mu_1 - \mu_2\}$. This set has the measure $\mathrm{mes}_{2n} \Gamma_{2n}(A) = \dfrac{1}{(2n)!}(1+\theta^2)^{1-2n}(\lambda_2 - \lambda_1)^{2n}$, $n \in N$.*

Theorem 5.9.2 [52, 53] *For any odd $n \geq 3$ and any parameters $\lambda_3 > \lambda_1$, $4\sigma \in (0, \lambda_3 - \lambda_1)$, there exists system (1_A) with $\lambda_1(A) = \lambda_1$ $\lambda_n(A) = \lambda_3$, $\sigma_0(A) = \sigma_1(A) = \sigma$ such that the set $\Gamma_n(A)$ has the inner Lebesgue measure $\underline{\mathrm{mes}}_n \Gamma_n(A) \geq [(\lambda_3 - \lambda_1)/\gamma]^n/n!$ with some number $\gamma > 4$.*

5.3.2 *Critical cases by Lyapunov $L_2(\sigma_L(A)) = 0$, Massera $L_m(\sigma_L(A)) = 0$, and Grobman $L_m(\sigma_G(A)) = 0$, $m > 1$.* The main question arising here is the following. Are the indicated equalities still sufficient for exponential stability of the zero solution of (2) under any perturbation $f \in F_m$, $m = 1 + \sigma/|\lambda_n(A)|$, or there exist systems (2) with these perturbations and with unstable zero solution?

The next theorem affirms the second alternative and establishes that the conditions of exponential stability by Lyapunov, Massera, and Grobman can not be improved if the whole set of linear systems (1_A) is considered.

Theorem 5.10 [25] *For any natural $n \geq 2$ and any constants $\lambda < 0, \sigma > 0$, there exist system (1_A) with $\lambda_n(A) = \lambda$, $\sigma_L(A) = \sigma_G(A) = \sigma$ and perturbation $f \in F_m$ with $m = 1 - \sigma/\lambda > 1$ such that the zero solution of (2) is unstable.*

Remark 5.4 [25] The m-perturbation f constructed in the proof of Theorem 5.10 when $\sigma = -\lambda$ and $m = 2$ is polynomial $f(t, y) = 3\sigma(y_2^2, y_1^2, 0, \ldots, 0) \in R^n$, $t \geq 0$. Hence, the equality $\lambda_n(A) + \sigma_L(A) = 0$ provides also the system (2) with holomorphic perturbation (3) and with unstable zero solution, i.e. the Lyapunov condition $\lambda_n(A) + \sigma_L(A) < 0$ can not be improved in the set of all linear systems (1_A) with holomorphic perturbations (3).

5.3.3 Critical case of the general Lyapunov Problem. As follows from Remark 5.2 and Theorems 5.5, 5.4, and 5.2, this case is determined by a single value $m_0 \equiv \inf\{m > 1: \Omega''_m(A) < 0\} > 1$, such that: 1) constructive m_0-exponent $\Omega''_{m_0}(A)$ is negative; 2) estimating m_0-exponent $\Omega'_{m_0}(A)$ is not defined. In this critical case for each $m = m_0 > 1$, we construct the following systems: 1) system (1_A) such that the zero solution of (2) with any perturbation $f \in F_m$ is exponentially stable; 2) system (1_A) such that the zero solution of (2) with some perturbation $f \in F_m$ is exponentially unstable. Simultaneously we state that the condition $\Omega'_m(A) < 0$ with estimating m-exponent is not necessary for exponential stability of the zero solution of (2) under any perturbation $f \in F_m$, and the condition $\Omega''_m(A) < 0$ with constructive m-exponent is not sufficient for stability of this kind. These results follow from the next two theorems.

Theorem 5.11 [25] *For any numbers $m > 1$ and $2 \leq n \in N$ there exists n-dimensional linear system (1_A) with undefined estimating m-exponent $\Omega'_m(A)$ such that the zero solution of (2) with any m-perturbation f is exponentially stable.*

Theorem 5.12 [25] *For any numbers $m > 1$ and $2 \leq n \in N$ there exists n-dimensional linear system (1_A) with negative constructive m-exponent $\Omega''_m(A)$ and a perturbation $f \in F_m$ such that the zero solution of the perturbed system (2) is unstable.*

Remark 5.5 [25] In the proof of Theorem 5.12 we use the so-called triangular m-perturbations. If we construct perturbations $f \in F_m$ of exponentially stable system (1_A) by means of rotations of solutions, we can obtain only exponentially stable systems (2) such that all their solutions with sufficiently small initial vectors have Lyapunov exponents $\leq \Omega''_m(A) < 0$.

5.4 Solution of general Lyapunov problem by diagonal approximation

In this Subsection we give the complete solution for the general Lyapunov problem on exponential stability of the zero solution of (2) with any perturbation $f \in F_m$ for linear approximation system (1_A) with diagonal coefficient matrix $A(t) = \mathrm{diag}[a_1(t), \ldots, a_n(t)]$. To this effect we introduce new m-exponent [25] $\gamma_m(A) \equiv \lim_{\alpha \to +0} \overline{\lim}_{k \to \infty} k^{-1} \ln \|\xi_m^{(\alpha)}(k)\|$ for diagonal system (1_A) where $\xi_m^{(\alpha)}(k)$ is n-dimensional vector function of $k \geq 0$ with consecutively computing components

$$\xi_{jm}^{(\alpha)}(k) = \max_{\substack{l \neq j; \\ i < k}} \left\{ x_j(k,i)\xi_{jm}^{(\alpha)}(i) + [\xi_{lm}^{(\alpha)}(i)]^m \int_i^k x_j(k,\tau)x_l^m(\tau,i)\,d\tau \right\},$$

where $\xi_{jm}^\alpha(0) = \alpha > 0$, $x_j(t,\tau) \equiv \exp \int_\tau^t a_j(s)\,ds$, $j,l = 1, \ldots, n$. This exponent satisfies the inequalities [25] $\Omega''_m(A) \leq \gamma_m(A)$ for all $m > 1$ and $\gamma_m(A) \leq \Omega'_m(A)$ on the domain of the estimating m-exponent $\Omega'_m(A)$.

Theorem 5.13 [25] *The zero solution of perturbed system (2) with any perturbation $f \in F_m$ is exponentially stable iff the m-exponent $\gamma_m(A)$ of linear diagonal approximation (1_A) is negative. In this case $\gamma_m(A)$ equals to the a priori m-exponent $\Omega_m(A)$ of (1_A).*

The m-exponent $\gamma_m(A)$ of diagonal (1_A) can be also computed via estimates [25]. Let $R_m(A)$ be the class of all piecewise continuous on semiaxis $t \geq 0$ n-dimensional vector functions $r(t)$ with components $r_j(t) > 0$ such that the following estimates are valid $r_j(s)x_j(t,s) + r_l^m(s)\int_s^t x_j(t,\tau)x_l^m(\tau,s)\,d\tau \leq r_j(t)$, $j,l = 1, \ldots, n$, $l \neq j$, for all $t - s \geq 1$. Then $\gamma_m(A) = \inf_{r \in R_m(A)} \lambda[r]$.

We also obtain the following generalization of the preceding theorem.

Theorem 5.14 [54] *The zero solution of (2) with diagonal approximation (1_A) and with any $f \in F_m$ is asymptotically stable and all solutions $y(t)$ of (1_A) with sufficiently small $y(0)$ satisfy the condition $\int_0^{+\infty} \|y(s)\|^{m-1}ds < +\infty$ iff there exists a piecewise continuous for $t \geq 0$ n-dimensional vector function $r(t)$ with components $r_j(t) > 0$ such that the above estimates are valid for all $t \geq s \geq 0$ and the condition $\int_0^{+\infty} \|r(s)\|^{m-1}ds < +\infty$ holds.*

The statement and the proof of Theorem 5.13 ensure

Corollary 5.1 [25] *The zero solution of (2) with any $f \in F_m$ is exponentially stable iff the a priori m-exponent $\Omega_m(A)$ of the diagonal approximation system (1_A) is negative.*

6 Problems for Investigation

1. The condition $\sigma_0(A) < \sigma_1(A)$ is sufficient for coincidence of characteristic aggregates $\lambda(A)$ and $\lambda(A + Q)$ of (1_A) and (1_{A+Q}) for each $Q \in E_{\sigma_1(A)}$. Are this condition exact in the following sense: for any vector $\lambda \in R^n$ with increasingly ordered components and any numbers $\sigma_1 > \sigma_0 > 0$ there exists system (1_A) with $\lambda(A) = \lambda$, $\sigma_0(A) = \sigma_0$, $\sigma_1(A) = \sigma_1$ such that for any sufficiently small $\varepsilon > 0$ one can find a matrix $Q_\varepsilon \in E_{\sigma_1(A)-\varepsilon}$, satisfying the inequality $\lambda(A + Q_\varepsilon) \neq \lambda(A)$?

2. Construct an algorithm to compute the lower σ-exponent $\Delta_\sigma(A) \equiv \inf_{Q \in E_\sigma} \lambda_1(A + Q)$, $\sigma > 0$, of system (1_A) and describe the properties of $\Delta_\sigma(A)$.

3. Investigate the structure of characteristic σ-sets $S_\sigma(A) \equiv \{\lambda(A + Q) \in R^n : Q \in E_\sigma\}$, $\sigma > 0$, of systems (1_A) including Grobman spectral sets $\Gamma_n(A)$. Are these sets convex or, at least, connected? Study the dependence of σ-set $S_\sigma(A)$ on parameter $\sigma > 0$.

4. For the critical case $\lambda_n(A) = 0$, create the similar theory for investigation of asymptotic stability of the zero solution of perturbed system (2) via characteristic degrees by B.P.Demidovich.

5. Are the conditions $(m-1)\lambda_n(A) + \sigma = 0$ and $\lambda_n(A) > \lambda_{n-1}(A)$, $n \geq 2$, sufficient for exponential stability of the zero solution of (2) with any perturbation $f \in F_m$, $m > 1$, when: 1) $\sigma = \sigma_G(A) > 0$; 2) $\sigma = \sigma_L(A) > 0$; 3) $\sigma = \sigma_L(A) > 0$, $m = 2$ and F_2 is the set of holomorphic perturbations (3)?

6. Obtain a criterion for exponential stability of the zero solution of (2) under any perturbation $f \in F_{m_0}$ in the critical case of the general Lyapunov problem on exponential stability when $m_0 > 1$, $\Omega''_{m_0}(A) < 0$, and $\Omega'_{m_0}(A)$ is undefined.

7. Is the following statement valid: $\lambda(A, f) < 0$, $\forall f \in F_m$, $m > 1 \Rightarrow \Omega_m(A) \equiv \sup_{f \in F_m} \lambda(A, f) < 0$?

8. By analogy with the higher exponent $\lambda(A, f)$ of (2), let us consider the lower exponent $\lambda_1(A, f) \equiv \lim_{\|y_0\| \to +0} \lambda[y(\cdot, y_0)]$. Compute the value $\omega_{1+0} \equiv \inf_{f \in F_{1+0}} \lambda_1(A, f)$ by means of the lower exponential index $\Delta_0(A)$ of linear approximation system (1_A). Study the properties of $\omega_m(A) \equiv \inf_{f \in F_m} \lambda_1(A, f)$ as function of $m > 1$.

9. Let us define the set of combined perturbations by $F_{m\sigma} \equiv \{f \colon \|f(t, y)\| \leq c_f e^{-\sigma t} \|y\|^m, \ y \in U_{\rho(f)}, \ t \geq 0\}$, $m > 1$, $\sigma > 0$ [14]. Investigate exponential stability of the zero solution of (2) with $f \in F_{m\sigma}$ in the critical case $(m-1)\lambda_n(A) + \sigma_G(A) - \sigma = 0$.

10. Give a complete description for functions of two variables $m > 1$ and $\sigma > 0$ representable by σm-exponents $\Omega_{m\sigma}(A) \equiv \sup_{f \in F_{m\sigma}} \lambda(A, f)$ of linear approximation system (1_A).

References

[1] Lyapunov, A.M. (1956). *Works*. Vol. 2. Publishing house of the Academy of Sciences of the USSR, Moscow, Leningrad (Russian).
[2] Bylov, B.F., Vinograd, R.E. et al. (1966). *Lyapunov Characteristic Exponents Theory and Its Applications to Stability Issues*. Nauka, Moscow (Russian).
[3] Izobov, N.A. (1974). Linear systems of ordinary differential equations. *Itogi Nauki i Tekhniki: Mathematical Analysis*. **12**, 71–146 (Russian).
[4] Izobov, N.A. (1980). Contribution to Lyapunov characteristic exponents theory for linear and quasilinear differential systems. *Mat. Zametki*, **28**(3), 459–476 (Russian).
[5] Izobov, N.A. (1993). Studies in Belorussia in the theory of characteristic Lyapunov exponents and Its applications. *Diff. Eqns.*, **29**(12), 2034–2055 (Russian).
[6] Grobman, D.M. (1952). Characteristic exponents of systems close to linear. *Mat. Sbornik*, **30**(1), 121–166 (Russian).
[7] Perron, O. (1929). Die Ordnungszahlen der Differentialgleichungs Systeme. *Math. Z.*, **31**, 748–766.
[8] Millionshchikov, V.M. (1965). Asymptotics for solutions of linear systems with small perturbations. *Doklady AN SSSR*, **162**(2), 266–268 (Russian).
[9] Izobov, N.A. (1969). On higher exponent of system with perturbations of higher order. *Vestnik Bel. Univ. Ser. 1*, **3**, 6–9 (Russian).
[10] Bogdanov, Yu.S. (1965). Generalized characteristic roots of nonautonomous systems. *Diff. Eqns.*, **1**(9), 1440–1148 (Russian).
[11] Izobov, N.A. (1982). Exponential indices and stability by the first approximation. *Vestsi AN BSSR. Ser. fiz.-mat. navuk*, **6**, 9–16 (Russian).
[12] Izobov, N.A. (1982). Upper bound for Lyapunov exponents of differential systems with perturbations of higher order. *Doklady AN BSSR*, **26**(5), 389–392 (Russian).
[13] Vinograd, R.E. and Izobov, N.A. (1970). Solution for Lyapunov problem on stability by the first approximation. *Transactions 5th Internat. Conf. on Nonlinear Oscillations*. 1969, Vol. 2. Kiev, 121–126 (Russian).
[14] Vinograd, R.E. and Izobov, N.A. (1970). Solution for Lyapunov problem on stability by the first approximation. *Diff. Eqns.*, **6**(2), 230–242 (Russian).
[15] Massera, J.L. (1956). Contribution to stability theory. *Ann. Math.*, **64**(4), 182–206.
[16] Malkin, I.G. (1966). *Motion Stability Theory*. Nauka, Moscow (Russian).
[17] Bol'shakov, N.E. (1973). The role of Perron's irregularity coefficient in the stability of a two-dimensional system with respect to a first approximation. *Diff. Eqns.*, **9**(2), 363–365 (Russian).
[18] Prokhorova, R.A. (1976). Stability with respect of a first approximation. *Diff. Eqns.*, **12**(4), 766–769 (Russian).
[19] Izobov, N.A. (1966). On stability in the first approximation. *Diff. Eqns.*, **2**(7), 898–907 (Russian).
[20] Millionshchikov, V.M. (1968). A criterion for small directional variation of solutions to linear system of differential equations under small perturbations of the system coefficients. *Mat. Zametki*, **4**(2), 173–180 (Russian).
[21] Millionshchikov, V.M. (1969). Proof of the central exponents attainability for linear systems. *Sib. Mat. Zhurnal*, **10**(1), 99–104 (Russian).
[22] Vinograd, R.E. (1969). Necessary and sufficient criterion and the exact asymptotic form for stability in the first approximation. *Diff. Eqns.*, **5**(5), 800–813 (Russian).
[23] Izobov, N.A. (1982). Exponential indices of linear system and how to compute them. *Doklady AN BSSR*, **26**(1), 5–8 (Russian).

[24] Izobov, N.A. (1969). The highest exponent of a linear system with exponential perturbations. *Diff. Eqns.*, **5**(7), 1186–1192 (Russian).

[25] Izobov, N.A. (1986). Stability with respect to a linear approximation. *Diff. Eqns.*, **22**(10), 1671–1688 (Russian).

[26] Nurmatov, A.M. (1987). Exponential indices of a triangular system and its diagonal approximation. *Diff. Eqns.*, **23**(5), 814–818 (Russian).

[27] Millionshchikov, V.M. (1992). The Baire class of Izobov indices. *Diff. Eqns.*, **28**(11), 2009 (Russian).

[28] Agafonov, V.G. (1993). The Baire class of Izobov indices. *Diff. Eqns.*, **29**(6), 1092–1093 (Russian).

[29] Agafonov, V.G. (1994). The Baire class of the Izobov upper exponent. *Diff. Eqns.*, **30**(6), 1089 (Russian).

[30] Bohl, P. (1913). Uber Differential gleichungen. *J. Reine und Angew. Math.*, **144**, 284–318.

[31] Izobov, N.A. (1992). Distribution of characteristic and other indices of two-dimensional linear systems. *Diff. Eqns.*, **28**(10), 1683–1698 (Russian).

[32] Izobov, N.A. (1998). Description of combined arrangement of exponents of two-dimensional linear differential systems. *Diff. Eqns.*, **34**(2), 166–174 (Russian).

[33] Izobov, N.A. (1994). Common distribution of characteristic, exponential, central and general indices of linear systems. *Uspekhi Mat. Nauk*, **49**(4), 96 (Russian).

[34] Izobov, N.A. and Barabanov, E.A. (1983). On the form of the leading σ-exponent of linear system. *Diff. Eqns.*, **19**(2), 359–362 (Russian).

[35] Fodor, J. (1979). Lyapunov problem on intermediate stability by the first approximation. *Szemelvények az ELTETTK Analizis II. Tanszék tudományos munkáibol*, Budapest (Russian).

[36] Barabanov, E.A. (1982). Properties of a leading σ-index. *Diff. Eqns.*, **18**(5), 739–744 (Russian).

[37] Barabanov, E.A. (1984). The leading σ-index of linear differential equations. *Diff. Eqns.*, **20**(2), 197–207 (Russian).

[38] Barabanov, E.A. and Nurmatov A.M. (1986). Leading σ-indices of linear Hamilton systems. *Diff. Eqns.*, **22**(9), 1491–1499 (Russian).

[39] Barabanov, E.A. (1989). Necessary conditions for the consistent behavior of the leading Sigma-indices of a triangular system and its diagonal-approximation system. *Diff. Eqns.*, **25**(10), 1662–1670 (Russian).

[40] Barabanov, E.A. (1990). Properties of the leading sigma-indices of a triangular system and of its diagonal approximation. *Diff. Eqns.*, **26**(2), 187–205 (Russian).

[41] Makarov, E.K. (1991). Contribution to the criterion of joint behavior of the leading sigma-exponents of a triangular system and its diagonal-approximation system. *Diff. Eqns.*, **27**(5), 910–911 (Russian).

[42] Bykov, V.V. (1997). Baire classification for Izobov σ-exponents. *Diff. Eqns.*, **33**(11), 1574 (Russian).

[43] Vetokhin, A.N. (1997). Contribution to Baire classification for Izobov σ-exponents. *Diff. Eqns.*, **33**(11), 1574 (Russian).

[44] Prokhorova, R.A. (1975). On some properties of lower exponent under Perron perturbations. *Diff. Eqns.*, **11**(6), 997–1004 (Russian).

[45] Izobov, N.A. (1985). On functions defined by central exponents of higher order. *Uspekhi Mat. Nauk*, **40**(4), 167–168 (Russian).

[46] Izobov, N.A. (1985). Properties of high-order central indices. *Diff. Eqns.*, **21**(11), 1867–1884 (Russian).

[47] Izobov, N.A. (1994). On polynomial representations of central high-order exponents. *Diff. Eqns.*, **30**(9), 1508–1515 (Russian).

[48] Izobov, N.A. and Stepanovich, O.P. (1990). On invariance of characteristic exponents under exponentially decreasing perturbations. *Archivum Mathematicum*, **2–3**, 107–114 (Russian).

[49] Izobov, N.A. and Stepanovich, O.P. (1990). Properties of the irregularity coefficients of linear systems. *Diff. Eqns.*, **26**(11), 1899–1906 (Russian).

[50] Izobov, N.A. and Stepanovich, O.P. (1990). On exponentially decreasing perturbations preserving characteristic exponents of linear diagonal system. *Diff. Eqns.*, **26**(6), 934–943 (Russian).

[51] Izobov, N.A. (1991). Characteristic indices of linear systems with Grobman perturbations. *Diff. Eqns.*, **27**(3), 428–437 (Russian).

[52] Izobov, N.A. (1991). Existence of Grobman spectral sets of positive measure for linear systems. *Diff. Eqns.*, **27**(6), 953–957 (Russian).

[53] Izobov, N.A. (1991). On Grobman spectral sets of characteristic exponents of linear systems. *Diff. Eqns.*, **27**(8), 1463–1464 (Russian).

[54] Izobov, N.A. (1995). Asymptotic stability and absolute integrability on the semiaxis of solutions to differential systems with higher-order perturbations. *Diff. Eqns.*, **31**(3), 417–421 (Russian).

Part 2
CONTEMPORARY DEVELOPMENT OF LYAPUNOV'S IDEAS OF DIRECT METHOD

2.1 VECTOR LYAPUNOV FUNCTIONS: NONLINEAR, TIME-VARYING, ORDINARY AND FUNCTIONAL DIFFERENTIAL EQUATIONS*

P. BORNE, M. DAMBRINE, W. PERRUQUETTI
and J.P. RICHARD

Ecole Centrale de Lille, LAIL (CNRS), Lille, France

1 Introduction

When dealing with the qualitative analysis of solutions of complex large scale functional differential equations (or ordinary differential equations) the seminal results of Lyapunov can be used and, more precisely, the second Lyapunov method introducing Lyapunov functions. Unfortunately, the more complex the considered dynamical system, the more difficult it is to find a Lyapunov function.

When faced with such a complex problem, the following general remarks can be made:

R1) Some of the complex aspects can be reduced to simpler connected problems. For example, this principle is used for the study of nonlinear dynamical systems: the linearized system provides some knowledge of the original system (under well-known conditions).

R2) Following Descartes' precepts, it can be relevant to split the problem into several components in order to extract an easier understanding of the initial problem through the understanding of each obtained components.

According to these remarks, it is interesting to apply the approach which consists in transforming a complex, large-scale dynamical system into several more simple dynamical systems of reduced dimensions. Let us examine these remarks by applying them to two examples: the first example will illustrate R1) and the second one R2).

---

*Advances in Stability Theory (Ed.: A.A. Martynyuk). Stability and Control: Theory, Methods and Applications, Taylor & Francis, London, **13** (2003) 49–73.

Example 1.1 Consider the system

$$\frac{dx}{dt} = 2x(-2 + \sin(t) + x), \quad t \in \mathbb{R}, \quad x \in \mathbb{R}. \tag{1}$$

Introducing the variable $z = \text{sign}(x)x$, this gives[1]

$$\frac{dz}{dt} = 2z(-2 + \sin(t) + x), \quad \text{if } x \neq 0, \tag{2}$$

$$\frac{dz}{dt} \leq 2z(-1 + z), \tag{3}$$

$$z(t) \leq \frac{z_0}{z_0 + (1 - z_0)\exp(2(t - t_0))}. \tag{4}$$

It can be deduced that the origin of system (1) is asymptotically stable and an estimate of its domain of attraction is $(-\infty, 1)$.

In this first example, a *comparison system* (CS) was implicitly used, since the solution of (3) has been upperbounded by the solution of the ordinary differential equation

$$\dot{y} = 2y(-1 + y).$$

Such a comparison system presents several interesting properties:

- Its solutions overestimate the actual system behaviour.
- It may infer a qualitative property $\mathcal{P}$ for the initial system, in this case the CS will be a *P-comparison system* (P-CS) with respect to the property $\mathcal{P}$. For instance, in Example 1.1, $\mathcal{P}$ is the exponential stability property.
- It may not depend on disturbances or time variable: this takes much of the hard work out of the study.
- It can be described by lower-order dynamical systems. In order to illustrate this last property, let us examine the following example:

Example 1.2 Consider the system

$$\frac{dx}{dt} = \begin{pmatrix} -2 + \sin(t) + \frac{1}{4}(x_1^2 + x_2^2) & -\sin(t) \\ \sin(t) & -2 + \sin(t) + \frac{1}{4}(x_1^2 + x_2^2) \end{pmatrix} x, \tag{5}$$

$$t \in \mathbb{R}, \quad x \in \mathbb{R}^2,$$

---

[1] As the derivative of the sign function at $x = 0$ has no meaning in the usual sense, we exclude this case in a first stage. In fact, using a more general notion of derivative (see [9]), a similar result can be directly obtained. Note that (4) is still valid for $x = 0$.

introducing the variable $v = \frac{1}{4}(x_1^2 + x_2^2)$, this gives

$$\frac{dv}{dt} = 2v(-2 + \sin(t) + v), \quad t \in \mathbb{R}, \quad v \in \mathbb{R}_+. \tag{6}$$

Then, using Example 1.1, one can conclude that the original system (5) is asymptotically stable and an estimate of its domain of attraction is $\{x \in \mathbb{R}^2 : (x_1^2 + x_2^2) < 4\}$.

However, this reduction of dimension may not be an advantage in all situations. In fact, using a standard Lyapunov function (v) always leads to a first order comparison system, which may represent a drastic cut-down. It thus seems interesting to reduce the loss of information by using not one but several Lyapunov functions: this leads to the notion of *vector Lyapunov function* (VLF) for which each component provides information about a part of the dynamics.

Finally, these introductory examples show that comparison techniques [2–4, 10, 25, 30–32, 37, 50] combined with VLF [1, 10, 20, 21, 23, 25, 32, 33, 37, 42] are certainly an interesting alternative for tackling the study of stability properties of dynamical systems.

Therefore the paper will be divided as follows:

- Section 2 briefly presents the notations used throughout the paper.
- Section 3 sets up the general framework for the considered dynamical systems, i.e. nonlinear, time-varying, ordinary and/or functional differential equations.
- Section 4 concerns the notion, the construction and the properties of CS. We shall see that, among VLF, the particular case of convex ones, illustrated by the *vector norms* (VN), provides a very convenient tool for the construction of CS.
- Section 5 presents different results concerning qualitative set properties for the considered dynamical systems: stability, attractivity, positive invariance. The definitions are recalled at the beginning of the section.
- Section 6 concerns the application of some results of the previous parts to the particular case of ordinary differential systems.

2 Notations

- $\mathcal{C} = C([-\tau, 0], \mathbb{R}^n)$, the set of continuous maps from $[-\tau, 0]$ into $\mathbb{R}^n$;
- $\mathcal{C}(\mathcal{D}) = C([-\tau, 0], \mathcal{D})$, the set of continuous maps from $[-\tau, 0]$ into $\mathcal{D}$;
- $\mathcal{C}_+ = \{y \in \mathcal{C} : y(s) \geq 0, \ s \in [-\tau, 0]\}$;
- x_t element of $\mathcal{C}$ associated with a map $x \colon \mathbb{R} \to \mathbb{R}^n$ by $x_t(s) = x(t+s)$, for all $s \in [-\tau, 0]$;
- φ_x element of $\mathcal{C}$, defined by $\varphi_x(s) = x$, for all $s \in [-\tau, 0]$;

- $B(x,\varepsilon)$ is an ε-ball centered at x in the metric space from which x is defined with the distance function $\rho(\cdot,\cdot)$;
- $\mu(A)$ is the matrix measure of the square matrix A;
- $a\mathcal{R}b$, elementwise relation $\mathcal{R}$ (a and b are vectors or matrices): for example $a < b$ (vectors) means $\forall i\colon a_i < b_i$.

3 Framework

A large number of processes can be modelled by a *functional differential equation*:

$$\dot{x}(t) = f(t, x(t), x_t, d), \tag{7}$$

$$x_{t_0} = \varphi \in \mathcal{C}, \tag{8}$$

where $t \in \mathbb{R}$ is the time variable, $d \in \mathcal{S}_d$ is either a vector or a function representing disturbances or parameter uncertainties of the system, $\mathcal{S}_d$ is a set of vector or functions for which some bounds are usually supposed to be known, $x(t) \in \mathbb{R}^n$ is a set of internal variables, x_t is the map defined by

$$x_t\colon [-\tau, 0] \to \mathbb{R}^n,$$

$$s \mapsto x(t+s). \tag{9}$$

In the paper, when the case $\tau = 0$ is considered, we shall use $x_t \triangleq x(t)$ for the sake of simplicity. In this case, (7) can be directly rewritten as

$$\dot{x}(t) = f(t, x(t), d), \tag{10}$$

$$x_{t_0} = x(t_0) = \varphi \in \mathbb{R}^n. \tag{11}$$

This represents a slight misuse of terminology since, in (7), f is a functional, x_{t_0} is a function and in (10), f is a function, x_{t_0} is a vector.

Assumption 3.1 *It is assumed that system (7) has solutions (for example f satisfies Carathéodory conditions: see [26]) defined over a maximal interval denoted by $\mathcal{I}_{(7)}(t_0, \varphi)$ where t_0 is the initial time and φ is the initial function defined over $[-\tau, 0]$ (most of the time φ is supposed to belong to $\mathcal{C}$).*

4 Comparison Systems

As seen in Example 1.1 of the introduction, it is interesting to obtain information about a complex system through a simpler one whose solutions overvalue the solutions of the initial system. Wazewski's contribution [50] is probably one of the most important in this field: it concerns differential inequalities and gives necessary and sufficient hypotheses ensuring that the solution of $\dot{x} = f(t,x)$, with initial state x_0 at time t_0 and function f satisfying the inequality $f(t,x) \leq g(t,x)$ is overvalued by the solution of the so-called "comparison system" $\dot{z} = g(t,z)$, with initial state $z_0 \geq x_0$ at time t_0, or, in other words, conditions on function g that ensure $x(t) \leq z(t)$ for $t \geq t_0$. These results were extended to many different classes of dynamical systems ([2, 10, 23, 30, 33, 48]).

In the next subsection, we first define the notions of *comparison system* (CS) and of *P-comparison system* (P-CS). The next subsection is then devoted to the construction of such CS.

4.1 Definitions

Focusing on the two systems:

$$\dot{x}(t) = f(t, x(t), x_t), \quad x(t) \in \mathbb{R}^n, \tag{12}$$

$$\dot{z}(t) = g(t, z(t), z_t), \quad z(t) \in \mathbb{R}^n, \tag{13}$$

we respectively note $z(t; t_0, \varphi_2)$ and $x(t; t_0, \varphi_1)$ the solutions of (13) with initial condition φ_2 and of (12) with initial condition φ_1.

Definition 4.1 System (13) is said to be a *comparison system* of (12) over $\Omega \subset \mathcal{C}$ if:

$$\forall (\varphi_1, \varphi_2) \in \Omega^2: \; \mathcal{I} = \mathcal{I}_{(12)}(t_0, \varphi_1) \cap \mathcal{I}_{(13)}(t_0, \varphi_2) \text{ is not reduced to } \{t_0\},$$

$$\varphi_2 \geq \varphi_1 \implies z(t; t_0, \varphi_2) \geq x(t; t_0, \varphi_1), \quad \forall t \in \mathcal{I}.$$

Obviously, one can go beyond this concept to derive a qualitative analysis for positive solutions. For example, if $z(t; t_0, \varphi_2) \geq x(t; t_0, \varphi_1) \geq 0$ and if solution $z(t)$ converges to zero, so does $x(t)$. For this reason, we introduce the following notion:

Definition 4.2 System (13) is a *P-comparison system* of (12) for property $(\mathcal{P})$ if

$$[(\mathcal{P}) \text{ holds for (13)} \Rightarrow (\mathcal{P}) \text{ holds for (12)}].$$

Example 4.1 If we consider a nonlinear system

$$\dot{x}(t) = f(x(t)), \tag{14}$$

then a P-comparison system for the property of uniform asymptotic stability of the zero solution is given by the first order approximation

$$\dot{z}(t) = Az(t), \tag{15}$$

where $A = \left(\frac{\partial f}{\partial x}\right)_{x=0}$.

A question naturally arises concerning properties of the function g ensuring that (13) is a comparison system of (12) over Ω. For this, the following notion is required:

Definition 4.3 A *functional*

$$g\colon \mathbb{R} \times \mathbb{R}^n \times \mathcal{C} \to \mathbb{R}^n$$
$$(t, x, y) \mapsto g(t, x, y)$$

(1) is *quasi-monotone non-decreasing in x* iff:

$$\forall t \in \mathbb{R}, \quad \forall y \in \mathcal{C}, \quad \forall (x, x') \in \mathbb{R}^n \times \mathbb{R}^n: \tag{16}$$

$$\forall i \in \{1, \ldots, n\}[(x_i = x'_i) \wedge (x \le x')$$
$$\Rightarrow g_i(t, x, y) \le g_i(t, x', y)], \tag{17}$$

(2) is *non-decreasing in y* iff:

$$\forall t \in \mathbb{R}, \ \forall x \in \mathbb{R}^n, \ \forall (y, y') \in \mathcal{C} \times \mathcal{C}: [y \le y'] \Rightarrow g(t, x, y) \le g(t, x, y'), \tag{18}$$

(3) is *mixed quasi-monotone non-decreasing in x, non-decreasing in y* iff:

$$\forall t \in \mathbb{R}, \quad \forall (x, x') \in \mathbb{R}^n \times \mathbb{R}^n, \quad \forall (y, y') \in \mathcal{C} \times \mathcal{C}: \tag{19}$$

$$\forall i \in \{1, \ldots, n\}[(x_i = x'_i) \wedge (x \le x') \wedge (y \le y')$$
$$\Rightarrow g_i(t, x, y) \le g_i(t, x', y')]. \tag{20}$$

Remark 4.1 The latter definition is a special case of mixed quasimonotonicity given in [30]. More general versions also exist (see [2, 3, 25, 32, 33]) and additional conditions are sometimes given (see [25, 23], or [50]).

The following result may easily be proved:

Lemma 4.1 *A functional* $g\colon (t, x, y) \mapsto g(t, x, y)$ *is quasi-monotone non-decreasing in x and non-decreasing in y iff it is mixed quasi-monotone non-decreasing in x, non-decreasing in y.*

Proof Omitted because obvious.

Example 4.2 If we consider

$$g\colon \mathbb{R} \times \mathbb{R}^n \times \mathcal{C} \to \mathbb{R}^n$$

$$(t, x, x_t) \mapsto \begin{pmatrix} a_{11}(x_1) & a_{12}(t) \\ a_{21}(t) & a_{22}(x_2) \end{pmatrix} x + \begin{pmatrix} b_{11}(t) & b_{12}(t) \\ b_{21}(t) & b_{22}(t) \end{pmatrix} x(t - \tau), \tag{21}$$

with $\forall t \in \mathbb{R}$, $a_{21}(t) \ge 0$, $a_{12}(t) \ge 0$, $\forall (i, j) \in \{1, 2\}^2: b_{ij}(t) \ge 0$, then g is quasi-monotone non-decreasing in x and non-decreasing in x_t.

Lemma 4.2 *If g is continuously differentiable with respect to x and y, and*

$$\forall t \in \mathbb{R}, \quad \forall x \in \mathbb{R}^n, \quad \forall y \in \mathcal{C}$$

$$\forall i \neq j \colon \frac{\partial g_i}{\partial x_j} \geq 0, \tag{22}$$

$$\forall (i,j) \colon \frac{\partial g_i}{\partial y_j} \geq 0, \tag{23}$$

then $g(t,x,y)$ is mixed quasi-monotone non-decreasing in x, non-decreasing in y.

Proof Using Lemma 4.1, let us first prove that g is quasi-monotone non-decreasing in x.

Let $y \in \mathcal{C}$, and consider any $(x, x') \in \mathbb{R}^n \times \mathbb{R}^n$ such that $\exists i \in \{1, \ldots, n\}$: $(x_i = x'_i) \wedge (x \leq x')$. Define Θ as:

$$\Theta \colon [0,1] \to \mathbb{R}^n$$
$$\theta \mapsto x + \theta(x' - x).$$

Naturally $g_i(t, \Theta(\theta), y)$ is continuously differentiable in its second variable, so

$$g_i(t, \Theta(1), y) - g_i(t, \Theta(0), y)$$

$$= \int_0^1 \sum_{j=1}^n \left(\frac{\partial g_i(t,x,y)}{\partial x_j} \right) \bigg|_{x = x + \theta(x'-x)} (x'_i - x_i)\, d\theta \leq 0. \tag{24}$$

The fact that g is non-decreasing in y may be proved in the same way.

Remark 4.2 In (23), y_j is a function and the map g_i is a functional.

It is now possible to state the main result of this subsection: a comparison principle for functional differential equations

Theorem 4.1 *Assume that:*

H1) $\forall t \in \mathbb{R}, \forall x \in \mathbb{R}^n, \forall y \in \mathcal{C} \colon f(t,x,y) \leq g(t,x,y)$,
H2) $g(t,x,y)$ is mixed quasi-monotone non-decreasing in x, non-decreasing in y,
H3) $g(t,x,y)$ is sufficiently smooth for (13) to possess, for every $z_{t_0} \in \Omega \subset \mathcal{C}$ and for every $t_0 \in \mathbb{R}$, a unique solution $z(t)$ for all $t \geq t_0$.

Then:

C1) *For any $z_{t_0} \in \Omega$, the inequality*

$$x(t) \leq z(t), \tag{25}$$

holds for every $t \geq t_0$ whenever it is satisfied for $t \in [t_0 - \tau, t_0]$. In other words, (13) is a comparison system of (12) over Ω.

C2) *Moreover, if $\forall t \geq t_0 \colon 0 \leq g(t, 0, \varphi_0)$ and $z_{t_0} \geq 0$, then $0 \leq z(t)$.*

Proof Let z^k be a solution of the system

$$\dot{z}^k(t) = g(t, z^k(t), z_t^k) + \delta_k, \tag{26}$$

with the initial condition

$$z_{t_0}^k = z_{t_0} + \varphi_{\delta_k}, \tag{27}$$

$$z_{t_0} \in \Omega \subset \mathcal{C}, \tag{28}$$

where δ_k is the n-vector each component of which is equal to $\left(\frac{1}{2}\right)^k$, and φ_{δ_k} is the constant function defined on $[-\tau, 0]$ of value δ_k.

By H3) $z^k(t)$ is defined and continuous with respect to time for all $t \geq t_0$, and naturally, the sequence $z^k(t)$ converges uniformly towards the solution $z(t)$ of (13) with the initial condition $z_{t_0} \in \Omega \subset \mathcal{C}$.

C1) $x(t) < z^k(t)$ holds for every $t \geq t_0$ if $z_{t_0} \geq x_{t_0}$ (both in $\Omega \subset \mathcal{C}$): otherwise, there would be a time $t_1 > t_0$ and an index $i_0 \in \{1, \ldots, n\}$ such that

$$x_{i_0}(t_1) = z_{i_0}^k(t_1). \tag{29}$$

Then, let $t_2 = \inf\{t \geq t_0 : \exists i \in \{1, ..., k\} : x_i(t) = z_i^k(t)\}$. By continuity and (27), we have $t_2 > t_0$. By denoting $\varepsilon(t) = z^k(t) - x(t)$, we then obtain from (26) and H1):

$$\dot{\varepsilon}_i(t_2) \geq g(t_2, z^k(t_2), z_{t_2}^k) - g(t_2, x(t_2), x_{t_2}) + \left(\frac{1}{2}\right)^k, \tag{30}$$

But, at time $t = t_2$, we have:

$$x_{i_0}(t_2) = z_{i_0}^k(t_2),$$

$$z_i^k(t_2) \geq x_i(t_2), \quad \forall i \neq i_0, \tag{31}$$

$$z_{t_2}^k \geq x_{t_2}.$$

According to H2), we have $g(t_2, z^k(t_2), z_{t_2}^k) - g(t_2, x(t_2), x_{t_2}) \geq 0$:

$$\dot{\varepsilon}_i(t_2) \geq \left(\frac{1}{2}\right)^k > 0. \tag{32}$$

It follows that function $\varepsilon_i(t)$ is increasing on a neighborhood of t_2. But, by definition of t_2, we have:

$$\dot{\varepsilon}_{i_0}(t_2^-) = \lim_{\theta \to 0^-} \frac{\varepsilon_{i_0}(t_2 + \theta) - \varepsilon_{i_0}(t_2)}{\theta} \leq 0, \tag{33}$$

which leads to a contradiction and completes the proof.

C2) In the same way: $\dot{z}_i^k(t_2) = g(t_2, z^k(t_2), z_{t_2}^k)) + \left(\frac{1}{2}\right)^k$ using H2) and $\forall t \geq t_0: 0 \leq g(t, 0, \varphi_0)$ leads to $\dot{z}_i^k(t_2) \geq \left(\frac{1}{2}\right)^k > 0$, which contradicts

$$\dot{z}_{i_0}^k(t_2^-) = \lim_{\theta \to 0^-} \frac{z_{i_0}^k(t_2 + \theta) - z_{i_0}^k(t_2)}{\theta} \leq 0.$$

Remark 4.3 One can refine the definitions given above by considering local comparison system and thus obtain a local version of this theorem (see [37, 39]).

Remark 4.4 Note that for ordinary differential systems, one can see that the given conditions are reduced to the well-known Wazewski conditions (see [50]).

Corollary 4.1 *Assume that:*

H1) $\forall t \in \mathbb{R}, \forall x \in \mathbb{R}_+^n, \forall y \in C_+ = \{y \in C: y(s) \geq 0, s \in [-\tau, 0]\}: 0 \leq g(t, 0, 0)$,

H2) $g(t, x, y)$ is mixed quasi-monotone non-decreasing in x, non-decreasing in y,

H3) $g(t, x, y)$ is sufficiently smooth for (13) to possess, for every $z_{t_0} \in \Omega \subset C$ and for every $t_0 \in \mathbb{R}$, a unique solution $z(t)$ for all $t \geq t_0$.

Then, for any $z_{t_0} \in C_+$, the inequality

$$0 \leq z(t), \tag{34}$$

holds for every $t \geq t_0$.

4.2 Vector Lyapunov functions

The concept of vector Lyapunov function (VLF) ([1, 31–33, 37]) and vector norms (VN) ([4, 6, 8]) have been widely developed in the literature. Here, we shall consider only the following definitions.

Definition 4.4 *V is said to be a vector Lyapunov function if:*

$$V: \mathbb{R}^n \to \mathbb{R}^k$$
$$x \mapsto V(x) = [v_1(x), \ldots, v_k(x)]^T, \tag{35}$$

where $v_i(x)$ are continuous, semi-positive definite functions and $[V(x) = 0 \Leftrightarrow x = 0]$.

If there is a scalar $c > 0$ such that, for all vector $x \in \mathbb{R}^n$, $\|V(x)\|_k \geq c\|x\|_n$, where $\|\cdot\|_k$ (respectively $\|\cdot\|_n$) denotes any given norm on $\mathbb{R}^k$ (resp. $\mathbb{R}^n$), then V is said to be *radially unbounded*.

Example 4.3 The map

$$V: \mathbb{R}^3 \to \mathbb{R}_+^2$$
$$x \mapsto V(x) = [x_1^2 + x_2^2,\ (x_2 - x_3)^2 + x_3^2]^T \qquad (36)$$

is a VLF whereas $[x_1^2 + 2x_1 x_2 + x_2^2,\ (x_2 - x_3)^2]^T$ is not a VLF.

In the following definitions, it is assumed that

$$\mathbb{R}^n = \bigoplus_{i=1}^{k} E_i, \qquad (37)$$

where E_i are vector subspaces of $\mathbb{R}^n$ with $\dim(E_i) = n_i$.

Definition 4.5 *P is said to be a regular vector norm if:*

$$P: \mathbb{R}^n \to \mathbb{R}_+^k$$
$$x \mapsto P(x) = [p_1(x^{[1]}), \ldots, p_k(x^{[k]})]^T, \qquad (38)$$

where $p_i(x^{[i]})$ are vector norms (in the classical sense) on E_i and $x^{[i]} = \text{Pr}_i(x)$ is the projection of x onto E_i. If the sum (37) is not direct, then P is said to be a *vector norm*.

Remark 4.5 If P is a vector norm, then it is also a vector Lyapunov function. Moreover, vector norms satisfy properties similar to those of classical scalar norms.

Example 4.4 Consider

$$P: \mathbb{R}^3 \to \mathbb{R}_+^2$$
$$x \mapsto V(x) = [|x_1| + |x_2| + |x_3|,\ |x_2|]^T, \qquad (39)$$

is a VN whereas $[|x_1| + |x_2|,\ |x_3|]^T$ is a regular VN.

Lemma 4.3 $P(x)$ *is radially unbounded and for* $c > 0$, $\bar{\mathcal{X}}_c = \{x \in \mathbb{R}^n : P(x) \leq c\}$ *are compact connected sets.*

4.3 Construction of comparison systems

One of the main advantages in using regular vector norms as particular vector Lyapunov functions is to allow for the computation of comparison systems in an easy and systematic way for a wide class of systems (7).

In this section, we assume that the mapping f appearing in (7) has the form:

$$f(t, x(t), x_t, d) = A(t, x(t), x_t, d)\, x(t) + B(t, x(t), x_t, d)\, x_t + c(t, x(t), x_t, d), \qquad (40)$$

where for any given $(t, x(t), x_t, d)$ (which will be shortened to (.) in the sequel), $A(.)$ is a $(n \times n)$-matrix with real entries, the map

$$B(.): C \to \mathbb{R}^n$$
$$\varphi \mapsto B(.)\varphi$$

is linear in φ and bounded, and $c(.)$ is a n-vector.

4.3.1 General expressions. Let us calculate the right-hand Dini derivative of $p_i(x^{[i]}(t))$ with respect to time for $1 \le i \le k$:

$$\begin{aligned}
D^+ p_i(x^{[i]}(t)) &= (\text{grad } p_i(x^{[i]}(t)))^T \dot{x}^{[i]}(t) = (\text{grad } p_i(x^{[i]}(t)))^T \text{Pr}_i \dot{x}(t) \\
&= (\text{grad } p_i(x^{[i]}(t)))^T \text{Pr}_i [A(.) x(t) + B(.) x_t + c(.)] \\
&= \sum_{j=1}^{k} (\text{grad } p_i(x^{[i]}(t)))^T \text{Pr}_i A(.) \text{Pr}_j x^{[j]}(t) \\
&\quad + \sum_{j=1}^{k} (\text{grad } p_i(x^{[i]}(t)))^T \text{Pr}_i B(.) \text{Pr}_j \circ x_t \\
&\quad + (\text{grad } p_i(x^{[i]}(t)))^T \text{Pr}_i c(.),
\end{aligned} \tag{41}$$

Let us consider each term separately.

First term: If $p_j(x^{[j]}(t)) \ne 0$, then

$$(\text{grad } p_i(x^{[i]}(t)))^T \text{Pr}_i A(.) \text{Pr}_j x^{[j]}(t)$$
$$= \frac{(\text{grad } p_i(x^{[i]}(t)))^T \text{Pr}_i A(.) \text{Pr}_j x^{[j]}(t)}{p_j(x_j(t))} p_j(x^{[j]}(t)) \le m_{ij}(.) p_j(x^{[j]}(t)), \tag{42}$$

where

$$m_{ij}(.) = \sup_{u \in \mathbb{R}^n} \left\{ \frac{\text{grad } p_i(u^{[i]})^T \text{Pr}_i A(.) \text{Pr}_j u^{[j]}}{p_j(u^{[j]})} \right\}. \tag{43}$$

Note that the previous inequality still holds if $p_j(x^{[j]}(t)) = 0$.

Second term: The mapping $\varphi \mapsto (\text{grad } p_i(x^{[i]}(t)))^T \text{Pr}_i B(.) \text{Pr}_j \circ \varphi$ is a bounded functional. So, using Riesz's representation of linear functional by Stieljes integral [44], we have

$$(\text{grad } p_i(x_i(t)))^T \text{Pr}_i B(.) \text{Pr}_j \circ \varphi = \sum_{l=1}^{n_j} \int_{-\tau}^{0} dk_{ijl}(.)(s) \, \varphi_{jl}(s), \tag{44}$$

where $k_{ijl}(.)$ are real-valued functions of bounded variation defined on $[-\tau, 0]$, and where $\varphi_j = [\varphi_{j1}, \ldots, \varphi_{jn_j}]^T$ denotes $\Pr_j \circ \varphi$. Each function $k_{ijl}(.)$ may be written as the difference between two non-decreasing functions [44]:

$$k_{ijl}(.) = \alpha_{ijl}(.) - \beta_{ijl}(.).$$

By defining $k'_{ijl}(.) = \alpha_{ijl}(.) + \beta_{ijl}(.)$, we have the inequalities

$$|(\operatorname{grad} p_i(x^{[i]}(t)))^T \Pr_i B(.) \Pr_j \circ \varphi| \leq \sum_{l=1}^{n_j} \int_{-\tau}^{0} dk'_{ijl}(.)(s) \, |\varphi_{jl}(s)|$$

$$\leq \int_{-\tau}^{0} \sum_{l=1}^{n_j} dk'_{ijl}(.)(s) \max_{1 \leq l \leq n_j} |\varphi_{jl}(s)|$$

and then, since norms on E_j are equivalent, for each j, there is a real number $\sigma_j > 0$ such that

$$|(\operatorname{grad} p_i(x^{[i]}(t)))^T \Pr_i B(.) \Pr_j \circ \varphi| \leq \int_{-\tau}^{0} \sum_{l=1}^{n_j} \sigma_j \, dk'_{ijl}(.)(s) \, p_j(\varphi_j(s)).$$

Denoting $n_{ij}(.)$ the functional defined on $C([-\tau, 0], \mathbb{R})$ by

$$n_{ij}(.): \psi \mapsto n_{ij}(.)\psi = \int_{-\tau}^{0} \sum_{l=1}^{n_j} \sigma_j \, dk'_{ijl}(.)(s) \, \psi(s), \tag{45}$$

we have

$$|(\operatorname{grad} p_i(x^{[i]}(t)))^T \Pr_i B(.) \Pr_j \circ \varphi| \leq n_{ij}(.) \, p_j \circ \varphi_j.$$

Third term: Let us denote

$$q_i(.) = |(\operatorname{grad} p_i(x^{[i]}(t)))^T \Pr_i c(.)|, \tag{46}$$

Then, summarizing the previous steps, the vector $D^+ P(x(t))$ satisfies the following inequality:

$$D^+ P(x(t)) \leq M(.) \, P(x(t)) + N(.) \, P \circ x_t + q(.),$$

where $M(.) = \{m_{ij}(.)\}$ is a $(k \times k)$ matrix, $N(.)$ is the mapping $\mathcal{C} \to \mathbb{R}^k$ given by $N(.)\varphi = \left[\sum_{j=1}^{k} n_{1j}(.) p_j \circ \varphi_j, \ldots, \sum_{j=1}^{k} n_{kj}(.) p_j \circ \varphi_j \right]^T$, and $q(.)$ denotes the k-vector $[q_1(.), \ldots, q_k(.)]^T$.

It only remains to be proved that the system

$$\dot{z}(t) = M(.)\, z(t) + N(.)\, z_t + q(.)$$

is a comparison system for the initial one: according to Theorem 4.1, it is enough to prove that the function $g(t,x,y) = M(.)\, x + N(.)\, y + q(.)$ is mixed quasi-monotone non-decreasing in x, non-decreasing in y. This is obvious if we observe that the off-diagonal elements of $M(.)$ are non-negative (in expression (43), the subvectors $u^{[i]}$ and $u^{[j]}$ can be chosen freely).

4.3.2 Particular cases for the construction of comparison systems. Simpler expressions may be found under stronger assumptions about the choice of the norms and the initial system.

Let us now consider a system of the form

$$\dot{x}(t) = A(t, x(t), x_t, d)\, x(t) + \sum_{k=1}^{m} B^k(t, x(t), x_t, d)\, x(t - \tau_k(t))$$

$$+ \int_{-\tau}^{0} K(s, t, x(t), x_t, d)\, x(t+s)\, ds + c(t, x(t), x_t, d), \quad (47)$$

where $A(.)$ and $B^k(.)$ (for $k = 1, \ldots, m$) and $K(s, t, x(t), x_t, d)$ are $n \times n$ matrices, and $\tau_k(t)$ are positive, bounded and piecewise-continuous functions such that

$$0 \leq \tau_k(t) \leq \tau, \quad \forall t \geq t_0, \quad \forall k = 1, \ldots, m.$$

To simplify, we consider here that the subspaces E_i of the partition (37) are given by

$$E_i = \text{span}\{e_j : j \in \mathcal{I}_i\}, \quad i = 1, \ldots, k,$$

where $\{e_1, e_2, \ldots, e_n\}$ denotes the canonical basis of $\mathbb{R}^n$ and where the $\mathcal{I}_i$'s are disjoint subsets of $\mathbb{N}_n = \{1, 2, \ldots, n\}$ such that

$$\mathbb{N}_n = \bigcup_i \mathcal{I}_i. \quad (48)$$

The norm p_i on the subspace E_i will be defined by $p_i(x^{[i]}) = \|x^{(i)}\|_i$, where $x^{(i)}$ is the n_i-vector of components x_j for $j \in \mathcal{I}_i$ and $\|\cdot\|_i$ is a given norm on $\mathbb{R}^{n_i}$.

According to the previous partition (48), we associate with a given matrix $A = \{a_{ij}\} \in \mathbb{R}^{n \times n}$, the k^2 matrices A_{ij} defined by

$$A_{ij} = \{a_{lm}\}_{\substack{l \in \mathcal{I}_i \\ m \in \mathcal{I}_j}} \quad \text{for } (i,j) \in (\mathbb{N}_k)^2.$$

Then, a comparison system for system (47) is

$$\dot{z}(t) = \Gamma(A(t,x(t),x_t,d))\,z(t) + \sum_{k=1}^{m} \Lambda(B^k(t,x(t),x_t,d))\,z(t-\tau_k(t)) \\
+ \int_{-\tau}^{0} \Lambda(K(s,t,x(t),x_t,d))\,z(t+s)\,\mathrm{d}s + P(c(t,x(t),x_t,d)), \qquad (49)$$

with the following notations

- $\Gamma(A)$ is the $(k \times k)$-matrix associated with $A = \{a_{ij}\} \in \mathbb{R}^{n \times n}$ by

$$\Gamma(A)_{lm} = \begin{cases} \mu_l(A_{ll}) & \text{for } l = m \\ \|A_{lm}\|_{lm} = \max_{0 \neq y \in R^{n_m}} \dfrac{\|A_{lm}y\|_l}{\|y\|_m} & \text{for } l \neq m, \end{cases} \qquad (50)$$

where $\mu_l(A_{ll})$ is the measure of the matrix A_{ll} associated with the norm $\|\cdot\|_l$ and is defined by the expression

$$\mu(A_{ll}) = \lim_{\varepsilon \to 0^+} \frac{\|I_{n_l} + \varepsilon A_{ll}\|_l - 1}{\varepsilon} \qquad (51)$$

(we refer to [15] for more details about properties of matrix measures).

- $\Lambda(A)$ is the $(k \times k)$-matrix associated with $A = \{a_{ij}\} \in \mathbb{R}^{n \times n}$ by

$$\Lambda(A)_{lm} = \|A_{lm}\|_{lm} \quad \text{for } l,m = 1,\ldots,k. \qquad (52)$$

This result may be found in another form in [46] or [47] and generalizes the formulas given in [13] (see also [12] and [10]).

We only outline the proof of this result. By continuity, we have

$$D^+ p_i(x^{[i]}(t)) = \lim_{\varepsilon \to 0^+} \frac{p_i(x^{[i]}(t+\varepsilon)) - p_i(x^{[i]}(t))}{\varepsilon} \\
= \lim_{\varepsilon \to 0^+} \frac{\|x^{(i)}(t) + \varepsilon \dot{x}^{(i)}(t)\|_i - \|x^{(i)}(t)\|_i}{\varepsilon}. \qquad (53)$$

Then, from (47) and the properties of the norm $\|\cdot\|_i$, this gives

$$D^+ p_i(x^{[i]}(t)) \leq \lim_{\varepsilon \to 0^+} \left(\frac{\|I_{n_i} + \varepsilon A_{ii}\| - 1}{\varepsilon} \right) \|x^{(i)}(t)\|_i \\
+ \sum_{j \neq i} \|A_{ij}(.)\|_{ij} \|x^{(j)}(t)\|_j \\
+ \sum_k \sum_j \|B_{ij}^k(.)\|_{ij} \|x^{(j)}(t-\tau_k(t))\|_j \\
+ \sum_j \int_{-\tau}^{0} \|K_{ij}(.)\|_{ij} \|x^{(j)}(t+s)\|_j \,\mathrm{d}s + \|c^{(i)}(.)\|_i. \qquad (54)$$

The result follows from definition (51).

There are simple and explicit expressions of the terms appearing in (50) and (52) in the case of usual norms. Let $A = \{a_{ij}\}$ denote a $(m \times n)$-matrix, $B = \{b_{ij}\}$ a $(n \times n)$-matrix, x a n-vector and y a m-vector.

With $p_1(x) = \|x\|_1 \triangleq \sum_{i=1}^{n} |x_i|$, $p_2(x) = \|x\|_\infty \triangleq \max_{1 \leq i \leq n} |x_i|$, we have:

$$\|A\|_{11} = \max_{j \in \{1,\ldots,n\}} \sum_{i=1}^{m} |a_{ij}|, \quad \|A\|_{12} \leq \sum_{i=1}^{m} \sum_{j=1}^{n} |a_{ij}|,$$

$$\|A\|_{22} = \max_{i \in \{1,\ldots,m\}} \sum_{j=1}^{n} |a_{ij}|, \quad \|A\|_{21} = \max_{\substack{1 \leq i \leq m \\ 1 \leq j \leq n}} |a_{ij}|, \quad (55)$$

$$\mu_1(B) = \max_j \left[\Re(b_{jj}) + \sum_{i \neq j} |b_{ij}|\right], \quad \mu_2(B) = \max_i \left[\Re(b_{ii}) + \sum_{j \neq i} |b_{ij}|\right].$$

It is possible to obtain other and simpler comparison systems at the expense of stronger estimations. Indeed, for any matrix $\tilde{M}(.)$, any mapping $\tilde{N}(.): \mathcal{C}(\mathbb{R}^k) \to \mathbb{R}^k$, and any k-vector $\tilde{q}(.)$ such that:

$$\tilde{M}(.) \geq M(.),$$
$$\tilde{N}(.)\varphi \geq N(.)\varphi, \quad \forall \varphi \in \mathcal{C}(\mathbb{R}^k), \quad (56)$$
$$\tilde{q}(.) \geq q(.),$$

the system

$$\dot{z}(t) = \tilde{M}(.) z(t) + \tilde{N}(.) z_t + \tilde{q}(.)$$

is also a comparison system of (47). For instance, when t does not appear explicitly, one can find a neighbourhood $\mathcal{N}$ of the origin $(0,0) \in \mathbb{R}^k \times \mathcal{C}(\mathbb{R}^k)$ and a map g mixed quasi-monotone non-decreasing in $P(x(t))$, non-decreasing in $P \circ x_t$, such that

$$g(P(x(t)), P \circ x_t) \geq M(x(t), x_t)P(x(t)) + N(x(t), x_t)P \circ x_t + q(x(t), x_t) \quad (57)$$

holds on $\mathcal{N}$, thus leading to a local CS $\dot{z}(t) = g(z(t), z_t)$.

Note that when M (constant) is Hurwitz (that is, all eigenvalues are in the open left-half plane), the opposite of M then belongs to the class of M-matrices (see [17]).

5 Qualitative Properties of Sets

In this part, $\mathcal{A}$ denotes a non-empty compact set which can be restricted to a point.

5.1 Stability, attractivity

5.1.1 Definitions. In this subsection, we formally define or recall different qualitative properties of sets such as stability, attractivity and so on. Moreover, as the dynamical description of the system is time-dependent, the same should be true for the definition (as mentioned and illustrated in [19]). Lastly, we will stress the importance of having information about initial conditions leading to a qualitative property for a solution: this is the notion of domain relative to a property.

Definition 5.1 The *set* $\mathcal{A} \subset \mathcal{C}$ is said to be *stable w.r.t.* $\{t_0\}$ for system (7) if

$$\mathcal{P}_s(t_0): \begin{cases} \forall \varepsilon > 0, \exists \delta(t_0, \varepsilon) > 0: \\ \varphi \in B(\mathcal{A}, \delta(t_0, \varepsilon)) \Rightarrow x_t(t_0, \varphi) \in B(\mathcal{A}, \varepsilon), \forall t \geq t_0, \end{cases} \quad (58)$$

holds, *stable w.r.t.* $\mathcal{T}$ if $\mathcal{P}_s(\mathcal{T})$: $[\forall t_0 \in \mathcal{T}, \mathcal{P}_s(t_0)]$ holds, *stable* if $\mathcal{P}_s(\mathbb{R})$: $[\forall t_0 \in \mathbb{R}, \mathcal{P}_s(t_0)]$ holds. These properties are said to be *uniform* if in $\mathcal{P}_s(t_0)$, $\delta(t_0, \varepsilon) = \delta(\varepsilon)$ does not depend on t_0, and the corresponding properties are denoted by $\mathcal{P}_{us}(.)$, where (.) is respectively $t_0, \mathcal{T}, \mathbb{R}$.

Definition 5.2 The *set* $\mathcal{A} \subset \mathcal{C}$, is said to be *exponentially stable w.r.t.* $\{t_0\}$ for system (7) if

$$\mathcal{P}_{es}(t_0): \begin{cases} \exists \delta(t_0) > 0, \exists \alpha(t_0) > 0, \exists \beta(t_0) > 0: \\ \varphi \in B(\mathcal{A}, \delta(t_0)) \Rightarrow x_t(t_0, \varphi) \in B(\mathcal{A}, \beta \exp(-\alpha(t-t_0))\rho(\varphi, \mathcal{A})), \\ \forall t \geq t_0 \end{cases} \quad (59)$$

holds, *exponentially stable w.r.t.* $\mathcal{T}$ if $\mathcal{P}_{es}(\mathcal{T})$: $[\forall t_0 \in \mathcal{T}, \mathcal{P}_{es}(t_0)]$ holds, *exponentially stable* if $\mathcal{P}_{es}(\mathbb{R})$: $[\forall t_0 \in \mathbb{R}, \mathcal{P}_s(t_0)]$ holds. These properties are said to be *uniform* if in $\mathcal{P}_s(t_0)$, $\delta(t_0) = \delta$, $\alpha(t_0) = \alpha$, and $\beta(t_0) = \beta$, i.e., they do not depend on t_0, the corresponding properties are denoted by $\mathcal{P}_{ue}(.)$, where (.) is respectively $t_0, \mathcal{T}, \mathbb{R}$.

Definition 5.3 The *set* $\mathcal{A} \subset \mathcal{C}$, is said to be *attractive w.r.t.* $\{t_0\}$ for system (7) if

$$\mathcal{P}_a(t_0): \begin{cases} \forall \varepsilon > 0, \exists \delta(t_0, \varepsilon) > 0, \exists T(t_0, \varepsilon) \geq t_0: \\ \varphi \in B(\mathcal{A}, \delta(t_0, \varepsilon)) \Rightarrow x_t(t_0, \varphi) \in B(\mathcal{A}, \varepsilon), \forall t \geq T(t_0, \varepsilon) \end{cases} \quad (60)$$

holds, *attractive w.r.t.* $\mathcal{T}$ if $\mathcal{P}_a(\mathcal{T})$: $[\forall t_0 \in \mathcal{T}, \mathcal{P}_a(t_0)]$ holds, *attractive* if $\mathcal{P}_a(\mathbb{R})$: $[\forall t_0 \in \mathbb{R}, \mathcal{P}_a(t_0)]$ holds. These properties are said to be *uniform* if in $\mathcal{P}_a(t_0)$, $\delta(t_0, \varepsilon) = \delta(\varepsilon)$ and $T(t_0, \varepsilon) = T(\varepsilon)$ does not depend on t_0, the corresponding properties are denoted by $\mathcal{P}_{ua}(.)$, where (.) is respectively $t_0, \mathcal{T}, \mathbb{R}$.

Definition 5.4 The set $\mathcal{A} \subset \mathcal{C}$ is said to be *asymptotically stable w.r.t.* $\{t_0\}$ (respectively *asymptotically stable w.r.t.* $\mathcal{T}$, *asymptotically stable*) for system (7) if $\mathcal{A}$ is stable w.r.t. $\{t_0\}$ and attractive w.r.t. $\{t_0\}$ (respectively stable w.r.t. $\mathcal{T}$ and attractive w.r.t. $\mathcal{T}$, stable and attractive). These properties are said to be *uniform* if attractivity and stability are uniform.

From a practical point of view, it is crucial in engineering sciences to have concrete knowledge of the initial sets implied in these definitions (in other words, to have information about the number δ appearing in the previous definitions). This induces the concept of domain relative to some property (see [19, 22]). The proposed definitions are inspired by those given for the equilibrium point of an ordinary differential equation ([19, 22]) and extended to sets in the case of ordinary differential equations [37]: here, we define them for the general class of functional differential equations.

Definition 5.5 The *domain of stability* $\mathcal{D}_s(t_0, \mathcal{A}) \subset \mathcal{C}$ of a non-empty compact set $\mathcal{A} \subset \mathcal{C}$ w.r.t. $\{t_0\}$ for system (7) is defined by:

(1) $\forall \varepsilon > 0, x_t(t_0, \varphi)) \in B(\mathcal{A}, \varepsilon), \forall t \geq t_0$, iff $\varphi_{x_0} \in \mathcal{D}_s(t_0, \mathcal{A})$,

(2) $\mathcal{D}_s(t_0, \mathcal{A})$ is a neighbourhood of $\mathcal{A}$.

The *domain of stability* $\mathcal{D}_s(\mathcal{T}, \mathcal{A}) \subset \mathcal{C}$ of a non-empty compact set $\mathcal{A} \subset \mathcal{C}$ w.r.t. $\mathcal{T}$ is defined by:

(1) $\forall t_0 \in \mathcal{T}, \mathcal{D}_s(t_0, \mathcal{A})$ exists,

(2) $\mathcal{D}_s(\mathcal{T}, \mathcal{A}) = \cap_{t_0 \in \mathcal{T}} \mathcal{D}_s(t_0, \mathcal{A})$.

The *domain of stability* of a non-empty compact set $\mathcal{A} \subset \mathcal{C}$ is $\mathcal{D}_s(\mathbb{R}, \mathcal{A})$.

The other *domains of uniform stability* $\mathcal{D}_{us}(., \mathcal{A})$ are similarly defined, where $(.)$ stands for t_0, $\mathcal{T}$, or $\mathbb{R}$.

The other domains corresponding to all the previously mentioned qualitative properties can be defined in a similar way and are omitted for sake of brevity. Note that $\mathcal{D}_{as}(., \mathcal{A}) = \mathcal{D}_s(., \mathcal{A}) \cap \mathcal{D}_a(., \mathcal{A})$, where $(.)$ is respectively $t_0, \mathcal{T}, \mathbb{R}$.

5.1.2 Results. This part concerns the qualitative properties defined in the previous part that can be deduced from the construction of a comparison system, as described in Section 4. Thus, we consider that the techniques described above lead to the following functional differential inequality

$$D^+ P(x) \leq g(P(x(t)), P(x_t)), \tag{61}$$

with which we associate

$$\dot{z}(t) = g(z(t), z_t). \tag{62}$$

Note that this is at least possible locally (see Section 4.3).

Let us recall that in that case if $g(0,0) \geq 0$, then Corollary 4.1 can be applied and the estimating relation $0 \leq P(x(t)) \leq z(t)$ holds for $t > t_0$ as soon as it is satisfied for $t \in [t_0 - \tau, t_0]$. Using this fact, we obtain:

Theorem 5.1 *Let $\mathcal{P}$ be one of the properties defined in the previous subsection. Assume that system (62) has a positive equilibrium point z_e with a non-empty domain $\mathcal{D}_\mathcal{P}(\varphi_{z_e})$, then $\mathcal{A} = \{\varphi \in \mathcal{C}\colon P(\varphi(s)) \leq z_e, \forall s \in [-\tau, 0]\}$ has property $\mathcal{P}$ with the associated domain estimated by $\mathcal{D}_\mathcal{P}(\mathcal{A}) = \{\varphi \in \mathcal{C}\colon P(\varphi(s)) \in \mathcal{D}_\mathcal{P}(z_e), \forall s \in [-\tau, 0]\}$.*

Remark 5.1 For VLF, one can obtain a result similar to Theorem 5.1, as soon as g in (61) is mixed quasi-monotone non-decreasing in $V(x(t))$, non-decreasing in $V(x_t)$ and V is radially unbounded.

Example 5.1 Consider the system defined by

$$\begin{cases} \dot{x}_1 = (1 - x_1^2 - x_2^2)x_1 + d_{12}(t)x_2 + \int_{t-\tau}^{t} d_{11}(s)x_1(s)\,ds, \\ \dot{x}_2 = (1 - x_1^2 - x_2^2)x_2 + d_{21}(t)x_1 + \int_{t-\tau}^{t} d_{22}(s)x_2(s)\,ds, \end{cases} \quad (63)$$

where $d_{ij}(t)$ are continuous functions, bounded in magnitude by 1, and let us consider the vector norm

$$P(x) = \begin{pmatrix} |x_1| \\ |x_2| \end{pmatrix}. \quad (64)$$

We then have

$$D^+P(x) \leq \begin{pmatrix} (1 - P_1^2(x) - P_2^2(x)) & 1 \\ 1 & (1 - P_1^2(x) - P_2^2(x)) \end{pmatrix} P(x) \\ + \int_{t-\tau}^{t} P(x(s))\,ds. \quad (65)$$

We can consider

$$\dot{z}(t) = g(z(t), z_t) = \begin{pmatrix} 1 - z_1^2(t) & 1 \\ 1 & 1 - z_2^2(t) \end{pmatrix} z(t) + \int_{t-\tau}^{t} z(s)\,ds, \quad (66)$$

to be a comparison system (g satisfies the hypotheses of Corollary 4.1). System (66) has an equilibrium point defined by

$$z_e = \sqrt{2 + \tau}\,(1, 1)^T. \quad (67)$$

A change of variable ($y + z_e = z$) leads to the system

$$\dot{y}(t) = \begin{pmatrix} -(5 + 3\tau) - 3\sqrt{2+\tau}\,y_1 - y_1^2(t) & 1 \\ 1 & -(5 + 3\tau) - 3\sqrt{2+\tau}\,y_1 - y_1^2(t) \end{pmatrix} y(t) \\ + \int_{t-\tau}^{t} y(s)\,ds. \quad (68)$$

Let us consider the function $v(y) = \max(|y_1|, |y_2|)$ and, following Razumikhin's method [41], assume that:

$$v(y(t+s)) < (1+\varepsilon)v(y(t)), \quad \forall s \in [-\tau, 0]$$

with $\varepsilon > 0$.

There is an index $i \in \{1, 2\}$, such that $v(y(t)) = |y_i(t)|$ and $\dot{v}(y(t)) = \text{sign}(y_i(t))\dot{y}_i(t)$. It follows that

$$\dot{v}(y(t)) \leq (-(4+2\tau) + \varepsilon - 3\sqrt{2+\tau}y_i - y_i^2(t))v(y(t)).$$

For any $y(t)$ such that $y_1 > -\sqrt{(2+\tau)}$ and $y_2 > -\sqrt{(2+\tau)}$, it is possible to choose ε such that $\dot{v}(y(t)) < 0$.

This proves that z_e is asymptotically stable with $\mathcal{D}_{as}(\varphi_{z_e}) = \mathcal{C}_+ \setminus \{0\}$. Then, according to Theorem 5.1, the set $\mathcal{A} = \{\varphi \in \mathcal{C} \colon P(\varphi(s)) \leq z_e, \forall s \in [-\tau, 0]\}$ is globally asymptotically stable.

Note that in the particular, ordinary, case $\tau = 0$ we find again the results proved in [37, 39], stating that $\mathcal{A} = \{x \in \mathbb{R}^n \colon P(x) \leq \sqrt{2}(1,1)^T\}$ is globally asymptotically stable for the corresponding ordinary differential system:

$$\begin{cases} \dot{x}_1 = (1 - x_1^2 - x_2^2)x_1 + d_{12}(t)x_2 \\ \dot{x}_2 = (1 - x_1^2 - x_2^2)x_2 + d_{21}(t)x_1 \end{cases} \quad (69)$$

Theorem 5.2 *Assume that:*

1) *there is a regular VN $P(x)$, such that*

$$\dot{z}(t) = Mz(t) + Nz_t + q, \quad (70)$$

(with $q > 0$), is a linear time-invariant comparison system of (7);
2) *there is a vector $d > 0$ such that $(Md + N\varphi_d) < 0$.*

Then $\mathcal{A} = \{\varphi \in \mathcal{C} \colon P(\varphi(s)) \leq z_e, \forall s \in [-\tau, 0]\}$, where $z_e \colon Mz_e + N\varphi_{z_e} + q = 0$, is globally uniformly exponentially stable.

Example 5.2 Considering system (63) and using (64) one can obtain:

$$D_t P(x) \leq \begin{pmatrix} -5 & 1 \\ 1 & -5 \end{pmatrix} P(x) + \int_{t-\tau}^{t} P(x(s))ds + 4\sqrt{2}\begin{pmatrix} 1 \\ 1 \end{pmatrix}. \quad (71)$$

The associated comparison system has an equilibrium point defined by

$$z_e = \frac{4\sqrt{2}}{4-\tau}(1,1)^T. \quad (72)$$

Let $d = \delta(1,1)^T$. One can easily check that condition 2) of Theorem 5.2 holds if $\tau < 4$. Considering any $\delta > \frac{4\sqrt{2}}{4-\tau}$ allows us to conclude that if $\tau < 4$, then the set

$$\mathcal{A} = \left\{\varphi \in \mathcal{C}\colon P(\varphi(s)) \leq \frac{4\sqrt{2}}{4-\tau}(1,1)^T,\ \forall s \in [-\tau,0]\right\}$$

is uniformly exponentially stable.

Note that in the particular case $\tau = 0$ we once more obtain the results proved in [37, 39]: the set $\mathcal{A} = \{x \in \mathbb{R}^n \colon P(x) \leq \sqrt{2}(1,1)^T\}$ is uniformly exponentially stable for the corresponding ordinary differential system (69).

5.2 Positive invariance

In this part, we define the notion of positive invariant sets for general, functional systems (7) and show how this property can be investigated by means of comparison systems.

Definition 5.6 A set $\mathcal{M} \subset \mathcal{C}$ is said to be *positively invariant* with respect to (7) and t_0 if, for any element $\varphi \in \mathcal{M}$, the solution $x(t,t_0,\varphi)$ of (7) is defined on $[t_0,\infty)$ and $x_t(t_0,\varphi) \in \mathcal{M}$ for all $t \geq t_0$.

Let us assume that

$$\dot{z}(t) = g(t, z(t), z_t) \qquad (73)$$

is a comparison system of (7) with respect to some vector Lyapunov function V. Then, the following result is obtained:

Theorem 5.3 *If there is a k-vector z_e with non-negative components such that*

$$g(t, z_e, \varphi_e) < 0, \quad \forall t \geq t_0,$$

where $\varphi_e \in C(\mathbb{R}^k)$ is defined by $\varphi_e(s) = z_e$ for all s in $[-\tau,0]$, then the set $\mathcal{C}(\mathcal{A})$, with $\mathcal{A} = \{x \in \mathbb{R}^n \colon V(x) < z_e\}$, is positively invariant for (7).

Proof The function g is non-decreasing in its third argument, so the system

$$\dot{y}(t) = g(t, y(t), \varphi_e) \qquad (74)$$

is a comparison system of (73) as long as $y(t) \leq z_e$. System (74) is a nonlinear, ordinary differential system, and applying Lemma 6 of [39] completes the proof.

Such a vector z_e may be proved to exist in some particular cases. For instance, consider a linear comparison system of the form

$$\dot{z}(t) = Mz(t) + \sum_{i=1}^{k} N^i z(t - \tau_i(t)) + \int_{-\tau}^{0} K(s) z(t+s)\,ds.$$

If the matrix $Z = M + \sum_{i=1}^{k} N^i + \int_{-\tau}^{0} K(s)\,ds$ is the opposite of an irreducible M-matrix, it is then possible to choose the vector z_e as a non-negative eigenvector associated with the importance eigenvalue of Z (i.e., the eigenvalue with the greatest real part). Proof of this result can be found in [10], and this reference also contains other results with less restrictive assumptions about the comparison system.

The positive invariance property is a standard tool in the problems of constrained control, we refer interested readers to the books [7, 43] and the papers [11, 14, 39].

6 Some Corollaries in the Ordinary Case

We recall here some results previously presented in [7], that appear as corollaries of the present study in the particular case of ordinary differential equations.

Let us consider the ordinary differential system (10) and suppose that the choice of some regular vector norm leads, for state vector in $\Omega \subset \mathbb{R}^n$, to an inequality such as

$$D^+ P(x(t)) \leq h(t, x(t), P(x(t))).$$

Several classes of functions h are investigated in the following results.

Theorem 6.1 *If it is possible to define a linear, time-invariant comparison system of (10) in a neighborhood $\Omega \subset \mathbb{R}^n$ of the equilibrium point $x = 0$ relative to a regular vector norm P:*

$$D^+ P(x(t)) \leq M\, P(x(t)), \quad \forall\, t \geq t_0, \quad \forall\, x \in \Omega$$

for which the (constant) matrix M is irreducible and of the Hurwitz type, then $x = 0$ is locally asymptotically stable and its domain of asymptotic stability $\mathcal{D}_{as}(0)$ includes the biggest domain $\mathcal{D}$ which is included in Ω and defined by one of the three following types of sets (or their union):

1. $\mathcal{D} = \mathcal{D}_1 = \{x \in \mathbb{R}^n : P^T(x)\, u_m(M^T) \leq \alpha\} \subset \Omega$,
2. $\mathcal{D} = \mathcal{D}_\infty = \{x \in \mathbb{R}^n : P(x) \leq \beta\, u_m(M)\} \subset \Omega$,
3. $\mathcal{D} = \mathcal{D}_c = \{x \in \mathbb{R}^n : P(x) \leq c\} \subset \Omega$, where $c \in \mathbb{R}_+^k$ is such that $Mc < 0$,

in which α and β are constant scalars and $u_m(A)$ denotes a positive importance eigenvector of the M-matrix A corresponding to its importance eigenvalue $\lambda_m(A)(< 0)$.

Moreover, the convergence in this domain is exponential (in $e^{\lambda_m(M)t}$).

If $\Omega \subset \mathbb{R}^n$, the property is global.

Theorem 6.2 *If it is possible to define a nonlinear and/or time-varying comparison system of (10) in a neighborhood $\Omega \subset \mathbb{R}^n$ of the equilibrium point $x = 0$, relative to a regular vector norm P:*

$$D^+ P(x(t)) \leq M(t, x(t)) \, P(x(t)), \quad \forall \, t \geq t_0, \quad \forall \, x \in \Omega$$

for which the matrix $M(t, x(t))$ verifies the following properties:

1. *$M(t, x(t))$ is irreducible $\forall \, t \geq t_0$, $\forall \, x \in \Omega$;*
2. *$\exists \, \varepsilon > 0$ such that $M(t, x(t)) + \varepsilon I_k$ is the opposite of an M-matrix;*
3. *the non-constant elements of $M(t, x(t))$ are grouped in one single column.*

Then $x = 0$ is locally exponentially stable for (10) (in $e^{-\varepsilon t}$) and its domain of asymptotic stability $\mathcal{D}_{as}(0)$ includes the biggest domain $\mathcal{D}$ of the type:

$$\mathcal{D} = \{x \in \mathbb{R}^n : P^T(x) \, u_m(M^T) \leq \alpha\}, \quad \text{with } \mathcal{D} \subset \Omega,$$

in which $u_m(M^T)$ is a common importance eigenvector of the matrices $M^T(t, x(t))$.

Theorem 6.3 *If it is possible to define a non-homogeneous, linear, time-invariant comparison system of (10) in the neighborhood $\Omega \subset \mathbb{R}^n$ of $x = 0$, relative to a regular vector norm P:*

$$D^+ P(x(t)) \leq M \, P(x(t)) + q, \quad \forall \, t \geq t_0, \quad \forall \, x \in \Omega$$

such that

1. *M is a constant matrix of Hurwitz type,*
2. *q is a constant, non-negative vector,*
3. *the set $\mathcal{L} \triangleq \{x \in \mathbb{R}^n : P(x) \leq -M^{-1} q\}$ is strictly included in Ω.*

Then, the set $\mathcal{L}$ is asymptotically stable for system (10). Moreover, the sets $\mathcal{D}_1, \mathcal{D}_\infty$ or $\mathcal{D}_c$ (Theorem 6.1) are estimates of the domain of asymptotic stability $\mathcal{D}_{as}(\mathcal{L})$ of $\mathcal{L}$ for any scalar α or vector c such that $\mathcal{L} \subset \mathcal{D}_i \subset \Omega$ (with $i = 1, \infty$ or c).

7 Conclusion

As shown in this paper, the comparison principle method is a general and useful tool in the investigation of the asymptotic behavior of many dynamical systems. It does not replace Lyapunov's second method but it may be seen as an intermediate step allowing for the construction of a Lyapunov function. In order to be efficient, tools have to be developed which facilitate the construction of comparison systems. One of the aims of this paper is to present such a tool, developed in the framework of stability study, by our team since the 1970's. Initial applications of this method concerned discrete or continuous nonlinear systems. The present chapter extends this works to a large family of systems — the class of functional differential equations.

Using the attractive topological properties of vector norms, and with slight restrictions on the structure of the obtained comparison systems, we have stated stability conditions that are easy to check (this is an important point for a practical use of the method). In addition, some means for estimating the domains of stability have been provided. Of course, there may be some cases which require arrangements the user to make some in order to reduce the conservatism of the methods. As a result, there are some works that have not been presented in this chapter and which may be useful: we refer the interested reader to the papers [5, 8]. Besides, analogous results are available for discrete-time systems (see [7]) and systems of neutral type ([[45]]).

References

[1] Bellman, R. (1962). Vector Lyapunov functions, *SIAM J. Control, Serie A*, **1**(1), 31–34.

[2] Bitsoris, G. (1978). *Principe de Comparaison et Stabilité des Systèmes Complexes*. Ph.D Thesis, Paul Sabatier University of Toulouse, France, no. 818.

[3] Bitsoris, G. (1983). Stability analysis of non linear dynamical systems. *Int. J. Control*, **38**(3), 699–711.

[4] Borne, P. and Gentina, J.C. (1974). On the stability of large non-linear systems. *Proc. Joint Automatic Control Conf.* Austin, Texas.

[5] Borne, P. and Richard, J.P. (1987). State-space modelling: state space transformation. *Systems and Control Encyclopedia*, 4540–4547. Pergamon Press.

[6] Borne, P. and Richard, J.P. (1990). Local and global stability of attractors by use of vector norms. *The Lyapunov Functions Method and Applications* (Eds.: P. Borne and V. Matrosov). J.C. Baltzer AG, Scientific publishing Co., 53–62.

[7] Borne, P., Richard, J.P. and Radhy, N.E. (1996). Stability, stabilization, regulation using vector norms. *Nonlinear Systems*, Vol. 2, *Stability and Stabilization* (Eds.: A. Fossard and D. Normand-Cyrot). Chapman & Hall, London, 45–90.

[8] Borne, P., Richard, J.P., Radhy, N.E. and Perruquetti, W. (1992). Estimation of attractors and stability region for nonlinear dynamical systems: improved results. *Handbook on Computational Systems Analysis*. Elsevier Science Publishers, Sydow edt.

[9] Clarke, F.H. (1983). *Optimization and Nonsmooth Analysis*, Canadian Mathematical Society Series of Monographs and Advanced Texts, Wiley-Interscience Publication.

[10] Dambrine, M. (1994). *Contribution à l'Étude des Systèmes à Retards*. Ph.D Thesis, University of Sciences and Technology of Lille, France, no. 1386.

[11] Dambrine, M., Goubet, A. and Richard, J.P. (1995). New results on constrained stabilizing control of time-delay systems. *Proc. 34th IEEE CDC*, 2052–2057. New Orleans.

[12] Dambrine, M. and Richard, J.P. (1993). Stability analysis of time-delay systems. *Dynamic Systems and Applications*, **2**, 405–414.

[13] Dambrine, M. and Richard, J.P. (1994). Stability and stability domains analysis for nonlinear differential-difference equations. *Dyn. Syst. and Applications*, **3**, 369–378.

[14] Dambrine, M., Richard, J.P. and Borne, P. (1995). Feedback control of time-delay systems with bounded control and state. *Mathematical Problems in Engineering*, **1**(1), 77–87.

[15] Desoer, C.A. and Vidyasagar, M. (1975). *Feedback Systems: Input-Output Properties*. Academic Press, New York.

[16] Driver, R. D. (1977). *Ordinary and Delay Differential Equations*. Springer Verlag, New York.

[17] Fiedler, M. and Ptak, V. (1962). On matrices with non-positive off-diagonal elements and positive principal minors. *Czec. Math. J.*, **12**(87), 382–400.

[18] Guckenheimer, J. and Holmes, P. (1983). Nonlinear oscillations, dynamical systems and bifurcations of vector fields. *Applied Mathematical Sciences*, **42**, Springer-Verlag, New York.

[19] Grujič, Lj.T. (1975). Novel Development of Lyapunov Stability of Motion. *Int. J. Control*, **22**(4), 525–549.

[20] Grujič, Lj.T., Gentina, J. C. and Borne, P. (1976). General aggregation of large-scale systems by vector Lyapunov functions and vector norms. *Int. J. Control*, **24**(4), 529–550.

[21] Grujič, Lj.T., Gentina, J.C., Borne, P., Burgat, C. and Bernussou, J. (1978). Sur la stabilité des systèmes de grande dimension, fonctions de Lyapunov vectorielles. *R.A.I.R.O Aut./Sys. Analysis and Control*, **12**(4), 319–348.

[22] Grujič, Lj.T. (1986). Stability domains of general and large scale systems. *Proceedings IMACS—IFACS, Modelling and Simulation for Control of Lumped and Distributed Parameter Systems (3-6 Juin 1986)*, 267–272.

[23] Grujič, Lj.T., Martynyuk, A.A. and Ribbens-Pavella, M. (1987). *Large Scale Systems Stability under Structural Perturbations. Lecture Notes in Control and Information Sciences*, Vol. 92. Springer Verlag, New York.

[24] Grujič, Lj.T. (1990). Solutions to Lyapunov stability problems: nonlinear systems with globally differentiable motions. *The Lyapunov functions method and applications* (Eds.: P. Borne and V. Matrosov). Scientific Publishing Co., J.C. Baltzer AG, 19–27.

[25] Habets, P. and Peiffer, K. (1972). Attractivity concepts and vector Lyapunov functions. *Nonlinear Vibration Problem (6th Int. Conference on Nonlinear Oscillations, Pozna)*, 35–52.

[26] Hale, J.K. (1977). *Theory of Functional Differential Equations*. Springer-Verlag, New York .

[27] Hahn, W. (1963). *Theory and Application of Liapunov's Direct Method*. Prentice-Hall, Englewood Cliffs, N.J.

[28] Hahn, W. (1967). *Stability of Motion*. Springer-Verlag, New York.

[29] Krasovskii, N. (1963). *Stability of Motion*. Stanford Univ. Press, Stanford, Calif.

[30] Laksmikantham, V. and Leela, S. (1969). *Differential and Integral Inequalities*, Vol. I and II. Academic Press, New York .

[31] Matrosov, V.M. (1963). On the theory of stability of motion. *Applied Math. Mech.*, **26**, 1506–1522.

[32] Matrosov, V.M. (1968; 1969). Comparison principle and vector Lyapunov functions I, II, III, IV. *Diff. Uravn.*, **4**(8), 1374–1386; **4**(10), 1739–1752; **5**(7), 1171–1185; **5**(12), 2129–2143 (Russian).

[33] Matrosov, V.M. (1971). Vector Lyapunov functions in the analysis of nonlinear interconnected system. *Symp. Math. Academic Press*, Vol. 6, 209–242. New York.
[34] Michel, A.N. and Miller, R.K. (1977). *Qualitative Analysis of Large Scale Dynamical Systems*. Academic Press, New York.
[35] Mori, T., Fukuma, N. and Kuwahara, M. (1982). On an estimate of the decay rate for stable linear delays systems. *Int. J. Control*, **36**(1), 95–97.
[36] Perruquetti, W. and Richard, J.P. (1993). Stability domains for differential inequalities systems. *Proc. IEEE/SMC'93*, **1**, 325–330, Le Touquet-France.
[37] Perruquetti, W. (1994). *Sur la Stabilité et l'Estimation des Comportements Non Linéaires, Non Stationnaires, Perturbés*. PhD Thesis, University of Sciences and Technology of Lille, France, no. 1286.
[38] Perruquetti, W. and Richard, J.P. (1995). Connecting Wazewski's condition with opposite of M-matrix: application to constrained stabilization. *Dynamic Systems and Applications*, **5**, 81–96.
[39] Perruquetti, W., Richard, J.P. and Borne, P. (1995). Vector Lyapunov functions: recent developments for stability, robustness, practical stability and constrained control. *Nonlinear Times & Digest*, **2**, 227–258.
[40] Perruquetti, W., Richard, J.P., Grujić, Lj.T. and Borne, P. (1995). On practical stability with the settling time via vector norms. *Int. J. Control*, **62**(1), 173–189.
[41] Razumikhin, B.S. (1960). The application of Lyapunov's method to problems in the stability of systems with delay. *Autom. i Telemekhanika*, **2**(6), 740–748.
[42] Richard, J.P., Borne, P. and Gentina, J.C. (1988). Estimation of stability domains by use of vector norms. *Information and Decision Technologies*, Vol. 14, 241–251. North-Holland.
[43] Richard, J.P., Goubet-Bartholomeus, A., Tchangani, A.P. and Dambrine, M. (1997). Nonlinear delay systems: tools for quantitative approach to stabilization. *Stability and Control of Time-Delay Systems*, Chapter 10, LNCIS, Springer-Verlag.
[44] Riesz, F. and Nagy, B. (1955). *Functional Analysis*. Ungar, New York.
[45] Tchangani, A.P., Dambrine, M. and Richard, J.P. (1998). Stability, attraction domains and ultimate boundedness for nonlinear neutral systems. *Mathematics and computers in simulation*, **45**(3–4), 291–298.
[46] Tchangani, A.P., Dambrine, M., Richard, J.P. and Kolmanovskii, V.B. (1998). Stability of nonlinear differential equations with distributed delay. *Nonlin. Anal.*, **34**, 1081–1095.
[47] Tchangani, A.P. (1999). *Sur la Stabilité des Systèmes Héréditaires non Linéaires*. PhD Thesis, University of Sciences and Technology of Lille, France.
[48] Tokumaru, H., Adachi, N. and Amemiya, T. (1975). Macroscopic stability of interconnected systems. *Proc. of IFAC 6th World Congress*, Paper 44.4, Boston.
[49] Wang, S.S., Chen, B.S. and Lin, T.P. (1987). Robust stability of uncertain time-delay systems. *Int. J. Control*, **46**(3), 963–976.
[50] Wazewski, T. (1950). Systèmes des équations et des inégalités différentielles ordinaires aux seconds membres monotones et leurs applications. *Ann. Soc. Polon. Math.*, **23**, 112–166.
[51] Xu, D. (1986). Stability criteria for time-varying delay systems. In: *Frequency Domain and State Space Methods for Linear Systems* (Eds.: C.I. Byrnes and A. Lindquist). Elsevier Science Publishers, 431–437.

2.2 SOME RESULTS ON TOTAL STABILITY PROPERTIES FOR SINGULAR SYSTEMS*

A. D'ANNA

Dipartimento di Matematica e Applicazioni R. Caccioppoli, Università di Napoli, Napoli, Italy

1 Introduction

The Liapunov's classical theorem on asymptotic stability of the zero solution of a differential equation $\dot{x} = f(x,t)$ in $\Re^n$ employs an auxiliary function $v(x,t)$ that satisfies some conditions. One of these obliges the derivative along the solutions $\dot{v}(x,t)$ to be definite negative, i.e. there exists a definite positive function $u(x)$ such that $\dot{v}(x,t) \leq -u(x)$.

Several authors have explored different ways to weaken this condition that happens rather restrictive; at first the set $E = \{x: u(x) = 0\}$ has been supposed composed by more than the unique point $x = 0$. Barbashin and Krasovskii, and later, La Salle were the first to investigate in this field assuming suitable conditions both on the set E and on the solutions of the differential equation [1, 2].

The comparison principle has been fruitfully employed in dealing with several qualitative problems [4].

Another direction many efforts have been directed to, was to consider in a neighbourhood of the set E one or several auxiliary functions endowed with suitable properties [3, 6 – 19].

This last method was used to deduce stability properties when the vector function f is not defined on a closed set M of $\Re^n$ [11, 15, 18, 19].

In this paper we come back to this last problem considering also the case of perturbations acting along the motions and we look for new conditions of stability and asymptotic behaviour for M. The results attained are illustrated by an example

---

*Advances in Stability Theory (Ed.: A.A. Martynyuk). Stability and Control: Theory, Methods and Applications, Taylor & Francis, London, **13** (2003) 75–88.

concerning the two-body problem when the motions take place in an atmosphere and the reference frame is non-inertial.

2 Notations, Definitions and Preliminary Results

2.1 Let $\mathfrak{R}^n$ be the usual Euclidean n-space and I denote the interval $0 \leq t < +\infty$. For $x \in \mathfrak{R}^n$, $|x|$ will indicate any norm in $\mathfrak{R}^n$; if $y \in \mathfrak{R}^n$, $\xi > 0$, $B(y,\xi)$ will be the open ball $\{x \in \mathfrak{R}^n, |x - y| < \xi\}$. For a set $G \subset \mathfrak{R}^n$, $\bar{G}$ and ∂G will denote respectively its closure and its frontier; further $d(x, G)$ is the distance of the point x from G.

In the following we refer to a closed set $M \subset \mathfrak{R}^n$ and, considering the numbers δ, λ, $0 < \delta < \lambda$, define

$$M_\lambda = \{x \in \mathfrak{R}^n, d(x, M) < \lambda\}, \quad M_{\delta,\lambda} = \{x \in \mathfrak{R}^n, \delta \leq d(x, M) < \lambda\}.$$

Consider now the systems of differential equations

$$\dot{x} = f(x, t), \qquad (2.1)$$

$$\dot{x} = f(x, t) + p(t). \qquad (2.2)$$

We call respectively *unperturbed* and *perturbed system*; we suppose f is an n-vector defined and continuous on the set $(\mathfrak{R}^n \setminus M) \times I$, while $p \colon I \to \mathfrak{R}^n$ is defined and continuous on I. As it is well known, for every point $(x_0, t_0) \in (\mathfrak{R}^n \setminus M) \times I$ there exists at least one non-continuable solution $x(t)$ of equation (2.1) or (2.2) defined on the interval $J(x_0, t_0) = (\omega', \omega)$ and such that $x(t_0) = x_0$; ω' and ω depend on x_0 and t_0. It may happen that $\omega' = -\infty$ and $\omega = +\infty$. We shall also sometimes write this solution as $x(t; t_0, x_0)$.

2.2 For a function $v(x, t) \colon (\mathfrak{R}^n \setminus M) \times I \to \mathfrak{R}$ we consider the upper right-hand derivative computed along the solutions of system (2.1) or (2.2), denoted by $D^+ v(x, t)$ or $D_p^+ v(x, t)$ respectively, and defined as $\limsup[v(x(t+h), t+h) - v(x(t), t)]/h$ for $h \to 0$. It is well known if v satisfies a local Lipschitz condition this derivative can be computed without knowing the solution $x(t)$ [5].

A function $a \colon I \to \mathfrak{R}$ is said to be of class $\mathcal{K}$ (in the sense of Hahn) if it is continuous, strictly increasing, and with $a(0) = 0$. For such a function we shall write $a \in \mathcal{K}$.

2.3 Let E be a non-empty set of M_λ, $E \cap \partial M \neq \emptyset$, and $\{W_y(x, t)\}$ be a family of functions, $W_y \colon B(y, \xi_y) \times I \to \mathfrak{R}$, $y \in \mathfrak{R}^n$, $\xi_y > 0$.

Definition 2.1 The *family of function* $\{W_y(x,t)\}$ is said to be *definitely negative on* E if there are two numbers $\xi > 0$ and $\beta > 0$ such that for $y \in E$ it is $\xi_y \geq \xi$ and moreover the function $W_y(x,t)$ satisfies

$$W_y(x,t) < -\beta \quad \text{for} \quad x \in B(y,\xi_y) \cap (M_\lambda \backslash M) \quad \text{and} \quad t \in I.$$

2.4 In the next sections we shall discuss some stability properties of the set M with respect to the solutions of system (2.2) and, therefore, we give the following definitions.

Definition 2.2 M is a *total-stable set* of equation (2.1) if there exists a point $x^* \in \partial M$ and for any $\varepsilon > 0$, $\rho > 0$ and $t_0 \in I$ there are $\delta(\varepsilon,\rho) > 0$, $\eta(\varepsilon,\rho) > 0$ such that if it is $d(x_0, M) < \delta$, $x_0 \in B(x^*,\rho)$ and $|p(t)| \leq \eta$, then the solution $x(t;t_0,x_0)$ of system (2.2) satisfies $x(t;t_0,x_0) \in M_\varepsilon$ for all $t \in J(x_0,t_0) \cap [t_0, +\infty)$.

Definition 2.3 M is a *total-asymptotically stable set* of equation (2.1) if it is total-stable and *total-attractor*, i.e. for any $\rho > 0$ and $\nu > 0$ there exist $\Delta(\rho) > 0$, $\eta(\rho,\nu) > 0$ such that if $d(x_0,M) < \Delta$, $x_0 \in B(x^*,\rho)$ and $|p(t)| \leq \eta$, then the solution $x(t;t_0,x_0)$ of system (2.2) satisfies $x(t;t_0,x_0) \in M_\nu$ for all $t \in [t_0+T,\omega)$, where $T = T(\rho,\nu,t_0,x_0)$.

At last we shall employ the following Lemma the proof of which may be found in [14]. Denoting by ∇ the gradient operator with respect to x, we have

Lemma 2.1 *Let $\xi > 0$ and E be a non-empty set of $\Re^n$. Then, there exist two numbers $\theta \in (0,\xi)$, $\psi \in (0,\theta)$, a countable family $\{B_i\}$ of open balls of $\Re^n$ having all the same radius θ and centres in points of E, and for each i a function $\alpha_i \colon \Re^n \to [0,1]$ of class $\mathbb{C}^\infty$ such that:*

(i) $\{B_i\}$ *is a covering of the set* $\bar{E}_\psi = \{x \in \Re^n : d(x,E) \leq \psi\}$;
(ii) *each ball B_i has a non-empty intersection with a number of other balls not greater than a fixed number n' (independent of i);*
(iii) $\operatorname{supp} \alpha_i \subset B_i$, $\sum_i \alpha_i(x) = 1$ *for* $x \in \bar{E}_\psi$, $0 \leq \sum_i \alpha_i(x) \leq 1$ *for* $x \in \cup B_i$, $\sum_i \alpha_i(x) = 0$ *for* $d(x,E) \geq \tau$, $\tau = \theta + \psi$;
(iv) *the gradients $\nabla \alpha_i$ are uniformly bounded.*

3 Two Lemmas

Lemma 3.1 *Let $c(d)$ and $a(d)$ be two functions of class $\mathcal{K}$, q and μ two positive parameters, h, m, β and φ positive constants. Admit the function $y(d) = -c(d) + a(d)$ satisfies:*

(i) $y(d) > 0$ *as* $d \in (0, d_1)$, $y(d_1) = 0$;
(ii) $y(d) < 0$ *as* $d > d_1$.

Then the function $g: I \to I$ defined by

$$g(d) = -hc(d) + qa(d) + \mu\left(-\tfrac{3}{4}\beta + mq\right) \quad \text{for} \quad d \in [0, \chi), \ \chi > 0,$$
$$g(d) = -hc(d) + qa(d) + \mu(\varphi + mq) \quad \text{for} \quad d \geq \chi, \tag{3.1}$$

satisfies

$$g(d) \leq -\gamma \quad \text{for} \quad d \geq 0 \quad \text{and} \quad q \in [0, \eta], \tag{3.2}$$

when

(a) $\phi = \varphi + \beta/4$,
(b) $\eta = \min[h/3, \mu\beta/4\hat{y}, \beta/4m]$,
(c) $\hat{y} = \max\{y(d), d \geq 0\}$, \hfill (3.3)
(d) $\mu = hc(\chi)/6\phi$,
(e) $\gamma = \mu\beta/4$.

Proof Given h, m, β and φ, we choose q satisfying $mq \leq \beta/4$ and define ϕ and $\hat{y}$ as in (3.3)(a) and (3.3)(c). Consequently, referring to the function $g(d)$ defined by (3.1), we note that for $d \in [0, \chi)$ it results

$$g(d) \leq q\hat{y} - (h-q)c(d) - \mu\beta/2.$$

Therefore, if we choose

$$q \leq \min[h, \mu\beta/4\hat{y}, \beta/4m], \tag{3.4}$$

we get $g(d) \leq -\mu\beta/4$ which implies (3.2). On the other hand, for $d \geq \chi$ it is

$$g(d) \leq q\hat{y} - (h-q)c(d) + \mu\phi$$

and hence, considering

$$q \leq \min[h/3, hc(\chi)/6\hat{y}], \tag{3.5}$$

and assuming μ verifying (3.3)(d) it follows $g(d) \leq -hc(\chi)/3$. Using (3.4) and (3.3)(d) we derive

$$hc(\chi)/6 = \mu\phi > \mu\beta/4 \tag{3.6}$$

and therefore (3.2) is verified under hypotheses (3.3).

Lemma 3.2 *Suppose for the system (2.1) there are a continuous function $v(x,t)\colon (M_\lambda\backslash M)\times I \to \Re$, a bounded set $E \subset M_\lambda\backslash M$, $E\cap \partial M \neq \emptyset$, and a family of functions $\{w_y(x,t)\}$, $w_y(x,t)\colon B(y,\xi_y)\times I \to \Re$, such that:*

(j) $D^+v(x,t) \leq -hc(d(x,E))$, $c \in \mathcal{K}$, $h > 0$ constant, $x \in M_\lambda\backslash M$, $t \in I$;
(jj) $D_p^+v(x,t) \leq D^+v(x,t) + a(d(x,E))|p(t)|$, $a \in \mathcal{K}$, $x \in M_\lambda\backslash M$, $t \in I$,

where the functions a and c verify hypotheses of Lemma 3.1;

(jjj) $w_y(x,t) \to 0$ as $d(x,E) \to 0$ uniformly for $y \in E$, $t \in I$;
(jv) $\nabla w_y(x,t)$ uniformly bounded for $y \in E$, $x \in B(y,\xi_y)$, $t \in I$;
(v) the family of functions $\{\dot{w}_y(x,t)\}$ is definitely negative on E;
(vj) for each $\sigma \in (0,\lambda)$ and for each compact set $S \subset M_{\sigma,\lambda}$ there is a positive constant K_σ such that $|f(x,t)| \leq K_\sigma$ for $x \in S$ and $t \in I$.

Then there exist the constant $H > 0$, $\tau > 0$ and for every $\sigma > 0$ three numbers $\mu(\sigma) = \mu_\sigma > 0$, $\eta(\sigma) = \eta_\sigma > 0$ $\gamma(\sigma) = \gamma_\sigma > 0$, $\mu_\sigma + \eta_\sigma + \gamma_\sigma \to 0$ as $\sigma \to 0$, and a function $V_\sigma(x,t)\colon (M_\lambda\backslash M)\times I \to \Re$ verifying

$$V_\sigma(x,t) \geq v(x,t) - \mu_\sigma H, \quad x \in M_\lambda\backslash M, \quad t \in I; \tag{3.7}$$

$$V_\sigma(x,t) \leq v(x,t) + \mu_\sigma H, \quad x \in M_\lambda\backslash M, \quad t \in I; \tag{3.8}$$

$$D_p^+ V_\sigma(x,t) \leq -\gamma_\sigma \quad \text{for} \quad |p(t)| \leq \eta_\sigma \quad \text{and} \quad x \in M_{\sigma,\lambda}, \quad t \in I, \tag{3.9}$$

$$\text{or} \quad d(x,E) \geq \tau, \quad x \in M_\sigma\backslash M, \quad t \in I.$$

Proof According to Definition 2.1, hypothesis (v) implies the existence of two numbers $\xi > 0$ and $\beta > 0$ such that if $y \in E$ the function $w_y(x,t)$ satisfies

$$\dot{w}_y(x,t) < -\beta \quad \text{for} \quad x \in B(y,\xi) \cap (M_\lambda\backslash M), \quad t \in I. \tag{3.10}$$

By employing Lemma 2.1 we get two positive numbers θ and ψ, $\xi > \theta > \psi$, a countable family $\{B_i\}$ of open balls of $\Re^n$ having all the same radius θ and centres $y_i \in \bar{E}_\psi$ and using hypotheses (jjj)–(jv) we can suppose the radius θ is such that the functions $w_i = w_{y_i}$ fulfil

$$|w_i(x,t)| + |\nabla w_i(x,t)| \leq w \quad \text{for all} \quad y_i \in \bar{E}_\psi, \quad x \in B_i, \quad t \in I, \tag{3.11}$$

where w is a positive constant.

For a point $x \in \cup B_i$ we denote by $I(x)$ the collection of the indices i satisfying the property $B_i \cap \{x\} \neq \emptyset$ whose number is at the most $n'+1$.

Let us now consider the function

$$h(x,t) = \sum_{i \in I(x)} \alpha_i(x) w_i(x,t), \quad x \in M_\lambda\backslash M, \quad t \in I. \tag{3.12}$$

Employing again Lemma 2.1 it is easy to verify $h(x,t)$ is of class C^1 and moreover fulfils

$$h(x,t) = 0 \quad \text{for} \quad x \in M_\lambda \setminus M, \quad d(x,E) \geq \tau, \quad \tau = \theta + \psi, \quad t \in I; \tag{3.13}$$

furthermore, from (3.11) we deduce that for the function (3.12) there exist two positive constants $H = (n'+1)w$ and m such that

$$|h(x,t)| \leq H, \quad x \in M_\lambda \setminus M, \quad t \in I; \tag{3.14}$$

$$|\nabla h(x,t)| \leq m, \quad x \in M_\lambda \setminus M, \quad t \in I. \tag{3.15}$$

It easy to show along the solutions of system (2.2) it results

$$D_p^+ h(x,t) = \sum_{i \in I(x)} \{\nabla \alpha_i \cdot f w_i + \alpha_i \dot{w}_i\} + \nabla h \cdot p. \tag{3.16}$$

Property (iv) of Lemma 2.1, hypotheses (jjj) and (vj) imply that for every $\sigma > 0$ there exists a number $\chi_\sigma \in (0, \min(\psi, \sigma))$ such that

$$\left|\sum_i \nabla \alpha_i \cdot f w_i\right| < \beta/4 \quad \text{for} \quad x \in M_{\sigma,\lambda} \quad \text{and} \quad d(x,E) < \chi_\sigma, \quad t \in I,$$

where β has been assigned in (3.10). Besides, owing to the conditions (iii) of Lemma 2.1 and (3.10), it results

$$\sum_i \alpha_i \dot{w}_i < -\sum_i \alpha_i \beta = -\beta, \quad x \in M_{\sigma,\lambda}, \quad d(x,E) < \chi_\sigma, \quad t \in I. \tag{3.17}$$

Employing (3.15), (3.17), formula (3.16) becomes

$$D_p^+ h(x,t) \leq -\tfrac{3}{4}\beta + m|p(t)|, \quad x \in M_{\sigma,\lambda}, \quad d(x,E) < \chi_\sigma, \quad t \in I. \tag{3.18}$$

Using again the properties of functions α_i, the hypothesis (vj) and (3.10), (3.11) we get a constant $\varphi_\sigma > 0$ satisfying

$$\sum_i \{\nabla \alpha_i \cdot f w_i + \alpha_i \dot{w}_i\} \leq \varphi_\sigma, \quad x \in M_{\sigma,\lambda}, \quad d(x,E) \geq \chi_\sigma, \quad t \in I. \tag{3.19}$$

Therefore, by means of (3.15) and (3.19), (3.16) implies

$$D_p^+ h(x,t) \leq \varphi_\sigma + m|p(t)|, x \in M_{\sigma,\lambda}, \quad d(x,E) \geq \chi_\sigma, \quad t \in I. \tag{3.20}$$

Now, choosing suitably the positive numbers $\mu(\sigma)$, $\gamma(\sigma)$ and $\eta(\sigma)$, we are able to show the function

$$V_\sigma(x,t) = v(x,t) + \mu(\sigma) h(x,t) \tag{3.21}$$

fulfils the conditions (3.7)–(3.9). In fact, the first two are a consequence of (3.14) for every $\mu > 0$. For the last one hypotheses (j) and (jj) imply that along the solutions of system (2.2) it is

$$D_p^+ V_\sigma(x,t) \leq -hc(d(x,E)) + a(d(x,E))|p(t)| + \mu(\sigma)D_p^+ h(x,t); \qquad (3.22)$$

from here, writing for simplicity $q = |p(t)|$, $d = d(x,E)$ and using (3.18) and (3.20), for every $x \in M_{\sigma,\lambda}$, $t \in I$ it follows

$$D_p^+ V_\sigma(x,t) \leq -hc(d) + qa(d) + \mu(\sigma)\left[-\tfrac{3}{4}\beta + mq\right] \qquad (3.23)$$
$$\text{as} \quad d \in (0, \chi_\sigma),$$

$$D_p^+ V_\sigma(x,t) \leq -hc(d) + qa(d) + \mu(\sigma)[\varphi_\sigma + mq] \qquad (3.24)$$
$$\text{as} \quad d \geq \chi_\sigma.$$

Formulas (3.23) and (3.24) show the second member of $D_p^+ V_\sigma(x,t)$ has the properties of the function $g(d)$ of Lemma 3.1. Therefore, considering (3.3), where $\chi = \chi_\sigma$ and $\varphi = \varphi_\sigma$, we get $\mu(\sigma)$, $\gamma(\sigma)$, $\eta(\sigma)$ defined as

(a) $6\mu(\sigma)[\varphi_\sigma + \beta/4] = hc(\chi_\sigma)$,

(b) $\gamma(\sigma) = \mu(\sigma)\beta/4$, $\qquad$ (3.25)

(c) $\eta(\sigma) = \min[h/3, \mu(\sigma)\beta/4\hat{y}, \beta/4m]$

and such that the function (3.21) verifies

$$D_p^+ V_\sigma(x,t) \leq -\gamma(\sigma) \quad \text{as} \quad x \in M_{\sigma,\lambda}, \quad t \in I \quad \text{and} \quad |p(t)| \leq \eta(\sigma). \qquad (3.26)$$

Moreover, when $d(x,E) \geq \tau$, $x \in M_\lambda$, $t \in I$, it is $V_\sigma(x,t) = v(x,t)$ and hence, considering the function $y(d) = -c(d) + a(d)$, formula (3.22) becomes

$$D_p^+ V_\sigma(x,t) \leq -(h-q)c(d) + qy(d).$$

Referring to the properties (i), (ii) of $y(d)$ quoted in Lemma 3.1, suppose $\tau < d_1$. As $\tau > \chi_\sigma$ and $q \leq h/3$ and moreover by (3.25)(c), (3.6) it results $qy \leq \mu\beta/4 < hc(\chi_\sigma)/6$ and therefore we derive

$$D_p^+ V_\sigma(x,t) \leq -2hc(\chi_\sigma)/3 + hc(\chi_\sigma)/6 = -hc(\chi_\sigma)/2.$$

If $\tau \geq d_1$, it is $y(d) < 0$ and therefore

$$D_p^+ V_\sigma(x,t) \leq -(h-q)c(d) \leq -2hc(\chi_\sigma)/3.$$

In both these possibilities (3.9) is obtained by virtue of (3.6) and by (3.25)(b). Finally, the property $\mu(\sigma) + \eta(\sigma) + \gamma(\sigma) \to 0$ as $\sigma \to 0$ is a direct consequence of definitions (3.25).

Thus the proof is achieved.

4 Stability Theorems

Theorem 4.1 *Suppose for the system (2.1) there exist the constants $H > 0$, $\tau > 0$, the functions $s \in \mathcal{K}$ and $\beta(d)$, $\alpha(d)$ strictly decreasing in $(0, \gamma)$ such that $0 < \alpha(d) < \beta(d)$, $\alpha \to +\infty$ for $d \to 0^+$, and a bounded set $E \subset M_\lambda$, $\partial M \cap E \neq \emptyset$. Moreover, admit for every $\sigma > 0$ there exist the numbers $\mu_\sigma > 0$, $\eta_\sigma > 0$, $\mu_\sigma + \eta_\sigma \to 0$ for $\sigma \to 0$, and a function $V_\sigma(x,t) : (M_\lambda \setminus M) \times I \to \Re$ satisfying the following conditions:*

(i) $V_\sigma(x,t) \geq -\beta(d(x,M)) - \mu_\sigma H$ *for* $x \in M_{\sigma,\lambda}$, $t \in I$;
(ii) $V_\sigma(x,t) \leq -\alpha(d(x,M)) + s(d(x,E)) + \mu_\sigma H$ *for* $x \in M_{\sigma,\lambda}$, $t \in I$;
(iii) $D_p^+ V_\sigma(x,t) \leq 0$ *for* $|p(t)| \leq \eta_\sigma$, *when* $t \in I$ *and further* $x \in M_{\sigma,\lambda}$ *or* $x \in M_\sigma \setminus M$ *and* $d(x,E) \geq \tau$.

Then M is uniform-total stable set of system (2.1).

Proof Consider a point $x^* \in \partial M \cap E$ and choose $\rho > 0$, $\varepsilon \in (0, \lambda)$ and $\delta \in \left(0, \alpha^{-1}(\beta(\varepsilon) + s(\rho + \tau))\right)$. Owing to the strict decreasing of $\alpha(d)$, it is

$$\alpha(\delta) > \beta(\varepsilon) + s(\rho + \tau)$$

and therefore for some $\mu > 0$ we get

$$\alpha(\delta) > \beta(\varepsilon) + s(\rho + \tau) + 2\mu H. \tag{4.1}$$

On the other hand, there exist a $\sigma \in (0, \delta)$ and a function $V_\sigma(x,t)$ verifying the hypotheses of theorem with $\mu_\sigma \leq \mu$. Therefore, from (4.1) it results $\alpha(\delta) > \beta(\varepsilon) + s(\rho + \tau) + 2\mu_\sigma H$ and also

$$-\beta(\varepsilon) - \mu_\sigma H > -\alpha(\delta) + s(\rho + \tau) + \mu_\sigma H. \tag{4.2}$$

Now we show that the solution $x(t)$ of system (2.2), corresponding to the perturbation term $|p(t)| \leq \eta_\sigma$ and to the initial condition $t_0 \in I$, $x_0 \in (M_\delta \setminus M) \cap B(x^*, \rho)$, fulfils

$$x(t) \in M_\varepsilon, \quad \forall t \in [t_0, \omega).$$

On the contrary, suppose there exist the instants $t_2 > t_1 > t_0$ satisfying the properties $d(x(t_1), M) = \delta$, $d(x(t_2), M) = \varepsilon$ and $\delta < d(x(t), M) < \varepsilon$ for $t_1 < t < t_2$. From hypothesis (i) we derive

$$V_\sigma(x(t_2), t_2) \geq -\beta(d(x(t_2), M)) - \mu_\sigma H = -\beta(\varepsilon) - \mu_\sigma H. \tag{4.3}$$

Let us set

$$t_3 = \sup\left\{t \leq t_1 : x(t) \in [E_\tau \cup B(x^*, \rho)] \cap M_\delta\right\}.$$

If it is $t_3 = t_1$, then $x(t_3) = x(t_1) \in \partial M_\delta$ and moreover $x(t_3) \in \bar{E}_\tau$ or $x(t_3) \in \bar{B}(x^*, \rho)$ and therefore $d(x(t_3), E) < \rho + \tau$. Hence, using hypotheses (iii), (ii) and (4.2), it follows

$$V_\sigma(x(t_2), t_2) \leq V_\sigma(x(t_1), t_1) \leq -\alpha(d(x(t_1), M)) + s(d(x(t_1), E)) + \mu_\sigma H$$
$$< -\alpha(\delta) + s(\rho + \tau) + \mu_\sigma H < -\beta(\varepsilon) - \mu_\sigma H$$

that contradicts (4.3).

In the case $t_3 < t_1$ it results $d(x(t_3), M) \leq \delta$ and further for $t_3 < t < t_1$ we get $x(t) \notin E_\tau \cap M_\delta$ and thus (iii) holds. Consequently, by (ii) and (4.2), we deduce

$$V_\sigma(x(t_2), t_2) \leq V_\sigma(x(t_1), t_1) \leq V_\sigma(x(t_3), t_3)$$
$$\leq -\alpha(d(x(t_3), M)) + s(d(x(t_3), E)) + \mu_\sigma H$$
$$< -\alpha(\delta) + s(\rho + \tau) + \mu_\sigma H < -\beta(\varepsilon) - \mu_\sigma H$$

and again we have a contradiction for (4.3).

Theorem 4.2 *Suppose for the system (2.1):*

(I) *there are a function $v: (M_\lambda \setminus M) \times I \to \Re$, a bounded set $E \subset M_\lambda \setminus M$ with $E \cap \partial M \neq \emptyset$, a family of functions $\{w_y\}$, $w_y: B(y, \xi_y) \times I \to \Re$, so that all the hypotheses of Lemma 3.2 are verified;*

(II) *there exist the functions $s \in \mathcal{K}$, $\alpha(d)$ and $\beta(d)$ strictly decreasing on $(0, \lambda)$, $0 < \alpha(d) < \beta(d)$ and $\alpha \to +\infty$ for $d \to 0^+$, such that for $v(x, t)$ it is*

$$v(x, t) \geq -\beta(d(x, M)),$$
$$v(x, t) \leq -\alpha(d(x, M)) + s(d(x, E));$$

(III) $v(x, t) \to +\infty$ *as $x \to \infty$, $x \in M_\lambda \setminus M_\sigma$, $t \in I$, where σ is any constant satisfying $0 < \sigma < \lambda$.*

Then M is a total-asymptotically stable set of (2.1).

Proof By Lemma 3.2 and hypotheses (I)–(II) for every σ we get a function V_σ satisfying the hypotheses of Theorem 4.1. Consequently M is a total-uniform-stable set of system (2.1). Therefore, given $\varepsilon_0 > 0$ and for any $\rho > 0$ there exist two positive numbers $\delta(\varepsilon_0, \rho) = \Delta$, $\eta(\varepsilon_0) = \eta_0$ such that if $t_0 \in I$, $x_0 \in M_\Delta \cap B(x^*, \rho)$ and $x^* \in \partial M \cap E$, $|p(t)| \leq \eta_0$, then the solution $x(t): J \to \Re$, $x(t_0) = x_0$, of equation (2.2) satisfies

$$d(x(t), M) < \varepsilon_0 \quad \text{for all} \quad t \in [t_0, +\infty) \cap J.$$

Owing to the properties of functions α, β, s quoted in (II), given $\nu > 0$, $\rho > 0$, $\tau > 0$ there is a $\nu' \in (0, \nu)$ satisfying the inequality $\alpha(\nu') > \beta(\nu) + s(\rho + \tau)$;

from here, corresponding to the number $H > 0$ obtained in Lemma 3.2, there is a $\mu > 0$ such that

$$\alpha(\nu') > \beta(\nu) + s(\rho + \tau) + 2\mu H. \tag{4.4}$$

Referring to the results and symbols of Lemma 3.2, we consider the property $\mu_\sigma \to 0$ for $\sigma \to 0^+$ of the function $\mu(\sigma)$ and therefore there is a $\sigma(\nu, \rho) < \nu'$ satisfying $\mu_\sigma < \mu$. As the function α is decreasing, from (4.4) we obtain

$$\alpha(\sigma) > \beta(\nu) + s(\rho + \tau) + 2\mu_\sigma H. \tag{4.5}$$

Furthermore, from (3.3) we derive the function $\eta(\sigma(\nu, \rho)) = \eta(\nu, \rho)$ and consequently we consider in equation (2.2) that $p(t)$ satisfies

$$|p(t)| \leq \min[\eta(\nu, \rho), \eta_0]. \tag{4.6}$$

We show for any $\nu > 0$, $\rho > 0$ and for $t_0 \in I$, $x_0 \in M_\Delta \cap B(x^*, \rho)$, there is a $T = T(x_0, t_0, \nu, \rho) > 0$ such that the solution $x(t)$ of (2.2) satisfying (4.6) and the initial conditions $x(t_0) = x_0$ fulfils

$$d(x(t), M) < \nu \quad \text{for all} \quad t \geq t_0 + T, \quad t \in J. \tag{4.7}$$

For this purpose, we discuss the following alternative involving $x(t)$.

A – There is a $\bar{t} \in J$ such that $x(\bar{t}) = \bar{x} \in (M_\sigma \setminus M) \cap (B(x^*, \rho) \cup E_\tau)$. We have $x(t; t_0, x_0) = x(t; x(\bar{t}; t_0, x_0), \bar{t}) = x(t; \bar{x}, \bar{t}) = \bar{x}(t)$. We claim $\bar{x}(t)$ verifies condition (4.7). Assume, on the contrary, that there are $t_2 > t_1 > \bar{t}$ satisfying

$$d(\bar{x}(t_1), M) = \sigma, \quad d(\bar{x}(t_2), M) = \nu, \quad \text{and}$$
$$\sigma < d(\bar{x}(t), M) < \nu \quad \text{for} \quad t_1 < t < t_2.$$

By Lemma 3.2 we get a number $H > 0$ and a function V_σ verifying conditions (3.7)–(3.9). In our hypotheses it results

$$V_\sigma(\bar{x}(t_2), t_2) \geq -\beta(d(\bar{x}(t_2), M)) - \mu_\sigma H. \tag{4.8}$$

Now we set $t_3 = \sup\{t \leq t_1 : \bar{x}(t) \in M_\sigma \cap [B(x^*, \rho) \cup E_\tau]\}$ and observe that the possibilities $t_3 = t_1$, $t_3 < t < t_1$ can be examined employing the same arguments used in the proof of Theorem 4.1. Thus (4.8) is violated and hence (4.7) holds.

B – For every $t \in J$ it results $x(t) \notin (M_\sigma \setminus M) \cap [B(x^*, \rho) \cup E_\tau]$. Given $\varepsilon \in [\sigma, \nu]$ suppose there is a $t' \in J$ such that

$$d(x(t), M) \geq \varepsilon, \quad \forall t \in [t', \omega). \tag{4.9}$$

We examine the following possibilities involving ω. If it is $\omega < +\infty$, $x(t)$ cannot be bounded. Suppose that $x(t)$ lies for $t \in [t_0,\omega)$ on a compact set of M_{ε_0}, then $x(t)$ is continuable to $t = \omega$ and this is clearly a contradiction. Then $x(t)$ might be unbounded. Consequently there is a sequence $\{t_j\} \to \omega$ as $j \to +\infty$ such that $\{x(t_j)\} \to \infty$. According to Lemma 3.2 we get a function $V_\sigma(x,t)$ satisfying conditions (3.7)–(3.9). By (III) and (4.9) for some $t_n > t'$ it results

$$V_\sigma(x(t_n), t_n) > V_\sigma(x(t'), t')$$

and this is a contradiction since $V_\sigma(x(t), t)$ is decreasing.

Suppose now $\omega = +\infty$. Again we are able to obtain a function $V_\sigma(x, t)$ inferiorly bounded in $M_{\varepsilon,\varepsilon_0}$ satisfying (3.9) and thus $V_\sigma(x(t), t) \to -\infty$ as $t \to +\infty$ and this again leads to a contradiction. Then there exists a $t'' \in [t', \omega)$ such that

$$d(x(t''), M) < \varepsilon. \tag{4.10}$$

As ε is arbitrary, we deduce M is a weak attractor.

If for $t > t''$ it is $d(x(t''), M) < \nu$, then the theorem is proved. Otherwise, according to (4.10), there exist two sequences $\{t'_j\}$ and $\{t''_j\}$ with $t'_j < t''_j < t'_{i+1} < \cdots < \omega$ such that

$$\sigma < d(x(t), M) < \nu \quad \text{for} \quad t'_i < t < t''_i \quad \text{and} \quad d(x(t'_i), M) = \sigma, \quad d(x(t''_i), M) = \nu.$$

Setting $\gamma^+ = \{x \in \mathfrak{R}^n / x = x(t), \ t \in [t_0, \omega)\}$, we show $\gamma^+ \cap M_{\sigma,\nu}$ is bounded. Suppose that it is not so. By (III) for a given $k > 0$ there is a $\delta_k > 0$ such that

$$d(x, E) > \delta_k, \quad x \in M_\sigma \backslash M_\nu, \quad t \in I \quad \Longrightarrow \quad v(x, t) > k.$$

Therefore, there is a sequence $\{t_j\} \to \omega$ as $j \to +\infty$ such that $\{x(t_j)\} \to \infty$. Consequently, for some j the function $V_\sigma(x,t)$ fulfils

$$V_\sigma(x(t_j), t_j) > k = V_\sigma(x(t_1), t_1)$$

and this inequality contradicts the decreasing of $V_\sigma(x(t), t)$. Thus there is a compact set S of $M_{\sigma,\lambda}$ such that $\gamma^+ \cap M_{\sigma,\nu} \subset S$. From (vj) of Lemma 3.2 there is a constant K_σ satisfying $|f(x,t)| \le K_\sigma$ for $(x,t) \in S \times I$, and therefore we have $t''_i - t'_i \ge (\nu - \sigma)/K_\sigma$ and hence $\omega = +\infty$. Consequently

$$V_\sigma(x(t''_j), t''_j) - V_\sigma(x(t'_j), t'_j) = \int_{t'_j}^{t''_j} D^+ V_\sigma(x(t), t)\, dt \le -\gamma(\sigma)(\nu - \sigma)/K_\sigma$$

and hence

$$\sum_{j=1,\ldots,n} [V_\sigma(x(t''_j), t''_j) - V_\sigma(x(t'_j), t'_j)] \le -n\gamma(\sigma)(\nu - \sigma)/K_\sigma.$$

From here, owing to the inequality $V_\sigma(x(t_j''), t_j'') > V_\sigma(x(t_{j+1}'), t_{j+1}')$ we get $V_\sigma(x(t_n''), t_n'') - V_\sigma(x(t_1'), t_1') < -n\gamma(\sigma)(\nu-\sigma)/K_\sigma$ and thus we deduce $V_\sigma(x(t_n''), t_n'') \to -\infty$ as $n \to +\infty$ which contradicts the boundedness of V_σ on $S \times I$ derived from (II). Thus the proof is complete.

5 Example

Consider in some reference frame $Oxyz$ a particle P of unitary mass subject to a friction force arising from an atmosphere at rest in $Oxyz$ and attracted towards the centre O by the Newton force. We denote by $\boldsymbol{r} = $ OP the vector joining O to the position P of P, by $\boldsymbol{v}$ the velocity of P and by r and v the respective Euclidean norms. Moreover, let $Oxyz$ be rotating uniformly in the inertial frame $O\xi\eta\zeta$ around the fixed axis $z \equiv \zeta$ with angular velocity Ω. In our model $-f(\boldsymbol{r},\boldsymbol{v})\boldsymbol{v}/v$ and $-g\mathcal{M}r^{-3}\boldsymbol{r}$ are respectively the dissipative and the Newton forces. We suppose $f(\boldsymbol{r},\boldsymbol{v}) \geq hc_1(v)$, where h is a positive constant and $c_1 \in \mathcal{K}$, g is the gravitational constant and $\mathcal{M}$ is the mass of centre O. Furthermore, on P are acting the forces Ω^2QP, QP is the vector joining the orthogonal projection Q of P on z to P, and $-2\boldsymbol{\Omega} \times \boldsymbol{v}$ because $Oxyz$ is a non inertial reference frame. The motions of P satisfy the differential equations

$$\dot{\boldsymbol{r}} = \boldsymbol{v},$$
$$\dot{\boldsymbol{v}} = -g\mathcal{M}r^{-3}\boldsymbol{r} + \Omega^2\text{QP} - 2\boldsymbol{\Omega} \times \boldsymbol{v} - f(\boldsymbol{r},\boldsymbol{v})\boldsymbol{v}. \tag{5.1}$$

We consider as Liapunov function the total energy

$$V = -\Omega^2(x^2 + y^2)/2 - g\mathcal{M}/r + v^2/2. \tag{5.2}$$

The system (5.1) and the function (5.2) are not defined on the set $M = \{(\boldsymbol{r}_M, \boldsymbol{v}_M) : \boldsymbol{r}_M = \boldsymbol{0}, \boldsymbol{v}_M = \boldsymbol{v}\}$. Let us introduce the set $E = \{(\boldsymbol{r}_E, \boldsymbol{v}_E) : \boldsymbol{r}_E = \boldsymbol{r}, r < \lambda, \boldsymbol{v}_E = \boldsymbol{0}\}$ and observe that it is $d[(\boldsymbol{r},\boldsymbol{v}), M] = \inf\left[|\boldsymbol{r} - \boldsymbol{r}_M|^2 + |\boldsymbol{v} - \boldsymbol{v}_M|^2\right]^{1/2} = r$ and $d[(\boldsymbol{r},\boldsymbol{v}), E] = \inf\left[|\boldsymbol{r} - \boldsymbol{r}_E|^2 + |\boldsymbol{v} - \boldsymbol{v}_E|^2\right]^{1/2} = v$; thus $E \cap M = \{\boldsymbol{0}, \boldsymbol{0}\}$. Now we suppose it is acting on P an additional force written, without loss of generality, in the form $\boldsymbol{p}(t) : I \to \Re^3$. Therefore, we associate to (5.1) the perturbed system

$$\dot{\boldsymbol{r}} = \boldsymbol{v},$$
$$\dot{\boldsymbol{v}} = -g\mathcal{M}r^{-3}\boldsymbol{r} + \Omega^2\text{QP} - 2\boldsymbol{\Omega} \times \boldsymbol{v} - f(\boldsymbol{r},\boldsymbol{v})\boldsymbol{v} + \boldsymbol{p}(t). \tag{5.3}$$

Along the solutions of (5.3) it results

$$\dot{V} = -f(\boldsymbol{r},\boldsymbol{v})v + \boldsymbol{p}(t) \cdot \boldsymbol{v}.$$

We choose a constant $\lambda \in (0, (g\mathcal{M}\Omega^{-2})^{1/3})$, introduce the function $w(r, v) = v$ and remark that by (5.1) it is $|\dot{w}| \geq g\mathcal{M} \; r^{-2} - \Omega^2 QP > 0$ for $v = 0$ and $r < \lambda$. Therefore, it is possible to associate to every point $(r_E, v_E) \in E$ a non vanishing component of w. Thus we obtain a family of (scalar) functions $\{w_{(r_E, v_E)}\}$ verifying conditions (jjj)–(v) of Lemma 3.2. Moreover, setting $\beta(d[(r,v), M]) = \Omega^2 r^2/2 + g\mathcal{M}/r$, $\alpha(d[(r,v), M]) = g\mathcal{M}/r$, $s(d[(r,v), E]) = v^2/2$, all the conditions of Theorem 4.2 are fulfilled. Therefore, the set M is total-asymptotically stable of (5.1).

If we compute λ supposing the Sun as point O we have $\lambda \cong 30 \times 10^6$ km. Therefore, the region of attraction of the Sun is of the same order of size of its distance from Mercury (58×10^6 km). This result may give a simple explanation of the absence of any planet within Mercury orbit.

References

[1] Barbashin, E.A. and Krasovskii, N.N. (1952). On the stability of a motion in the large. *Dokl. Akad. Nauk SSSR*, **86**, 453–456 (Russian).

[2] La Salle, J.P. (1962). Asymptotic stability criteria. *Proc. Symp. Appl. Math.*, **13**, 299–303.

[3] Matrosov, V.M. (1962). On the stability of motion. *Prikl. Mat. Mekh.*, **26**, 885–895 (Russian).

[4] Corduneanu, C. (1960). Application of differential inequalities to theory of stability. *An. Sti. Univ. "Al. I. Cuza" Iasi, Sect. Ia Mat.*, **6**, 47–58.

[5] Yoshizawa, T. (1969). *Stability Theory by Liapunov's Second Method*. The Math. Society of Japan.

[6] Rouche, N. (1968). On the stability of motion. *Int. J. Non-Linear Mech.*, **3**, 295–306.

[7] Salvadori, L. (1968). Sulla stabilità del movimento. *Le Matematiche*, **XXIV**, 218–239.

[8] Salvadori, L. (1971). Famiglie ad un parametro di funzioni di Liapunov nello studio della stabilita. *Ist. Naz. Alta Mat.*, **VI**, 309–330.

[9] Matrosov, V.M. (1971). Vector Liapunov functions in the analysis of nonlinear interconnected systems. *Istituto Naz. Alta. Mat.*, **VI**, 209–242.

[10] Gambardella, L. and Tenneriello, L. (1971). On a theorem of N. Rouche. *Rend. Acc. Scienze Fis. e Mat. Napoli, serie 4*, **XXXVIII**, 145–150.

[11] Rouche, N. (1971). Attractivity of certain sets proved by using several Liapunov functions. *Istituto Naz. Alta. Mat.*, **VI**, 331–343.

[12] Fergola, P. and Moauro, V. (1972). On the stability of a set with respect to one of its subsets. *Ricerche di Matematica*, **XXI**, 161–175.

[13] D'Anna, A. (1973). Asymptotic stability proved by using vector Liapunov functions. *Ann. Soc. Sci. Bruxelles, II*, **87**, 119–139.

[14] Salvadori, L. (1974). Some contributions to asymptotic stability theory. *Ann. Soc. Sci. Bruxelles, I*, **88**, 183–194.

[15] Corne, J.L. and Rouche, N. (1973). Attractivity of closed sets proved by using a family of Liapunov functions. *J. Diff. Eqns*, **13**, 231–246.

[16] Habets, P. and Risito, C. (1973). Stability criteria for systems with first integrals generalizing theorems of Routh and Salvadori. *Equa-Diff. 73*, Bruxelles Louvain-La-Neuve Hermann, 569–580.

[17] Habets, P. and Peiffer, K. (1975). Attractivity concepts and vector Liapunov functions. *Nonlin. Vibr. Problems*, **16**, 35–52.

[18] D'Anna, A. (1977). Proving conditional attractivity of a close set with a family of Liapunov functions. *Int. J. Non-Linear Mech.*, **12**, 103–111.

[19] D'Anna, A. (1993). Stability properties for perturbed singular systems. *7th Conf. on Waves and Stability in Continuous Media, Bologna 4–9 ottobre 1993, Series on Adv. for Appl. Sci.*, **23**, 110–117.

2.3 STABILITY THEORY OF VOLTERRA DIFFERENCE EQUATIONS*

F. DANNAN[1†], S. ELAYDI[1] and P. LI[2]

[1] *Department of Mathematics, Trinity University, San Antonio, USA*
[2] *Department of Mathematics and Statistics, The Flinders University of South Australia, Adelaide, Australia*

1 Introduction

Stability theory of Volterra differential and integro-differential equations has been extensively investigated in the literature [1, 3, 4, 10, 21, 22, 23, 26, 28, 29]. A parallel theory for Volterra difference equations is still under development. In this survey, we present the current state of affairs of this theory. In a series of papers, the first author and his collaborators made an earnest effort to build a solid foundation for Volterra difference equations [12–18]. Among other things, the use of the z-transform methods for equations with convolutions has been very fruitful and has produced some of the most beautiful results in the subject. A resolvent matrix has been defined and a variation of constants formula has been developed. Significant results were produced by Kolmanovskii and his collaborators in [6–9]. Their results include interesting applications to numerical methods of Volterra integral equations. In [5] the authors developed a resolvent matrix different from ours and used it to construct a variation of constants formula. Relevant results may also be found in [19, 20, 30–33]. A readable introductory account on the subject may be found in [12, 24, 25]. Although applications to numerical methods are important, we chose not to include them in this article. We refer the interested reader to the references [2, 26, 27]. We end the paper with some open problems (see Section 8).

Advances in Stability Theory (Ed.: A.A. Martynyuk). Stability and Control: Theory, Methods and Applications, Taylor & Francis, London, **13** (2003) 89–106.
†On leave from University of Damascus and partially supported by a Fulbright Grant.

2 Z-transform Methods

Consider the Volterra difference equation of convolution type

$$x(n+1) = Ax(n) + \sum_{r=0}^{n} B(n-r)x(r), \quad n \in \mathbb{Z}^+ \tag{2.1}$$

and its perturbation

$$y(n+1) = Ay(n) + \sum_{r=0}^{n} B(n-r)y(r) + g(n), \quad n \in \mathbb{Z}^+, \tag{2.2}$$

where A is a $k \times k$ nonsingular matrix, $B(n) \in \ell^1(\mathbb{Z}^+)$ is a $k \times k$ matrix function, and $g(n)$ is a vector function in $\mathbb{R}^k$.

To obtain a variation of constants formula for equation (2.2), we define the resolvent matrix $R(n)$ of equation (2.1) as the unique solution of the matrix equation

$$R(n+1) = AR(n) + \sum_{j=0}^{n-1} B(n-j)R(j), \quad n \in \mathbb{Z}^+ \tag{2.3}$$

with $R(0) = I$, the identity matrix.

Note that $R(n)$ is the counterpart of the fundamental matrix in ordinary difference and differential equations; its columns are the vector solutions $x_i(n)$, $1 \leq i \leq k$ of equation (2.1) with $x_i(0) = e_i$, where e_i is the standard i-th unit vector in $\mathbb{R}^k$. The Z-transform [12] of a sequence $x(n)$ is defined as

$$\tilde{x}(n) = z[x(n)] = \sum_{n=0}^{\infty} x(n)z^{-n}, \quad |z| > d, \tag{2.4}$$

where $z \in \mathbb{C}$, and d is the radius of convergence of $\tilde{x}$.

Taking the z-transform of both sides of equation (2.3) yields

$$\left[zI - A - \tilde{B}(z)\right]\tilde{R}(z) = zI, \quad |z| > d.$$

Since zI is nonsingular, it follows that

$$\tilde{R}(z) = z[zI - A - \tilde{B}(z)]^{-1}, \quad |z| > d. \tag{2.5}$$

Now taking the Z transform of both sides of equation (2.2) and using equation (2.5) yields

$$\tilde{y}(z) = \tilde{R}(z)y_0 + \frac{1}{z}\tilde{R}(z)\tilde{g}(z), \quad |z| > d.$$

And by taking the inverse z-transform we obtain the promised variation of constants formula

$$y(n, 0, y_0) = R(n)y_0 + \sum_{j=0}^{n-1} R(n-j-1)g(j). \tag{2.6}$$

We are now ready to present the Fundamental Theorem of Stability of Equations of Convolution type.

Theorem 2.1 [16, 17] *The following statements are equivalent for equation (2.1).*

1. $\det(zI - A - \tilde{B}(z)) \neq 0$, for $|z| \geq 1$;
2. $R(n) \in \ell^1(\mathbb{Z}^+)$;
3. *The zero solution is uniformly asymptotically stable* [12];
4. *Both $R(n)$ and $h(n)$ tend to zero as n approaches infinity, where*

$$h(n) := \sum_{i=0}^{\infty} \left| \sum_{j=0}^{n-1} R(n-j-1)B(j+i+1) \right|.$$

A word of caution is now in order. Although the resolvent matrix $R(n)$ shares many nice features of fundamental matrices, it fails however in satisfying an important property namely, the semigroups property. It was pointed out in Lemma 3 of [17] that the statement $R(n-s)R(s) = R(n)$ is in general false, and it is true if and only if $B(n) \equiv 0$ (which makes equation (2.1) an ordinary difference equation).

Such a failure of the semigroups property paved the way to conclude that for Equation (2.1) uniform asymptotic stability does not imply exponential stability. The question now is when are these latter notions equivalent. For integro-differential equations, the same question was raised by Corduneanu and Lakshmikanthan [4] and was successfully answered for the scalar case by Murakami [29]. The following result is the discrete analogue.

Theorem 2.2 [17] *Suppose that the zero solution of equation (2.1) is uniformly asymptotically stable. Then it is exponentially stable if and only if $B(n)$ decays exponentially, i.e., $\|B(n)\| \leq M\nu^n$, $n \in \mathbb{Z}^+$, for some $M > 0$ and $\nu \in (0,1)$.*

Explicit criteria for stability using Theorem 2.1(1) may be found in [12, 14]. Here is one of the main results in this direction.

Theorem 2.3 *Let $A = (a_{ij})$ and $B(n) = (b_{ij}(n))$ such that*

$$\beta_{ij} = \sum_{n=0}^{\infty} |b_{ij}(n)| < \infty.$$

Then the zero solution of equation (2.1) is uniformly asymptotically stable if either one of the following conditions hold.

$$\text{(a)} \quad \sum_{j=1}^{k}(|a_{ij}| + \beta_{ij}) < 1, \quad 1 \leq i \leq k, \tag{2.7}$$

$$\text{(b)} \quad \sum_{i=1}^{k}(|a_{ji}| + \beta_{ji}) < 1, \quad 1 \leq j \leq k. \tag{2.8}$$

Note that if A and $B(n)$ are scalar, then condition (2.7) reads as

$$|A| + \sum_{n=0}^{\infty} |B(n)| < 1. \tag{2.9}$$

Criterion (2.9) provides us with a useful tool to illustrate the intriguing Theorem 2.2. Consider the scalar difference equation

$$x(n+1) = \frac{1}{3} x(n) + \sum_{j=0}^{n} \frac{x(j)}{(n-j+1)(n-j+3)}. \tag{2.10}$$

Then $A = \frac{1}{3}$, $B(n) = \frac{1}{(n+1)(n+3)} \in \ell^1(\mathbb{Z}^+)$. Moreover, $A + \sum_{n=0}^{\infty} B(n) = \frac{1}{3} + \frac{1}{2} < 1$, which implies by criterion (2.9) that the zero solution of equation (2.10) is uniformly asymptotically stable.

But by Theorem 2.1, the zero solution is not exponentially stable. For a negative statement about stability we have the following result from [14]. Let us assume that

$$\nu_{ij} = \sum_{n=0}^{\infty} b_{ij}(n) < \infty.$$

Theorem 2.4 *Suppose that the following statements hold:*

1. $a_{ii} + \nu_{ii} > 1$, $1 \leq i \leq k$,

2. $(a_{ii} + \nu_{ii} - 1)(a_{jj} + \nu_{jj} - 1) > \sum_{r}' |a_{ir} + \nu_{ir}| \sum_{r}' |a_{jr} + \nu_{jr}|$,

   where

$$\sum_{r}' a_{ir} = \sum_{r=1}^{k} a_{ir} - a_{ii}.$$

Then if k is odd, the zero solution of equation (2.1) is not asymptotically stable. If k is even, then the zero solution of equation (2.1) may or may not be asymptotically stable.

In the scalar case there are much more powerful results which we now report.

Theorem 2.5 [14] *Suppose that A and $B(n)$ are scalar. Then the zero solution of equation (2.1) is not asymptotically stable if any one of the following conditions hold:*

1. $A + \sum_{n=0}^{\infty} B(n) \geq 1$,

2. $A + \sum_{n=0}^{\infty} B(n) \leq -1$ and $\sum_{n=0}^{\infty} B(n) > 0$,

3. $A + \sum_{n=0}^{\infty} B(n) < -1$, $\sum_{n=0}^{\infty} B(n) < 0$, and $\sum_{n=0}^{\infty} B(n)$ is sufficiently small.

3 Renewal Equations

Consider a sequence of repeated trials with possible outcomes E_j ($j = 1, 2, \ldots$). We agree that the expression "ξ occurs at the n-th place in the sequence $E_{j1}, E_{j2}, \ldots$" is an abbreviation for "the subsequence $E_{j1}, E_{j2}, \ldots$ has the attribute ξ." This convention implies that the occurrence of ξ at the n-th trial depends solely on the outcome of the first n trials. We adopt the following notation from Feller [20]. Let

$$x(n) = P\{\xi \text{ occurs at the } n\text{-th trial}\},$$
$$b(n) = P\{\xi \text{ occurs for the first time at the } n\text{-th trial}\}.$$

It is convenient to set $b(0) = 0$ and $x(0) = 1$. Then the probability that ξ occurs for the first time at trial number and then again at a later trial $n > j$ is, by definition, $b(j)x(n-j)$. Moreover, the probability that ξ occurs at the n-th trial for the first time is $b(n) = b(n)x(0)$. Since these cases are mutually exclusive,

$$x(n) = \sum_{j=1}^{n} b(j)x(n-j), \quad n \geq 1. \tag{3.1}$$

To proceed further, we need a couple of definitions from Feller [20].

Definition 3.1 A *recurrent event* ξ will be called *persistent* if $b = \sum_{n=1}^{\infty} b(n) = 1$ and *transient* if $b < 1$. It is periodic if there exists an integer $m > 1$ such that ξ can occur only at trials number $m, 2m, 3m, \ldots$, i.e. $x(n) = 0$ whenever n is not divisible by m. The greatest m with this property is called the period of ξ.

Theorem 3.1 [20] *The following statements hold true.*

(a) *For ξ to be transient, it is necessary and sufficient that*

$$x = \sum_{n=0}^{\infty} x(n)$$

is finite.

(b) *Let ξ be persistent and not periodic and denote it by $\mu = \sum_{j=1}^{\infty} jb(j)$. Then*

$$x(n) \to \mu^{-1} \quad \text{as} \quad n \to \infty.$$

(c) *If ξ is persistent and has period s, then $x(ns) = \frac{s}{\mu}$ while $x(k) = 0$ for every k not divisible by s.*

Proof The proof by Feller was accomplished via the use of generating functions. Since we have already used the z-transform in Section 2, we found it more appropriate to adapt the proof to our methods. Notice that $x(n)$ in the left side of equation (3.1) misses $x(0)$. Hence $\tilde{x}(z) = \sum_{j=1}^{\infty} x(j)z^{-j} + x(0) = \sum_{j=1}^{\infty} x(j)z^{-j} + 1$. Now taking the z-transform of both sides of equation (3.1) yields

$$\tilde{x}(z) = \frac{1}{1 - \tilde{b}(z)}. \tag{3.2}$$

Observe that $\tilde{x}(z)$ increases monotonically as z decreases to one. Hence for each N

$$\sum_{n=0}^{\infty} x(n) \leq \lim_{z \to 1} \tilde{x}(z) \leq \sum_{n=0}^{\infty} x(n) = x.$$

Now if ξ is transient, $\tilde{x}(z)$ approaches $(1-b)^{-1}$ as z approaches one. Hence $x = \sum_{n=0}^{\infty} x(n) = \frac{1}{1-b}$. On the other hand, if ξ is persistent, i.e., $b = 1$, then $\tilde{x}(z)$ approaches infinity as z approaches one. This completes the proof of part (a). The proofs of pats (b) and (c) will be omitted.

A more general equation was also introduced by Feller [20]. This equation takes the form

$$y(n) = g(n) + \sum_{j=1}^{n} b(j)y(n-j) \tag{3.3}$$

with the assumption

$$\begin{aligned} b(n) \geq 0, \quad b = \sum_{n=1}^{\infty} b(n) < \infty; \\ g(n) \geq 0, \quad g = \sum_{n=0}^{\infty} g(n) < \infty. \end{aligned} \tag{3.4}$$

Theorem 3.2 [20] *Suppose that equation (3.3) holds and that $b(n)$ is not periodic.*

1. *If $b < 1$, then $y(n) \to 0$ as $n \to \infty$ and*

$$\sum_{n=0}^{\infty} y(n) = \frac{g}{1-b}.$$

2. *If $b = 1$, then $y(n) \to g\mu^{-1}$.*
3. *If $b > 1$, there exists a unique positive root z_0 of the equation $\tilde{b}(z) = 1$, and*

$$z_0^{-n} y(n) \to \frac{-\tilde{g}(z_0)}{z_0 \tilde{b}'(z_0)} \quad \text{as} \quad n \to \infty.$$

(*Clearly* $z_0 > 1$ *and hence* $\tilde{b}'(z_0)$ *is finite. Then* $y(n)$ *behaves like a geometric sequence with ration* $\frac{1}{z_0} > 1$ *and consequently,* $y(n) \to \infty$ *as* $n \to \infty$).

Proof See [20].

4 Liapunov Functions

For equations of nonconvolution type or when the matrix A is not constant, z-transform techniques are not effective in detecting stability. In this situation, the method of Liapunov functions comes to the rescue. A hybrid of Liapunov and z-transform techniques was used successfully by Raffoul [30].

Consider the following difference system of nonconvolution type

$$x(n+1) = A(n)x(n) + \sum_{j=0}^{n} B(n,j)x(j) \qquad (4.1)$$

and its perturbation

$$y(n+1) = A(n)y(n) + \sum_{j=0}^{n-1} B(n,j)y(j) + g(n). \qquad (4.2)$$

We now define the resolvent matrix of equation (3.1) as the unique solution $R(n,m)$ of the matrix equation

$$R(n+1,m) = A(n)R(n,m) + \sum_{r=m}^{n} B(n,r)R(r,m), \qquad n \geq m, \qquad (4.3)$$

with $R(m,m) = I$. Using equation (4.3) yields the following variation of constants formula.

Lemma 4.1 [13] *The unique solution* $y(n, n_0, y_0)$ *of equation (4.2) with* $y(n_0) = y_0$ *is given by*

$$y(n, n_0, y_0) = R(n, n_0)y_0 + \sum_{j=n_0}^{n-1} R(n, j+1)g(j). \qquad (4.4)$$

It is a good exercise to show that $y(n, n_0, y_0)$ in equation (4.4) is indeed a solution of equation (4.2).

In our first application of Liapunov method we assume that

$$\beta_{ij}(n) = \sum_{s=n}^{\infty} |b_{ij}(s,n)| < \infty, \qquad (4.5)$$

and for any $n_0 \geq 0$

$$\sup \sum_{r=0}^{n_0-1} \sum_{s=n_0}^{\infty} |b_{ij}(s,r)| < \infty, \quad 1 \leq i, \quad j \leq k. \tag{4.6}$$

We use here the vector norm $|x| = \sum_{i=1}^{k} |x_i|$. For any $n_0 \in \mathbb{Z}^+$ and initial function $\phi \colon [0, n_0] \to \mathbb{R}^k$, there is a unique solution $y(n, n_0, \phi) \equiv y(n)$ which satisfies equation (4.1) on $[n_0, \infty)$ and $y(n) = \phi(n)$ on $[0, n_0]$. Note that all of our intervals are discrete.

Theorem 4.1 [13] *Suppose that for $1 \leq i \leq k$, $n \geq n_0$,*

$$\sum_{j=1}^{k} [|a_{ji}(n)| + \beta_{ji}(n)] \leq 1 - c \tag{4.7}$$

for some $c \in (0,1)$. If in addition, conditions (4.5) and (4.6) hold, then the zero solution of equation (4.1) is globally uniformly asymptotically stable, and in fact, is exponentially stable.

Proof Define the Liapunov functional as

$$V(n, x(\cdot)) = \sum_{i=1}^{k} \left[|x_i(n)| + \sum_{j=1}^{k} \sum_{r=0}^{n-1} \sum_{s=n}^{\infty} |b_{ij}(s,r)| |x_j(r)| \right].$$

Then we show that

$$\Delta V_{(3.1)}(n, x(\cdot)) \leq -c|x(n)| \leq -cV(n, x(\cdot)).$$

Hence

$$|x(n)| \leq V(n, x(\cdot)) \leq (1-c)^n V(n_0, \phi(\cdot)) \leq M(1-c)^n \|\phi\|,$$

where

$$\|\phi\| = \sup\{|\phi(s)| \colon s \in [0, n_0]\}.$$

A second approach to study stability is through the use of vector Liapunov functions. To simplify our notation let us rewrite equation (4.1) in the form

$$x(n+1) = \sum_{j=0}^{n} C(n,j) x(j), \tag{4.8}$$

where $C(n,m) = A(n) + B(n,m)$ and $C(n,j) = B(n,j)$ for $n \neq j$. We define the absolute value of a matrix $A = (a_{ij})$ as the matrix $|A| = (|a_{ij}|)$, where entries of

$|A|$ are the absolute values of the entries of A. We say that $A \leq C$ if $a_{ij} \leq c_{ij}$ for $1 \leq i, j \leq k$.

We make the following assumption:

(H) For each $n \in \mathbb{Z}^+$, $\sum_{i=0}^{\infty} C(i,n)$ is absolutely convergent, and if

$$C = \sup_{n \geq 0} \left\{ \sum_{i=n}^{\infty} |C(i,n)| \right\},$$

then $|\lambda_j| < 1$ for all eigenvalues λ_j of C.

Theorem 4.2 [15] *Assume that assumption (H) holds. Then the zero solution of equation (3.1) is globally uniformly asymptotically stable.*

Proof Use the vector Liapunov functional

$$V(n, x(\cdot)) = (I - C)^{-1} \left[|x(n)| + \sum_{r=0}^{n-1} \sum_{s=n}^{\infty} |C(s,r)| |x(r)| \right].$$

We observe that the matrix series $\sum_{n=0}^{\infty} C^n$ converges to $(I - C)^{-1}$ if and only if $|\lambda_j| < 1$ for all eigenvalues λ_j of C. Moreover, if $C \geq 0$, then $(I - C)^{-1} \geq 0$.

We now turn our attention to the method of Kolmanovskii and his collaborators [6–9]. The authors consider the scalar equation

$$x(n+1) = -\sum_{i=0}^{n} a(n,i) x(i), \quad n \in \mathbb{Z}^+. \tag{4.9}$$

They defined a $n \times n$ matrix M to be positive definite if

$$z^T M z \geq c z_n^2 \quad \text{for all} \quad z = (z_1, z_2, \ldots, z_n)^T \in \mathbb{R}^n. \tag{4.10}$$

Let $B(n) = (b_{ij}(n))$, $C(n) = (c_{ij}(n))$, $0 \leq i, j \leq n$ be defined by

$$b_{ij}(n) = a(n-1, i+j-n),$$
$$c_{ij}(n) = a(n, i+j-n+1) - a(n-1, i+j-n).$$

Theorem 4.3 [9] *Suppose that the coefficients $a(n,i)$ can be extended in such a way that*

(1) $0 \leq a(n, n+1)$, and $\sup_{n \leq 0} a(n, n+1) < 2$.

(2) The matrix $B(n)$ is positive definite as defined by equation (4.10).

(3) The matrix $C(n)$ is negative semidefinite.

Then the zero solution on equation (4.9) is asymptotically stable. Moreover, if in (1), $\sup_{n \leq 0} a(n, n+1) \leq 2$, *then the zero solution of equation (4.9) is stable.*

Proof Use the Liapunov functional

$$V(n, x(\cdot)) = \sum_{i=0}^{n} \sum_{j=0}^{n} a(n-1, n-i-j) x(n-i) x(n-j).$$

Observe that if a matrix M is positive definite in the traditional sense, then it is positive definite in the sense of equation (4.10).

5 A Comparison of Stability Notions

Let us consider the linear Volterra difference equations

$$x(n+1) = A(n)x(n) + \sum_{j=0}^{n} B(n,j)x(j), \quad n \in \mathbb{Z}^+, \tag{5.1}$$

$$u(n+1) = A(n)u(n) + \sum_{j=0}^{n} B(n,j)u(j) + p(n), \tag{5.2}$$

$$y(n+1) = A(n)y(n) + \sum_{j=-\infty}^{n} B(n,j)y(j), \quad n \in \mathbb{Z}^+. \tag{5.3}$$

We make the following assumptions on the $k \times k$ matrices $A(n)$ and $B(n,j)$.

(C_1) $\sup_{n \in \mathbb{Z}} \{\|A(n)\| + \sum_{j=-\infty}^{n} \|B(n,j)\|\} < \infty.$

(C_2) For any $\varepsilon > 0$ there exists $j_0 = j_0(\varepsilon) \in \mathbb{Z}^+$ such that

$$\sum_{j=-\infty}^{n-j_0} \|B(n,j)\| < \varepsilon \quad \text{for all} \quad n \in \mathbb{Z}.$$

Define $\Omega(A, B) := \{(\tilde{A}, \tilde{B}):$ there exists a sequence $\{n_j\}$ in $\mathbb{Z}^+$ with $n_j \to \infty$ as $j \to \infty$ such that $(A(n+n_j), B(n+n_j, s+n_j)) \to (\tilde{A}(n), \tilde{B}(n, s))\}$ as $n_j \to \infty$.

It is not hard to see that conditions (C_1) and (C_2) hold true for any $(\tilde{A}, \tilde{B}) \in \Omega(A, B)$. Associated with $(\tilde{A}, \tilde{B}) \in \Omega(A, B)$ is the limiting equation

$$z(n+1) = \tilde{A}(n)z(n) + \sum_{j=-\infty}^{n} \tilde{B}(n,j)z(j). \tag{5.4}$$

Definition 5.1 The *zero solution* of equation (5.3) is said to be

(a) *collectively uniformly stable* if for any $\varepsilon > 0$ there exists $\delta = \delta(\varepsilon) > 0$ such that if $\tau \in \mathbb{Z}$ and ϕ is an initial function on $(-\infty, \tau]$ with $\|\phi\|_{(-\infty,\tau]} < \delta$ and if $(\tilde{A}, \tilde{B}) \in \Omega(A, B)$, then $|z(n, \tau, \phi)| < \varepsilon$ for all $n \geq \tau$, where $z(n, \tau, \phi)$ is the solution of equation (5.4).

(b) *collectively uniformly asymptotically stable* if it is collectively uniformly stable, and if in addition there exists $\mu > 0$ such that for any $\varepsilon > 0$ there exists $N = N(\varepsilon) \in \mathbb{Z}^+$ such that if $\tau \in \mathbb{Z}$ and ϕ is an initial function on $(-\infty, \tau]$ with $\|\phi\|_{(-\infty,\tau]} < \mu$ and if $(\tilde{A}, \tilde{B}) \in \Omega(A, B)$, then $|z(n, \tau, \phi)| < \varepsilon$ for all $n \geq \tau + N$.

Definition 5.2 The *zero solution* of equation (5.1) is said to be *totally stable* (TS) if for any $\varepsilon > 0$ there exists $\delta = \delta(\varepsilon) > 0$ such that if $s \in \mathbb{Z}^+$ and ϕ is an initial function on $[0, s]$ with $\|\phi\|_{[0,s]} < \delta$ and if $p: [s, \infty) \to \mathbb{R}^k$ is any function with $\|p\|_{[s,\infty)} < \delta(\varepsilon)$, then $|u(n, s, \phi, p)| < \varepsilon$ for all $n \geq s$, where $u(n, s, \phi, p)$ is a solution of equation (5.2).

With some obvious modification of Definition 5.2 we extend the notion of total stability to equation (5.3).

We are now ready to present the main theorem of this section. We adopt the notation "US" for uniformly stable and "UAS" for uniformly asymptotically stable.

Theorem 5.1 [17] *The following statements are equivalent:*

1. *The zero solution of equation (5.1) is UAS.*
2. *The zero solution of equation (5.3) is UAS.*
3. *The zero solution of equation (5.1) is TS.*
4. *The zero solution of equation (5.3) is TS.*
5. *The zero solution of equation (5.3) is collectively UAS.*

The next theorem utilizes the resolvent matrix of equation (5.1) to provide a criterion for UAS.

Theorem 5.2 [17] *The zero solution of equation (5.1) is UAS if and only if*

$$\sup_{n \in \mathbb{Z}^+} \sum_{j=0}^{n} \|R(n, j)\| < \infty. \tag{5.5}$$

6 The Methods of Kolmanovskii

Consider the general Volterra difference equation

$$y(n+1) = F(n, y(n-\ell), y(n-\ell+1), \ldots, y(n), f(n, y(0), y(1), \ldots, y(n))), \tag{6.1}$$

with $\ell, n \in \mathbb{Z}^+$, $y(n) \in \mathbb{R}^k$, where $F: \mathbb{Z}^+ \times \mathbb{R}^{(\ell+2)k} \to \mathbb{R}^k$, $f(n,0,\ldots,0) = 0$ for all $n \in \mathbb{Z}^+$.

Kolmanovskii's procedure [7] for constructing a Liapunov function is as follows.

1. We write the function F in either the form

$$F = F_1(n, y(n-\tau), y(n-\tau+1), \ldots, y(n)) \\ + F_2(n, y(n-\ell), y(n-\ell+1), \ldots, y(n)), \quad (6.2)$$

with $F_1(n,0,\ldots,0) = F_2(n,0,\ldots,0) = 0$ or the form

$$F = F_1(n, y(n-\tau), y(n-\tau+1), \ldots, y(n)) \\ + \Delta F_2(n, y(-\ell), y(-\ell+1), \ldots, y(n)), \quad (6.3)$$

where $\tau \geq 0$ is any fixed integer.

2. For the associated difference equation

$$x(n+1) = F_1(n, x(n-\tau), x(n-\tau+1), \ldots, x(n)). \quad (6.4)$$

We construct a Liapunov functional $v(n, x(n-\tau), \ldots, x(n))$ satisfying certain stability conditions to be specified.

3. The sought for Liapunov functional V is now written in the form $V = V_1 + V_2$, where V_1 depends on the representation used for F. For if we use the representation (6.2), then

$$v_1(n, y(-\ell), \ldots, y(n)) = v(n, y(n-\tau), \ldots, y(n)) \quad (6.5)$$

and if we use the representation (6.3), then

$$V_1(n, y(-\ell), \ldots, y(n)) = v(n, y(n-\tau), \ldots, y(n)) \\ - F_2(n, y(-\ell), \ldots, y(n)). \quad (6.6)$$

4. We choose V_2 in such a way that the functional $V = V_1 + V_2$ will satisfy sufficient conditions for asymptotic stability.

Let us now apply the preceding procedure to the following Volterra difference equation

$$y(n+1) = \sum_{j=0}^{n} b(n-j)y(j), \quad n \in \mathbb{Z}^+. \quad (6.7)$$

Using $\tau = 1$ and representation (6.2) we obtain the following result [7].

Theorem 6.1 *Suppose that the following conditions hold.*
1. $|b(1)| < 1$, $|b(0)| < -b(1) + 1$,
2. $\alpha_2|d_{12} + b(0)d_{22}| < 1$, *and* $\alpha_2|d_{22}| + |b_1| + \gamma_1| < 1$, *where*

$$\alpha_2 = \sum_{j=2}^{\infty} |b(j)|, \quad \gamma_1 = |d_{12} + b(0)d_{22}| + d_{22}|b(1)| + \alpha_2,$$

with

$$d_{11} = 1 + b^2(1)d_{22}, \quad d_{21} = b(1)b(0)(1-b(1))^{-1}d_{22},$$
$$d_{22} = 2(1-b(1))(1+b(1))^{-1}[(1-b(1))^2 - b^2(0)]^{-1}.$$

Then the zero solution of equation (6.7) is asymptotically stable.

Proof We let $F_1(n, y(n-1), y(n)) = b(0)y(n) + b(1)y(n-1)$, and

$$F_2(n, y(0), \ldots, y(n)) = \sum_{j=2}^{n} b(j)y(n-j).$$

Put $V_1 = [y(n-1), y(n)]D[y(n-1), y(n)]^T$, where D is a positive definite 2×2 matrix that satisfies

$$A^T D A - D = -I_2, \quad A = \begin{pmatrix} 0 & 1 \\ b(1) & b(0) \end{pmatrix}.$$

We take

$$V_2(n, y(0), \ldots, y(n)) = \gamma_1 \sum_{\ell=2}^{n} y^2(n-\ell) \sum_{j=\ell}^{\infty} |b(j)|.$$

Another result may be obtained if one uses $\tau = i$ and representation (6.3). We omit the details and refer the interested reader to [7].

The above procedure has been applied to the more general equation

$$y(n+1) = \sum_{j=0}^{n} b(n, j)y(n-j). \tag{6.8}$$

For more details, see [7].

7 Nonlinear Equations

Consider the nonlinear system

$$x(n+1) = \sum_{j=0}^{n} K(n,j,x(j)), \qquad (7.1)$$

where $K(n,j,x)$ is a function from $\mathbb{Z}^+ \times \mathbb{Z}^+ \times \mathbb{R}^k \to \mathbb{R}^k$ continuous in x and such that $K(n,j,0) = 0$ for all $n, j \geq 0$. For the first result we make the following assumptions.

(A_1) For every $x \in \mathbb{R}^k$, $j, n \geq 0$,

$$|K(n,j,x)| \leq B(n,j)|x|, \quad B(n,j) > 0, \qquad (7.2)$$

where $|\cdot|$ is any vector norm on $\mathbb{R}^k$.

(A_2)

$$\beta(n) = \sum_{s=n}^{\infty} B(s,n) \leq 1 - c, \quad c \in (0,1), \quad n \in \mathbb{Z}^+. \qquad (7.3)$$

We associate with equation (6.1) the scalar equation

$$y(n+1) = \sum_{j=0}^{n} B(n,j)y(j). \qquad (7.4)$$

Lemma 7.1 *Suppose that assumptions (A_1) and (A_2) hold. If $x(n)$ and $y(n)$ are solutions of equations (7.1) and (7.4), respectively, with $|x(0)| \leq y(0)$, then $|x(n)| \leq y(n)$ for all $n \in \mathbb{Z}^+$.*

Using this lemma, Elaydi [12] proved the following result.

Theorem 7.1 *If assumptions (A_1) and (A_2) hold, then the zero solution of equation (7.1) is globally asymptotically stable.*

Proof Use Lemma 7.1 and Theorem 4.1.

An alternative approach to determine global stability is the use of vector Liapunov functions. This approach produced Theorem 4.2 which we will now exploit to accomplish the task at hand. In the sequel we will make the following assumptions.

(H_1) For $j, n \geq 0$, there exists nonnegative $k \times k$ matrices $B(n,j)$ such that

$$|K(n,j,x)| \leq B(n,j)|x|,$$

where we adopt the notation and terminology used in Theorem 4.2.

(H_2) $\sum_{i=0}^{\infty} B(i,n)$ converges for all $n \in \mathbb{Z}^+$. Furthermore, all the eigenvalues of the matrix

$$B = \sup_{n \geq 0} \left(\sum_{i=n}^{\infty} B(i,n) \right)$$

lie inside the complex unit circle.

We now associate to equation (7.1) the linear system

$$y(n+1) = \sum_{j=0}^{n} B(n,j) y(j). \tag{7.5}$$

Theorem 7.2 [15] *Suppose that assumptions (H_1) and (H_2) hold. Then the zero solution of equation (7.1) is globally asymptotically stable.*

Proof The proof is based on a comparison between solutions of equations (7.1) and (7.5) and an application of Theorem 4.2.

If one specializes Theorem 7.2 to scalar equations of convolution type, then assumptions (H_1) and (H_2) may be greatly simplified.

Consider the scalar equation

$$x(n+1) = \sum_{j=0}^{n} B(n-j) f(x(j)) \tag{7.6}$$

such that the following assumptions hold.

(H_3) $B = \sum_{n=0}^{\infty} |B(n)| < \infty$, and

(H_4) $f \in C[\mathbb{R}, \mathbb{R}]$, $f(0) = 0$, $|f(u)| \leq |u|$ for all $u \in \mathbb{R}$.

Theorem 7.3 [15] *If assumptions (H_3) and (H_4) hold, then the zero solution of equation (7.6) is globally asymptotically stable provided that $B < 1$.*

8 Open Problems

We now present some open problems pertaining to both scalar and vector difference equations.

8.1 Scalar difference equations

Consider the scalar equation

$$x(n+1) = Ax(n) + \sum_{j=0}^{n} B(n-j) x(j). \tag{8.1}$$

Open Problem 1 Determine the stability of equation (8.1) when

$$A + \sum_{n=0}^{\infty} B(n) = -1 \quad \text{and} \quad \sum_{n=0}^{\infty} B(n) < 0.$$

Open Problem 2 Determine the stability of the zero solution of equation (8.1) when

$$-1 < A + \sum_{n=0}^{\infty} B(n) < 1.$$

Open Problem 3 In Theorem 2.5 (3), can we omit the assumption that $\sum_{n=0}^{\infty} B(n)$ is sufficiently small?

Open Problem 4 If in Theorem 2.4 $a_{ii} + \nu_{ii} < 1$, for $1 \leq i \leq k$, what can we conclude about the stability of the zero solution of equation (8.1)?

8.2 Vector difference equations

Here we assume that A and $B(n)$ are $k \times k$ matrices.

Open Problem 5 Suppose that any one of the conditions in Theorem 2.1 holds. Then by Theorem 2.2 the zero solution of equation (8.1) is exponentially stable if and only if $B(n)$ is of exponential decay. Find an estimate of the rate of decay of solutions of equation (8.1) if $B(n)$ is not of exponential decay.

References

[1] Brauer, F. (1978). Asymptotic stability of a class of integrodifferential equations. *J. Diff. Eqns*, **28**, 180–188.

[2] Brunner, H. and Van der Houwen, P.J. (1985). *The Numerical Solution of Volterra Equations*. SIAM, Philadelphia.

[3] Burton, T.A. (1983). *Volterra Integral and Differential Equations*. Academic Press, Orlando, Florida.

[4] Corduneanu, C. and Lakshmikantham, V. (1980). Equations with unbounded delay. *Nonlin. Anal.: TMA*, **4**, 831–877.

[5] Crisci, M.R., Jackiewicz, Z., Russo, E. and Vecchio, A. (1991). Stability analysis of discrete recurrence equations of Volterra type with degenerate kernels. *J. Math. Anal. Applic.*, **161**, 49–62.

[6] Crisci, M.R., Kolmanovskii, V.B., Russo, E. and Vecchio, A. (1997). Boundedness of discrete Volterra equations. *J. Math. Anal. Applic.*, **211**(1), 106–130.

[7] Crisci, M.R., Kolmanovskii, V.B., Russo, E. and Vecchio, A. (1995). Stability of continuous and discrete Volterra integro-differential equations by Liapunov approach. *J. Integral Eqns*, **7**, 393–411.

[8] Crisci, M.R., Kolmanovskii, V.B., Russo, E. and Vecchio, A. (1998). Stability of difference Volterra equations: Direct Liapunov method and numerical procedure. *Comp. Math. Appl.*, **36**(10–12), 77–97.

[9] Crisci, M.R., Kolmanovskii, V.B., Russo, E. and Vecchio, A. (2000). Stability of discrete Volterra equations of Hammerstein type. *J. Diff. Eqns Applic.*, **6**(2), 127–145.

[10] Cushing, J.M. (1977). *Integrodifferential Equations and Delay Models in Population Dynamics*. Springer-Verlag, Berlin.

[11] Elaydi, S. (1999). *An Introduction to Difference Equations*. Second Edition, Springer-Verlag, New York.

[12] Elaydi, S. (1996). Global stability of difference equations. *Proc. of the First World Congress of Nonlinear Analysis* (Ed.: V. Lakshmikantham). Walter de Gruyter, Berlin, 1131–1138.

[13] Elaydi, S. (1994). Periodicity and stability of linear Volterra difference systems. *J. Math. Anal. Applic.*, **181**, 483–492.

[14] Elaydi, S. (1993). Stability of Volterra difference equations of convolution type. *Proc. of the Special Program at Nankai Institute of Mathematics* (Ed.: L. Shan-Tao). World Scientific, Singapore, 66–73.

[15] Elaydi, S. and Kocic, V. (1994). Global stability of a nonlinear Volterra difference system. *Diff. Eqns Dyn. Sys.*, **2**, 337–345.

[16] Elaydi, S. and Murakami, S. (1996). Asymptotic stability versus exponential stability in linear Volterra difference equations of convolution type. *J. Diff. Eqns Applic.*, **2**, 401–410.

[17] Elaydi, S. and Murakami, S. (1998). Uniform asymptotic stability in linear Volterra difference equations. *J. Diff. Eqns Applic.*, **3**, 203–218.

[18] Elaydi, S., Murakami, S. and Kamiyama, E. (1999). Asymptotic equivalence for difference equations with infinite delay. *J. Diff. Eqns Applic.*, **5**, 1–23.

[19] Elaydi, S. and Zhang, S. (1994). Stability and periodicity of difference equations with finite delay. *Funkcialaj Ekvacioj*, **37**, 401–413.

[20] Feller, W. (1970). *An Introduction to Probability Theory and Its Applications*, Vol. 1. John Wiley, New York.

[21] Hino, Y. and Murakami, S. (1991). Stability properties of linear Volterra equations, *J. Diff. Eqns*, **89**, 121–137.

[22] Hino, Y. and Murakami, S. (1991). Total stability and uniform asymptotic stability for linear Volterra equations. *J. London Math. Soc.*, **43**, 305–312.

[23] Jordan, G.S. (1979). Asymptotic stability of a class of integrodifferential systems. *J. Diff. Eqns*, **31**, 359–365.

[24] Kelley, W.G. and Peterson, A.C. (1991). *Difference Equations: An Introduction with Applications*. Academic Press, New York.

[25] Li, P. (1997). *Stability of Nonlinear Discrete Systems with Applications to Population Dynamics*, PhD Thesis. The Flinders University of South Australia.

[26] Linz, P. (1985). *Analytical and Numerical Methods for Volterra Equations*. SIAM, Philadelphia.

[27] Lubich, C. (1983). On the stability of linear multistep methods for Volterra convolution equations. *IMA J. Num. Anal.*, 439–465.

[28] Miller, R.K. (1971). Asymptotic stability properties of linear Volterra integrodifferential equations. *J. Diff. Eqns*, **10**, 485–506.

[29] Murakami, S. (1991). Exponential asymptotic stability for scalar linear Volterra equations. *Diff. and Integral Eqns*, **4**, 519–525.

[30] Raffoul, Y. (1998). Boundedness and periodicity of Volterra systems of difference equations. *J. Diff. Eqns Applic.*, **4**(4), 381–393.

[31] Zhang, S. (1994). Boundedness of infinite delay difference systems. *Nonlin. Anal. TMA*, **22**, 1209–1219.

[32] Zhang, S. (1994). Stability of infinite delay difference systems. *Nonlin. Anal. TMA*, **22**, 1121–1129.

[33] Zouyousefain, M. and Leela, S. (1990). Stability results for difference equations of Volterra type. *Appl. Math. Comp.*, **36**, 51–61.

2.4 CONSISTENT LYAPUNOV METHODOLOGY FOR EXPONENTIAL STABILITY: PCUP APPROACH*

Ly.T. GRUYITCH

University of Technology Belfort-Montbeliard, Belfort, France

1 Introduction

Exponential stability (of an equilibrium, of a motion or of a set) ensures a higher system motions quality than asymptotic stability (of an equilibrium, of a motion or of a set, respectively). The former provides information about both an upper bound of an overshoot of an accepted norm of any system motion and the rate of motions convergence that is exponential, while the latter fails to provide such information. Like asymptotic stability, exponential stability properties of nonlinear systems were studied by employing Lyapunov's original methodology for nonlinear systems [1] to be called the *classical Lyapunov methodology (for nonlinear systems)*. It is characterized by stability conditions expressed in terms of existence of a positive definite function with appropriate properties. In this framework, Krasovskii resolved the problem of the necessary and sufficient conditions for exponential stability [2].

The problem of an exact single-step direct construction of a system Lyapunov function, which has been a fundamental problem of the stability theory and its effective applications, has been solved for different classes of systems as reviewed and originally contributed in [3]. The solution reflects a new Lyapunov stability methodology called the *consistent Lyapunov methodology (for nonlinear systems)* [3], which was developed also for exponential stability of an equilibrium state of a time-varying nonlinear system with differentiable motions [4] or with continuous motions [5], and for global exponential stability of sets of the systems with differentiable motions [6].

---

*Advances in Stability Theory (Ed.: A.A. Martynyuk). Stability and Control: Theory, Methods and Applications, Taylor & Francis, London, **13** (2003) 107–120.

The purpose of this paper is to further develop the consistent Lyapunov methodology to (non-global and global) exponential stability of sets of time-varying nonlinear systems obeying a recently discovered *Physical Continuity and Uniqueness Principle* [7–9].

The consistent Lyapunov methodology applied to exponential stability starts with an arbitrary choice of a function $e(.)$ from a well defined functional family $E(.)$. It continues with a test of the necessary and sufficient conditions for exponential stability, which are imposed on a solution function $v(.)$ to $D^+v(t,x) = -e(t,x)$. The result is decisive. If it is positive, then the system possesses the requested exponential stability property. Otherwise, it does not.

2 Notation

The letter A will denote a compact connected invariant set, exponential stability of which is studied. In the special case of a study of exponential stability of a state x^*, the set A is singleton: $A = \{x^*\}$. In the case of exponential stability of a closed trajectory (of an orbit) T the set $A = T$. A neighborhood of A will be denoted by $N(A)$ or by $S(A)$. A distance of a point x from the set A is denoted by $d(x, A) = \inf\{\|x - y\|: y \in A\}$ with $\|.\|: R^n \to R_+$ being a norm on R^n, where $R_+ = [0, \infty) = \{\gamma: \gamma \in R, \infty > \gamma \geq 0\}$. The norm can be Euclidean norm. Besides $R^+ = (0, \infty) = \{\gamma: \gamma \in R, \infty > \gamma > 0\}$. Let $R_0 = [t_0, \infty)$. Notice that $A \subset N(A)$. The boundary, closure and interior of the set A are designated by ∂A, $\mathrm{Cl}\, A$ and $\mathrm{In}\, A$, respectively.

A motion of a system, which passes through an initial state $x_0 \in R^n$ at an initial moment t_0 is denoted by $x(.; t_0; x_0)$. Its vector value at a moment $t \in R$ is $x(t; t_0; x_0)$, $x(t) \equiv x(t; t_0; x_0)$. If t_0 is fixed then $x(t; t_0; x_0) \equiv x(t; x_0)$. If $v(.): R \times R^n \to R$ is continuous then its total time right-hand upper derivative along $x(.)$ is its Dini derivative $D^+v(t,x) = \lim \sup\{[v[t+\theta, x(t+\theta; t, x)] - v(t, x)]\theta^{-1}: \theta \to 0^+\}$. Other notation is explained in the sequel.

3 Physical Continuity and Uniqueness Principle [7–9]

Newton's explication [10] of the nature of time has been recently broadened by leading to a new time-space co-ordinate transformation that generalizes the Lorentz transformation and leads to new results on the time relativity [11]. In order to avoid any ambiguity, time is understood in the sequel in the following sense:

Time is an independent physical variable the value of which is strictly monotonously continuously increasing independently of all other (physical and mathematical) variables, processes and events, which is used to uniquely determine the order of events happening. The *value of time* is called *moment* or *instant* and it is denoted by t or τ and a subscript: t_2 or τ_b. An arbitrary instant will be denoted as time itself by t or by τ.

A physical variable can take only unique value at a fixed point at one moment and can change its value at the fixed point only continuously in terms of time. These facts have led to the following principle:

Physical Continuity and Uniqueness Principle (PCUP):

    I Scalar form [7–9]
        a) Physical Continuity Principle
            A physical variable can change its value from one value to another one only by passing through all intermediate values.
        b) Physical Uniqueness Principle
            A physical variable possesses a unique local instantaneous real value at any place (in any being or in any object) at any moment.
    II Vector form
        A vector variable obeys the Physical Continuity and Uniqueness Principle if and only if all its entries obey the Principle.

In view of this principle and the nature of time we may conclude that a necessary condition (but not sufficient) for a mathematical model of a physical system to be an adequate description of the physical system is that all its variables satisfy the PCUP.

4 System Description

Large classes of overall control systems and of other dynamical systems are described by (1),

$$\frac{dx}{dt} = f(t, x), \qquad (1)$$

and by their property to obey the PCUP as explained in what follows.

Smoothness Property:

(i) There is an open connected neighbourhood $S(A)$ of a compact connected invariant set A of system (1), $S(A) \subseteq R^n$, such that for every $x_0 \in S(A)$:
    a) system (1) has a unique solution $x(.; t_0, x_0)$ through x_0 at $t = t_0 \in R$ on a largest interval I_0, $I_0 = I_0(t_0, x_0)$, $I_0 \subseteq R$, $I_0 \neq \emptyset$, $\emptyset$ is the empty set, and
    b) $x(t; t_0, x_0)$ obeys the PCUP on $I_0 \times R \times S(A)$: $x(t; t_0, x_0) \in C[I_0 \times R \times S(A)]$.

(ii) For every $(t_0, x_0) \in R \times [R^n - S(A)]$ and every motion $x(.; t_0, x_0)$ of system (1), $x(t; t_0, x_0)$ obeys the PCUP on I_0, $I_0 = I_0[t_0, x_0; x(.)]$, where $I_0 = \emptyset$ is permitted.

5 Problems Statement

The following problems are considered and will be solved:
What are the necessary and sufficient conditions, which are not expressed in terms of existence of a Lyapunov function, for:

a) the invariant compact connected nonempty set A to be exponential stable,
b) for system motions to obey the following estimates for some numbers $a_i \in R^+$ and $b_i \in R^+$, $i = 1,2$, and on some open connected neighborhood $N(A)$ of the set A, $N(A) \subseteq S(A)$:

$$a_1 d(x_0, A) \exp[-b_1(t - t_0)] \leq d[x(t; t_0, x_0), A] \leq a_2 d(x_0, A) \exp[-b_2(t - t_0)], \quad (2)$$
$$\forall (t, t_0, x_0) \in R_0 \times R \times N(A),$$

c) for a direct single step exact construction of a system Lyapunov function for exponential stability of the set A?
d) What are possible conceptual applications?

It will be said that the estimates (2) hold globally if and only if they hold for $N(A) = R^n$.

6 Family $E(A; S; f)$ of Functions $e(.)$. Lyapunov Functions Generation

A characteristic advantage of the consistent Lyapunov methodology compared with the classical one is a well determination of a family $E(A; S; f)$ of functions denoted by $e(.)$, each of which can be selected arbitrarily to generate a system Lyapunov function.

Definition 6.1 A *function* $e(.)\colon R \times R^n \to R_+$ *belongs to the functional family* $E(A; S; f)$ *if and only if:*

1) $e(.)$ obeys the PCUP on $S(A)$: $e(t,x) \in C(R \times S, R_+)$, and
2) there are numbers $\eta_i \in R^+$, $\eta_i = \eta_i(e; A; S)$, $i = 1,2$, and a natural number k such that

$$\eta_1 d^k(x, A) \leq e(t, x) \leq \eta_2 d^k(x, A), \quad \forall (t, x) \in R \times S(A). \quad (3)$$

Functions belonging to the functional family $E(A; S; f)$ are inherent for constructing a system Lyapunov function $v(.)$ via the equations (4) with (4a) taken along motions of system (1)

$$D^+ v(t, x) = -e(t, x), \quad e(.) \in E(A; S; f), \quad (4a)$$

and with (4b) as the boundary condition,

$$v(t, x) = 0, \quad \forall (t, x) \in R \times \partial A. \quad (4b)$$

6.1 The basic solution

At first a general result will be presented. It will be used to deduce specific criteria.

Theorem 6.1 *Let system (1) possess the smoothness property.*

(a) *In order for the solutions of the system to obey the exponential estimates (2) on some neighborhood $N(A)$ of A, $N(A) \subseteq S(A)$, of the invariant compact connected nonempty set A, for the set A to be exponential stable, and for a direct single exact construction of a system Lyapunov function $v(.)$ it is necessary and sufficient that*

$$f(t, x) \neq 0 \quad \text{for all} \quad (t, x) \in R \times [N(A) - A],$$

and that for an arbitrary function $e(.) \in E(A; S; f)$ there exists a unique solution function $v(.)$ to Lyapunov's like equations (4), which belongs to the functional family $E(A; N; f)$.

(b) *In order for the solutions of the system to obey globally the exponential estimates (2), for the invariant compact connected nonempty set A to be global exponential stable, and for a direct single exact construction of a system Lyapunov function $v(.)$ it is necessary and sufficient that*

$$f(t, x) = 0 \quad \text{only if} \quad (t, x) \in R \times A,$$

and that for an arbitrary function $e(.) \in E(A; R^n; f)$ there exists a unique solution function $v(.)$ to Lyapunov's like equations (4), which belongs to the functional family $E(A; R^n; f)$.

Proof Let system (1) possess the smoothness property.

(a) At first the statement under (a) will be proved.

Necessity. Let the solutions of the system obey the exponential estimates (2) on some neighborhood $N(A)$, $N(A) \subseteq S(A)$, of the invariant compact connected nonempty set A. Hence, the set A is exponential stable. The inequalities (2) show that there is not a system equilibrium state in $N(A)$, which implies

$$f(t, x) \neq 0 \quad \text{for all} \quad (t, x) \in R \times [N(A) - A].$$

Let the solutions of the system obey the exponential estimates (2) on some neighborhood $N(A)$ of A. Hence, the equation (4a) integrated from $t = t_0$ to $t = \infty$ becomes

$$v[\infty, x(\infty; t_0, x_0)] - v(t_0, x_0) = -\int_{t_0}^{\infty} e[t, x(t; t_0, x_0)]\, dt,$$

or,

$$v(t_0, x_0) = \int_{t_0}^{\infty} e[t, x(t; t_0, x_0)]\, dt,$$

where $e(.) \in E(A; S; f)$ that together with (i) of the smoothness property ensures uniqueness and continuity of $v(.)$ and

$$\eta_{e1} d^k(x, A) \leq e(t, x) \leq \eta_{e2} d^k(x, A), \quad \forall (t, x) \in R \times S(A).$$

The last equation, inequalities (2) and the preceding inequalities prove that:

$$\eta_{v1} d^k(x, A) \leq v(t, x) \leq \eta_{v2} d^k(x, A), \quad \forall (t, x) \in R_+ \times N(A),$$

in view of arbitrariness of $(t, x) \in R \times N(A)$ and for

$$\eta_{vi} = b_i^{-1} \eta_{ei} a_i, \quad i = 1, 2.$$

Hence, all the conditions of the theorem statement are necessary.

Sufficiency. Let all the conditions of the theorem statement hold. Hence, there are:

$$k \in \{1, 2, \ldots, n, \ldots\} \quad \text{and} \quad \eta_{ei}, \eta_{vi} \in R^+, \ i = 1, 2,$$

such that both

$$\eta_{e1} d^k(x, A) \leq e(t, x) \leq \eta_{e2} d^k(x, A), \quad \forall (t, x) \in R \times S(A)$$

and

$$\eta_{v1} d^k(x, A) \leq v(t, x) \leq \eta_{v2} d^k(x, A), \quad \forall (t, x) \in R_+ \times N(A)$$

hold. These inequalities and the equations (4) imply (2) for

$$a_1 = \eta_{v1} \eta_{v2}^{-1}, \quad a_2 = \eta_{v2} \eta_{v1}^{-1}, \quad b_1 = \eta_{e2} \eta_{v1}^{-1}, \quad b_2 = \eta_{e1} \eta_{v2}^{-1},$$

$$a_i, b_i \in R^+, \ i = 1, 2,$$

which completes the proof.

b) The statement under b) results from $N(A) = S(A) = R^n$.

6.2 Conceptual applications: *E*-functions $f(.)$

A function $f(.)$ is said to be *E-function* (to belongs to the *E*-class of functions) if and only if $\|f(.)\|^2 \in E(A; M; f)$, where M is some open connected neighborhood of A, $M(A) \subseteq S(A)$. It is *global E-function* if and only if $M = S(A) = R^n$.

Theorem 6.2 *Let system (1) possess the smoothness property.*

(a) *Let $A = \{x^*\}$. If $f(.)$ is an E-function then in order for the solutions of the system to obey the exponential estimates (2) and for the set A to be exponential stable it is necessary and sufficient that both*

$$f(t,x) \neq 0 \quad \text{for all} \quad (t,x) \in R \times [N(A) - A],$$

and the function $v(.)$ defined by:

$$v(t,x) = -\int_{x^*}^{x} f^T(t,x)\,dx = -\frac{1}{2}\int_{x^*}^{x}\left[(dx)^T f(t,x) + f^T(t,x)\,dx\right],$$

$$x = x(t),$$

is a unique solution to Lyapunov's like equations (4), which is also an E-function.

(b) *Let $A = \{x^*\}$. If $f(.)$ is a global E-function then in order for the solutions of the system to obey globally the exponential estimates (2) and for the set A to be global exponential stable it is necessary and sufficient that both*

$$f(t,x) = 0 \quad \text{only if} \quad (t,x) \in R \times A,$$

and the function $v(.)$ defined by:

$$v(t,x) = -\int_{x^*}^{x} f^T(t,x)\,dx = -\frac{1}{2}\int_{x^*}^{x}\left[(dx)^T f(t,x) + f^T(t,x)\,dx\right],$$

$$x = x(t),$$

is a unique solution to Lyapunov's like equations (4), which is also a global E-function.

Proof Theorem 6.2 results from Theorem 6.1 for $e(.,.) = \|f(.,.)\|^2 = f^T(.)f(.) = f^T(.)\frac{dx}{dt}$.

6.3 Conceptual application: power and energy approach

It is now possible to present the necessary and sufficient exponential stability conditions in terms of the system power $P(.)$ and energy $E(.)$.

Theorem 6.3 *Let system (1) possess the smoothness property.*

(a) *Let $A = \{x^*\}$. Let the system power $P(.) \in E(A; S; f)$ and be negative definite with respect to A on $S(A)$. In order for the set A to be global exponential stable it is necessary and sufficient that both*

$$f(t, x) \neq 0 \quad \text{for all} \quad (t, x) \in R \times [N(A) - A],$$

and the system energy $E(.,.)$ obeys that $[E(.,.) - E(., x^*)]$ is positive definite function with respect to A on $N(A)$ and it belongs to $E(A; N; f)$ for some neighborhood $N(A)$ of A.

(b) *Let $A = \{x^*\}$. Let the system power $P(.) \in E(A; R^n; f)$ and be global negative definite with respect to A. In order for the set A to be global exponential stable it is necessary and sufficient that both*

$$f(t, x) = 0 \quad \text{only if} \quad (t, x) \in R \times A,$$

and the system energy $E(.,.)$ obeys that $[E(.,.) - E(., x^*)]$ is global positive definite function with respect to A and it belongs to $E(A; R^n; f)$.

Proof Theorem 6.3 results from Theorem 6.1 for $e(.,.) = P(.,.)$ and $v(.,.) = [E(.,.) - E(., x^*)]$.

6.4 Conceptual application: exponential absolute stability

Absolute stability has attracted a lot of interest since 1944 [12] in general [13–17], and in particular as exponential absolute stability [18, 19]. Let the system (1) take the form of a Lur'e-Postnykov (for short: Lur'e) system:

$$\frac{dx}{dt} = A(t)x + B(t)g(t, \Sigma), \quad A(t) \colon R \to R^{n \times n}, \quad B(t) \colon R \to R^{n \times m},$$

$$C(t) \colon R \to R^{m \times n}, \quad \Sigma = C(t)x, \quad \Sigma = (\sigma_1, \sigma_2, \ldots, \sigma_m)^T,$$

where the vector nonlinearity $g(.)$,

$$g(.) = [g_1(.) g_2(.) \ldots g_m(.)]^T \colon R \times R^m \to R^m,$$

satisfies the Lur'e conditions on the Lurie sector L, $L = L_1 \times L_2 \times \cdots \times L_m$:

a) $g(.)$ fulfils the PCUP, i.e. $g(t, \Sigma) \in C(R \times R^m)$,

b) $g(t, 0) = 0$ for all $t \in R$,

c) $\dfrac{g_i(t, \Sigma)}{\sigma_i} \in L_i, \quad \forall (t, \Sigma) \in R \times R^m, \quad \sigma_i \neq 0, \quad \forall i = 1, 2, \ldots, m,$

d) $\|A(t)x + B(t)g[t, C(t)x]\| = 0$ if and only if $x = 0$, $t \in R$.

Theorem 6.4 *Let system (1) be of the Lur'e form and let it possess the smoothness property.*

Let $\|A(.) + B(.)g[., C(.)]\|^2$ be a global E-function for every Lur'e vector nonlinearity $g(.)$ on the Lur'e sector L, and let $A = \{0\}$. In order for the set A to be exponential absolute stable on L it is necessary and sufficient that the solution function $v(.)$ to Lyapunov's like equations, which is defined by:

$$v(t,x) = -\int_0^x [A(t)x + B(t)g[t, C(t)x]]^T \, dx$$

$$= -\frac{1}{2}\int_0^x \left\{ [A(t)x + B(t)g[t, C(t)x]]^T \, dx + dx^T [A(t)x + B(t)g[t, C(t)x]] \right\},$$

$$x = x(t),$$

is global positive definite and global E-function for every Lur'e vector nonlinearity $g(.)$ in the Lur'e sector L.

Proof Theorem 6.4 results from Theorem 6.3 for $f(t,x) = A(t)x + D(t)g[t, C(t)x]$

7 Example

Let system (1) take the next specific form of a third nonlinear dynamic system obeying the PCUP globally, i.e. $S = R^3$:

$$\frac{dx}{dt} = \begin{bmatrix} \dot{x}_1 \\ \dot{x}_2 \\ \dot{x}_3 \end{bmatrix} = f(x) = \begin{bmatrix} f_1(x) \\ f_2(x) \\ f_3(x) \end{bmatrix},$$

$$\dot{x}_1 \triangleq \frac{dx_1}{dt} = (-x_1 + 4x_2^3 + 4x_3^7)\frac{h(x) - 10}{h(x)}, \quad h(x) = 6x_1^2 + 12x_2^4 + x_3^8,$$

$$\dot{x}_2 \triangleq \frac{dx_2}{dt} = (-x_1 - \frac{1}{2}x_2 + 2x_3^7)\frac{h(x) - 10}{h(x)},$$

$$\dot{x}_3 \triangleq \frac{dx_3}{dt} = (-6x_1 - 12x_2^3 - \frac{1}{4}x_3^7)\frac{h(x) - 10}{h(x)}.$$

The system has the next invariant connected compact set:

$$A = \{x \colon h(x) = 6x_1^2 + 12x_2^4 + x_3^8 \leq 10\}.$$

The distance function is defined by

$$d(x, A) = [h(x) - 10].$$

Let

$$e(x) = 2[h(x) - 10]$$

so that

$$e(x) = 2d(x, A),$$

which yields

$$\eta_1 = \eta_2 = 2, \quad k = 1$$

in

$$\eta_1 d^k(x, A) \leq e(x) \leq \eta_2 d^k(x, A), \quad \forall x \in R^3.$$

It follows now that

$$e(.) \in E(A; S; f).$$

Theorem 6.1 application continues with solving the equations (4):

$$\frac{dv(x)}{dt} = -e(x) = -2[h(x) - 10]$$

and

$$v(x) = 0 \quad \forall x \in \partial A.$$

It is easy to verify that the function $v(.)$:

$$v(x) = [h(x) - 10] = [6x_1^2 + 12x_2^4 + x_3^8 - 10]$$

obeys both

$$v(x) = 0 \iff x \in \partial A$$

and

$$\eta_3 d^k(x, A) \leq v(x) \leq \eta_4 d^k(x, A), \quad \forall x \in R^n, \quad \eta_1 = \eta_2 = 1, \quad k = 1.$$

Theorem 6.1, all the conditions of which are satisfied, enables us to conclude that the set A,

$$A = \{x \colon h(x) = 6x_1^2 + 12x_2^4 + x_3^8 \leq 10\}$$

is global exponential stable.

Simulation results are presented in Figure 7.1 through Figure 7.4.

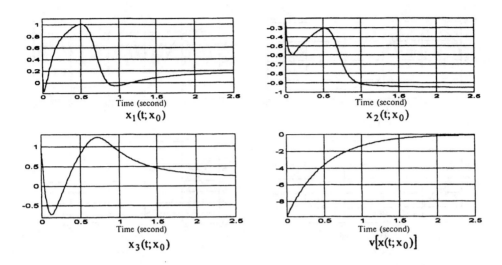

Figure 7.1. $x_0 = [0.1 \ -0.2 \ 0.8]^T \in \text{Int } A$.

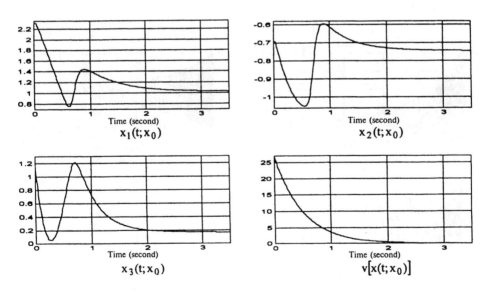

Figure 7.2. $x_0 = [2.3 \ -0.7 \ 1.1]^T \notin A$.

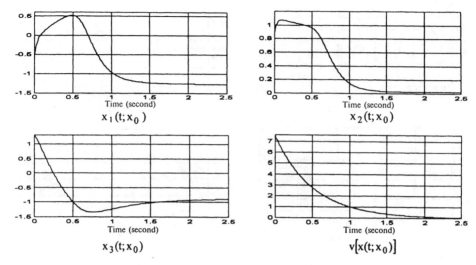

Figure 7.3. $x_0 = [-0.5 \ \ 0.9 \ \ 1.3]^T \notin A$.

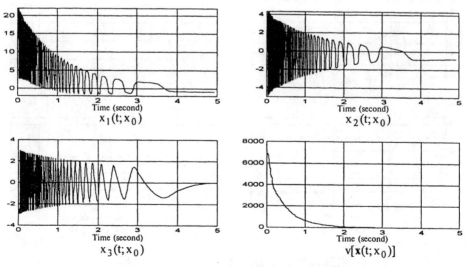

Figure 7.4. $x_0 = [4 \ \ -5 \ \ -2]^T \notin A$.

8 Conclusion

A new physical principle called the Physical Continuity and Uniqueness Principle has been explained. It reflects the essential, but simple (obvious) physical properties of physical variables.

A new understanding of the real nature and meaning of time, which slightly generalizes that by Newton and which led to new results in the relativity theory incorporating those by Lorentz and Einstein as special cases, has been also explained.

The Physical Continuity and Uniqueness Principle shows it full usefulness when it is linked with the new understanding of time as shown in the paper.

The paper presents a development of the consistent Lyapunov methodology (for nonlinear systems) to non-global and global exponential stability of invariant compact connected nonempty sets of time varying nonlinear systems obeying the Physical Continuity and Uniqueness Principle.

The main theorem presents the complete solution to the problem of the necessary and sufficient conditions for exponential stability of the sets and for a direct single exact construction of a system Lyapunov function. Its conceptual applications are shown to particular classes of the systems described by (1) including time-varying Lur'e-Postnykov systems. Besides, the conditions are presented in terms of the system power and energy.

References

[1] Lyapunov A.M. (1956). *The General Problem of the Stability of Motion.* (Kharkov Mathematical Society, Kharkov, 1892) published in: *Academician A.M. Lyapunov. Collected Works. Vol. II*, Academy of Sciences of USSR, Moscow, 5–263 (Russian). French translation: *Annales de la Faculté des Sciences de l'Université de Toulouse*, 2nd Ser., Vol. 9, 203–474 (1907). English translation: *Int. J.Control*, **55**, 531–773 (1992); also the book: Taylor and Francis, London (1992).

[2] Krasovskii, N.N. (1963). *Stability of Motion.* Stanford University Press, Stanford.

[3] Gruyitch, Ly.T. (1997) Consistent Lyapunov methodology for time-invariant nonlinear systems. *Avtomatika i Telemekhanika*, **12**, 35–73 (Russian).

[4] Grujić, Lj.T. (1995). New Lyapunov methodology and exact construction of a Lyapunov function: exponential stability. *Problems of Nonlinear Analysis in Engineering Systems*, Kazan, **1**, 9–14.

[5] Grujić, Lj.T. (1997). Exponential quality of time-varying dynamical systems: stability and tracking. *Stability and Control: Theory, Methods and Applications. Advances in Nonlinear Dynamics* (Eds.: S. Sivasundaram and A.A. Martynyuk). Gordon and Breach Science Publishers, Amsterdam, Vol. 5, 51–61.

[6] Gruyitch, Ly.T. (1998). Consistent Lyapunov methodology: Global exponential stability of sets. *Proc. 5th International Seminar on Stability and Oscillations of Nonlinear Control Systems*, Moscow, June 3–5, 5–8.

[7] Gruyitch, Ly.T. (in print) Physical Continuity and Uniqueness Principle. Exponential Natural Tracking Control. *Int. J. on Neural, Parallel and Scientific Computing*, Atlanta, Georgia.

[8] Gruyitch, Ly.T. (1998). Exponential stabilizing natural tracking control of robots: theory. *Proc. of the Third ASCE Speciality Conference on Robotics for Challenging Environments*, Albuquerque, New Mexico, USA, April 26–30, 286–292.

[9] Gruyitch, Ly.T. (in print). Vector Lyapunov function synthesis of aircraft control. *Proc. 2nd Int. Conference on Nonlinear Problems in Aviation and Aerospace*, Daytona Beach, Florida, USA, April 29—May 1.

[10] Newton, I. (1952, the first addition 1687). *Mathematical Principles of Natural Philosophy. Book I. The motion of bodies*. William Benton, Publisher. Encyclopaedia Britannica, Inc., Chicago.

[11] Grujić, Lj.T. (1997). Multiple time scale systems, time and modelling. *Proc. of the IFAC Conference on Control of Industrial Systems*, Elsevier Science Ltd., Oxford, 189–194.

[12] Lur'e, A.I. and Postnykov, V.N. (1944). On the stability theory of control systems. *Prikl. Mat. Mekh.*, **VIII**(3), 246–248 (Russian).

[13] Popov, V.M. (1973). *Hyperstability of Control Systems*. Springer-Verlag, New York.

[14] Yakubovich, V.A. (1964). Method of nonlinear matrix inequalities in the stability theory of nonlinear control systems. 1. Absolute stability of forced oscillations. *Avtomatika i Telemekhanika*, **XXV**(7), 1017–1029 (Russian).

[15] Kalman, R.E. (1963). Liapunov functions for the problem of Lur'e in automatic control. *Proc. Nat. Acad. Sci. U.S.A.*, **49**(2), 201–205.

[16] Pyatnitskiy, E.S. (1970). Absolute stability of time-varying second order control systems with one time varying element. *Avtomatika i Telemekhanika*, **1**, 5–16 (Russian).

[17] Pyatnitskiy, E.S. (1970). Absolute stability criteria for nonlinear systems. Free and forced motions. Automatika i Telemekhanika, **3**, 5–15 (Russian).

[18] Šiljak, D.D. (1971). On Exponential Absolute Stability. *Publications of the Faculty of Electrical Engineering of the University of Belgrade, Series: Electronics, Telecommunications and Automatics*, Belgrade, Serbia, Yugoslavia, **53–58**, 25–30.

[19] Šiljak, D.D. and Sun, C.K. (1972). On Exponential Absolute Stability. *Int. J. Control*, **16**, 1003–1018.

2.5 ADVANCES IN STABILITY THEORY OF LYAPUNOV: OLD AND NEW*

V. LAKSHMIKANTHAM[1] and S. LEELA[2]

[1] Department of Applied Mathematics, Florida Institute of Technology, Melbourne, USA
[2] Department of Mathematics, SUNY at Geneseo, Geneseo, USA

1 Introduction

It is well known that Lyapunov's second method is an interesting and fruitful technique that has gained increasing significance and has given decisive impetus for modern development of stability theory of dynamic systems. A manifest advantage of this method is that it does not require the knowledge of solutions of differential equations and thus has exhibited a great power in real world applications. There are several books available expounding the main ideas of Lyapunov's second method.

It is now well recognized that the concept of Lyapunov functions can be utilized to investigate various qualitative and quantitative properties of nonlinear differential equations. Lyapunov functions serve as vehicles to transform a given complicated differential system into a relatively simpler system, and therefore, it is enough to investigate the properties of this simpler system [8, 10, 13, 14, 16, 18, 25, 26, 27, 29]. It is also being realized that the same versatile tools are adaptable to study entirely different nonlinear systems, and these effective methods offer an exciting prospect for further advancement.

2 Lyapunov Stability

Consider the differential system

$$\frac{dx}{dt} = f(t,x), \quad x(t_0) = x_0, \quad t_0 \geq 0, \tag{1}$$

Advances in Stability Theory (Ed.: A.A. Martynyuk). Stability and Control: Theory, Methods and Applications, Taylor & Francis, London, **13** (2003) 121–134.

where $f \in C[\mathbb{R}_+ \times S_\rho, \mathbb{R}^n]$ and $S_\rho = [x \in \mathbb{R}^n : |x| < \rho]$. Assume, for convenience, that the solutions $x(t) = x(t, t_0, x_0)$ of (1) exist, and are unique for $t \geq t_0$.

The well known original theorems of Lyapunov for stability and asymptotic stability of the trivial solution of (1) have been refined, extended and generalized in various aspects. We shall discuss below some important trends that have occurred in recent years.

2.1 Loss of decrescentness

Let us first consider loss of decrescentness, which needs boundedness of $f(t, x)$ to yield asymptotic stability and is due to Marachkov. An interesting generalization of this result due to Salvadori uses two Lyapunov functions, see [11, 13]. The first Lyapunov function V serves to derive stability and the second relates suitably to the first one. The advantage is that one can utilize the monotone character of $V(t, x(t))$.

Theorem 2.1 *Assume that*

(i) $V \in C'[\mathbb{R}_+ \times S_\rho, \mathbb{R}_+]$, V *is positive definite,* $V(t, 0) \equiv 0$ *and* $V'(t, x) \leq -c(W(t, x))$ *on* $\mathbb{R}_+ \times S_\rho$, *where* $c \in \mathcal{K}$;

(ii) $W \in C'[\mathbb{R}_+ \times S_\rho, \mathbb{R}_+]$, W *is positive definite and* $W'(t, x)$ *is bounded from above or from below on* $\mathbb{R}_+ \times S_\rho$.

Then $x = 0$ of (1) is asymptotically stable.

Here and later, the class $\mathcal{K} = [a \in C[\mathbb{R}_+, \mathbb{R}_+] : a(u)$ is strictly increasing in u and $a(0) = 0]$.

Theorem 2.1 may be generalized in different ways. The following is a typical result which shows the ideas involved.

Theorem 2.2 *Suppose that*

(i) $V_1, V_2 \in C[R_+ \times S(\rho), R_+]$, V_1, V_2 *are locally Lipschitzian in* x, $V_1(t, 0) = V_2(t, 0) \equiv 0$, V_1 *is positive definite and* $D^+V(t,x) \leq -\lambda(t)C(V_1(t,x))$ *on* $R_+ \times S(\rho)$, *where* $V = V_1 + V_2$ *and* $\lambda \in C[R_+, R_+]$ *is integrally positive, that is* $\int_I \lambda(s)\, ds = \infty$ *whenever* $I = \bigcup_{i=1}^\infty [\alpha_i, \beta_i]$, $\alpha_i < \beta_i < \alpha_{i+1}$ *and* $\beta_i - \alpha_i \geq \delta > 0$;

(ii) *for every* α, $\alpha_1 > 0$ *and for every* $y \in C[R_+, S(\rho)]$, *the inequalities* $V(t, y(t)) \leq \alpha$, $V_1(t, y(t)) \geq \alpha_1$, *imply that the function*

$$\int_0^t [D^+V_2(s, y(s))]_{+-}\, ds \quad \text{is uniformly continuous on} \quad R_+,$$

where $[\cdot]_{+-}$ means that either the positive part $[\cdot]_+$ or the negative part $[\cdot]_-$ is considered for all $s \in R_+$.

Then $x = 0$ asymptotically stable and $V_2(t, x(t))$ has a finite limit as $t \to \infty$.

Setting $W = V_1$ and $V = V_1 + V_2$ so that $V_2 = V - V_1$, we see that V_2 satisfies condition (ii) and hence if $\lambda(t) \equiv 1$, Theorem 2.2 reduces to Theorem 2.1. Furthermore, one can modify Theorem 2.2 to yield partial asymptotic stability with the additional gain of showing the rest of the components of $x(t)$ tend to a finite limit as $t \to \infty$. This has a good application in mechanical systems. See Hatvani [3, 4] for details.

2.2 Loss of negative definiteness

The loss of negative definiteness in Lyapunov's second theorem needs compensating conditions to guarantee asymptotic stability. If $f(t, x)$ is autonomous one has the invariance principle of LaSalle [18, 27] which followed Krasovskii's result of periodic systems [10]. For extensions to general systems, see Matrosov [10, 18, 27].

A stability property may be considered as a family of properties depending on some parameters. As a result, when we employ a single Lyapunov function to prove a given stability property, the Lyapunov function used is assumed to play the role for every choice of these parameters. Consequently, if we utilize a family of Lyapunov functions instead of a single one, it is natural to expect that each member of the family has to satisfy weaker requirements. This is precisely the idea of using a perturbing family of Lyapunov functions.

Theorem 2.3 *Assume that*

(i) $V_1 \in C[R_+ \times S(\rho), R_+]$, V_1 *is locally Lipschitzian in* x, $V_1(t, 0) \equiv 0$ *and* $D^+V_1(t, x) \leq 0$ *on* $R_+ \times S(\rho)$;

(ii) *for every* $0 < \eta < \rho$, *there exist a family* $V_{2\eta} \in C[R_+ \times S(\rho) \cap S^c(\eta), R_+]$, $V_{2\eta}$ *is locally Lipschitzian in* x *and for* $(t, x) \in R_+ \times S(\rho) \cap S^c(\eta)$, $b(\|x\|) \leq V_{2\eta}(t, x) \leq a(\|x\|)$, $a, b \in \mathcal{K}$, $D^+V_1(t, x) + D^+V_{2\eta}(t, x) \leq 0$.

Then $x = 0$ is equistable.

Of course if $V_1 \equiv 0$, Theorem 2.3 yields uniform stability showing the advantage of utilizing a family of Lyapunov functions in proving uniform stability properties. For details, see [7, 13].

2.3 Practical stability

Complete stability (global asymptotic stability) is a more desirable feature in applications. Sometimes, even instability may be good enough. Since the desired state of a system may be mathematically unstable but the system may oscillate sufficiently near this state so that its performance is considered acceptable. For example, aircrafts, missiles, chemical processes or space vehicles.

Definition 2.1 The *system* (1) is said to be *practically stable*, if given (λ, A) with $0 < \lambda < A$, we have $|x_0| < \lambda$ implies $|x(t)| < A$, for $t \geq t_0$. See [14] for further details.

2.4 Eventual stability

When the trivial solution of (1) does not exist, we may still have stability eventually, which generalizes Lyapunov stability. See [10].

Definition 2.2[1] The *system* (1) is said to be *eventually stable*, if, for each $\epsilon > 0$, there exist $\delta(\epsilon) > 0$, $\tau(\epsilon) > 0$ such that

$$|x_0| < \delta \quad \text{implies} \quad |x(t)| < \epsilon, \quad t \geq \tau(\epsilon).$$

A generalization of eventual stability is known as M_0-stability discussed in [13]. Here we have an estimate of solutions as $|x(t, t_0, x_0)| \leq |x_0| + \lambda(t_0)$, $t \geq t_0$, where $\int_{t_0}^{t_0+1} \lambda(s)\, ds \to 0$ as $t_0 \to \infty$. If $\lambda(t_0) \to 0$ as $t_0 \to \infty$, we get eventual stability.

3 Stability in Terms of Two Measures

The concept of Lyapunov stability has given rise to several other new notions of stability that are important in applications. For example, partial stability, conditional stability orbital stability, stability of invariant sets, Lagrange stability and boundedness concepts to name a few. In order to unify a variety known concepts of stability and boundedness, it is found beneficial to employ two different measures and obtain criteria in terms of these distinct measures [16]. Let us define such a concept.

Definition 3.1 Let $\Gamma = [h \in C[\mathbb{R}_+ \times \mathbb{R}^n, \mathbb{R}_+] : \inf h(t, x) = 0$ for $(t, x) \in \mathbb{R}_+ \times \mathbb{R}^n]$ and $h_0, h \in \Gamma$. Then the differential *system* (1) is said to be (h_0, h)-*stable*, if for each $\epsilon > 0$ and $t_0 \in \mathbb{R}_+$, there exists a $\delta = \delta(t_0, \epsilon) > 0$ such that $h_0(t_0, x_0) < \delta$ implies $h(t, x(t)) < \epsilon$, $t \geq t_0$.

For this concept to make sense, we need to have a relationship between the two measures. We say that h_0 is finer than h if there exists a $\rho > 0$ and a function $\varphi \in \mathcal{K}$ such that $h(t, x) < \varphi(h_0(t, x))$ whenever $h_0(t, x) < \rho$.

A few choices of the two measures given below will demonstrate the generality of Definition 3.1, which reduces to

(1) the stability of the trivial solution of (1) if $h(t, x) = h_0(t, x) = |x|$;

---

[1]Remark of Editors. See T.Yoshizawa, *Stability Theory by Liapunov's Second Method*, The Math. Soc. of Japan, 1966 for details.

(2) the partial stability of the trivial solution of (1) if $h(t,x) = |x|_s$, $1 \leq s < n$ and $h_0(t,x) = |x|$;

(3) the eventual stability of (1) if $h(t,x) = |x|$ and $h_0(t,x) = |x| + \sigma(t)$, where $\sigma \in \mathcal{L} = [\sigma \in C[\mathbb{R}_+, \mathbb{R}_+] \colon \sigma(t)$ is strictly decreasing in t with $\sigma(t) \to 0$ as $t \to \infty]$;

(4) the stability of the invariant set $A \subset \mathbb{R}^n$, if $h(t,x) = h_0(t,x) = d(x,A)$, where d is the distance function;

(5) the conditional stability of the trivial solution of (1) if $h_0(t,x) = |x| + d(x,M)$, $h(t,x) = |x|$, where M is the k-dimensional manifold containing origin;

(6) the stability of conditionally invariant set B with respect to A, $A \subset B \subset \mathbb{R}^n$, if $h(t,x) = d(x,B)$, $h_0(t,x) = d(x,A)$.

To find sufficient conditions for any type of stability concepts to hold, the use of comparison principle offers the most general setup. It is well known that one can achieve this by employing a single or a vector Lyapunov function [10, 18, 27]. The method of vector Lyapunov functions offers a very flexible and effective mechanism to investigate qualitative properties of nonlinear differential equations, including large scale systems [30, 31].

We have seen that using the technique of perturbing Lyapunov functions and employing a family of Lyapunov functions are helpful in discussing nonuniform properties of solutions of differential systems under weaker assumptions.

In [15], a new approach is initiated to the method of vector Lyapunov functions by combining the ideas involved in the foregoing techniques and this helps in distributing the burden between groups of components of the vector Lyapunov function and the comparison function. As a result, this approach contributes to the enrichment of the method of vector Lyapunov functions by including and improving earlier known results and enhancing the applicability of the method.

Since the method of variation of parameters and the comparison principles via Lyapunov-like functions are both extremely useful, it is natural to combine these two approaches in order to exploit the benefits of the two important methods.

Consider the known differential system

$$\frac{dy}{dt} = F(t,y), \quad y(t_0) = x_0, \qquad (2)$$

where $F \in C[\mathbb{R}_+ \times \mathbb{R}^n, \mathbb{R}^n]$. Let $y(t, t_0, x_0)$ be the unique solution of (2) existing for $t \geq t_0$. Then we can formulate the comparison principle in terms of the variational Lyapunov method with inequalities between vectors being component-wise [17].

Theorem 3.1 *Assume that*

(A_1) $V \in C[\mathbb{R}_+ \times \mathbb{R}^n, \mathbb{R}^N]$, $V(t,x)$ and $|y(t,s,x)|$ are locally Lipschitzian in x for each (t,s);

(A_2) for $t_0 \leq s \leq t$,

$$D_-V(t,s,x) = \liminf_{h \to 0^-} \frac{1}{h}\big[V(s+h, y(t,s+h, x+hf(s,x))) - V(s, y(t,s,x))\big]$$
$$\leq g(t, s, V(s, y(t,s,x)));$$

(A_3) $g \in C[\mathbb{R}_+^2 \times \mathbb{R}_+^N, \mathbb{R}^N]$, $g(t,s,u)$ is quasimonotone nondecreasing in u for each (t,s) and $r(t,s,t_0,u_0)$ is the maximal solution of

$$\frac{du(s)}{ds} = g(t,s,u(s)), \quad u(t_0) = u_0 \geq 0,$$

existing on $t_0 \leq s \leq t < \infty$.
Then $V(t_0, y(t, t_0, x_0)) = u_0$ implies

$$V(t, x(t, t_0, x_0)) \leq r_0((t, t_0, V(t_0, y(t, t_0, x_0)))), \quad t \geq t_0,$$

where $r_0(t, t_0, u_0) = r(t, t_0, u_0)$.

For various choices of (2), one gets several special results. For example, $F \equiv 0$ in (2) yields the well known comparison theorem in terms of vector Lyapunov function. $N = 1$ gives the comparison result with a single Lyapunov function.

Once we have this comparison result, it is not difficult to investigate stability properties of (1), see [17]. However, in this set up, to prove a result analogous to the second theorem of Lyapunov, one needs the concept of strict stability which we define now.

Definition 3.2 The *trivial solution* of (1) is said to be *strictly stable* if given $\epsilon_1 > 0$ and $t_0 \in \mathbb{R}_+$, there exists a $\delta_1 > 0$ such that $|x_0| < \delta_1$ implies $|x(t, t_0, x_0)| < \epsilon_1$, $t_1 \geq t_0$, and for every $\delta_2 \leq \delta_1$, there exists an $\epsilon_2 < \delta_2$ such that $\delta_2 < |x_0|$ implies $\epsilon_2 < |x(t, t_0, x_0)|$ for $t \geq t_0$.

For further discussion on strict stability, see [19].

4 Cone Valued Lyapunov Functions

An unpleasant fact in the method of vector Lyapunov functions is the requirement of a quasimonotone nondecreasing property of the comparison system, which is defined as $x \leq y$ and $x_i = y_i$ for some $1 \leq i \leq N$ implies $g_i(t,x) \leq g_i(t,y)$. Since the comparison systems with the desired property like stability exist without satisfying the quasimonotone property, the limitation of the method of vector Lyapunov functions is obvious. To circumvent this unpleasant requirement, one needs to employ arbitrary cones rather than the standard cone $\mathbb{R}_+^N$, which is used in the method of vector Lyapunov functions. We shall develop the method of cone valued functions now [5, 8, 20]. See for matrix-valued Lyapunov functions [25, 26].

Let $K \subseteq \mathbb{R}^N$ be a cone, that is, K is closed, convex with $\lambda K \subset K$ for all $\lambda \geq 0$ and $K \cap \{-K\} = \{0\}$ with interior $K^0 \neq \emptyset$. For any $x, y \in \mathbb{R}^N$, we let $x \leq y$ iff $y - x \in K$. Let $K^* = [\varphi \in \mathbb{R}^N : \varphi(x) \geq 0$ for all $x \in K]$ and $K_0^* = K^* - \{0\}$. Then we can define the quasimonotone property of a function $F : \mathbb{R}^N \to \mathbb{R}^N$ relative to K as follows:

$$x \leq y \quad \text{and} \quad \varphi(y - x) = 0 \quad \text{for}$$
$$\varphi \in K_0^* \implies \varphi(F(y) - F(x)) \geq 0.$$

Then we can state a general set of criteria for (h_0, h)-stability unifying several concepts.

Theorem 4.1 *Assume that*

(A_0) $h_0, h \in \Gamma$ *and* h_0 *is finer than* h;

(A_1) $V \in C[\mathbb{R}_+ \times \mathbb{R}^n, K]$ *and* $V(t, x)$ *is locally Lipschitzian in* x *relative to* K;

(A_2) $Q_0, Q \in \Sigma = [Q \in C[K, \mathbb{R}_+] : Q(0) = 0$ *and* $Q(w)$ *is increasing in* w *relative to* $K]$ *and* Q_0 *is finer than* Q;

(A_3) $g \in C[\mathbb{R}_+ \times K, \mathbb{R}^N]$ *and for* $(t, x) \in S(h, \rho)$, $D^+ V(t, x) \leq g(t, V(t, x))$, *where* $g(t, w)$ *is quasimonotone in* w *relative to* K;

(A_4) $b(h(t, x)) \leq Q(V(t, x))$ *if* $h(t, x) < \rho$ *and* $Q_0(V(t, x)) \leq a(h_0(t, x))$ *if* $h_0(t, x) < \rho_0$, *where* $a, b \in \mathcal{K}$.

Then the (Q_0, Q)-*stability properties of*

$$\frac{dw}{dt} = g(t, w), \quad w(t_0) = w_0 \geq 0,$$

imply the corresponding (h_0, h)-*stability properties of (1) respectively.*

The special case $K = \mathbb{R}_+^N$, $Q(w) = Q_0(w) = \sum_{i=1}^{N} w_i$, $h(t, x) = h_0(t, x) = |x|$ is the method of vector Lyapunov functions.

Consider the example $g(w) = Aw$, where $A = \begin{pmatrix} 0 & 1 \\ -k^2 & -2\beta \end{pmatrix}$, $k, \beta > 0$ so that Aw is not quasimonotone relative to cone $\mathbb{R}_+^2$. But the cone $K = [u : u = d_1 w_1 + d_2 w_2, w_i \geq 0, i = 1, 2]$, where $d_1 = \begin{pmatrix} 1 \\ -b \end{pmatrix}$, $d_2 = \begin{pmatrix} 0 \\ 1 \end{pmatrix}$, we find, if $b = \beta > k$, will satisfy the conditions of the method of cone valued Lyapunov functions. See [5, 8, 20] for details.

5 Analysis of Invariant Sets

In the notion of eventual stability, we note that, although $\{0\}$ is not an invariant set of (1) (the trivial solution), it is so asymptotically. Hence eventual stability is nothing but Lyapunov stability of asymptotically invariant set $\{0\}$. See [10].

Definition 5.1 The *set $x = 0$* is *asymptotically invariant* relative to (1) if given any decreasing sequence $\{\epsilon_p\}$, $\epsilon_p \to 0$ as $p \to \infty$, there exists an increasing sequence $\{\tau_p\}$, $\tau_p \to \infty$ as $p \to \infty$ such that $|x(t, t_0, 0)| < \epsilon_p$, $t \geq \tau_p$, $t_0 \geq \tau_p$.

This leads to the idea that once a different kind of invariant set is introduced, the consideration of its stability properties runs parallel to the known results of standard invariant sets. Thus one can think of analyzing various invariant sets whose study blends the qualitative and quantitative properties of systems yielding several different stability concepts [6].

Let $A, B \in C[\mathbb{R}_+, \Omega]$ such that $A(t) \subset B(t)$ for $t \in \mathbb{R}_+$, where Ω denotes all nonempty closed subsets of $\mathbb{R}^n$, which together with the Hausdorff distance becomes a metric space.

Definition 5.2 The *set $B(t)$* is said to be

(i) *conditionally invariant* relative to $A(t)$ and (1) if for certain $t_0 \geq 0$, $x_0 \in A(t_0)$ implies $x(t) \in B(t)$, $t \geq t_0$;

(ii) *conditionally quasi-invariant* relative to $A(t)$ and (1) if for certain $t_0 \geq 0$, $x_0 \in A(t_0)$, there exists a $T > 0$ such that $x(t) \in B(t)$ for $t \geq t_0 + T$.

In this terminology, we can have uniform concepts as well if they don't depend on t_0. Moreover, if $A(t) \equiv B(t)$, one gets self-invariant sets. What we consider in the theory of Lyapunov stability can be called uniform self-invariant sets.

One can consider moving invariant sets when the system (1) depends on a parameter, see [1, 2, 9, 21]. Consider the differential system

$$\frac{dx}{dt} = f(t, x, \lambda), \quad x(t_0) = x_0, \quad t_0 \geq 0, \tag{3}$$

where $f \in C[R_+ \times R^n \times R^d, R^n]$ and $\lambda \in R^d$ is an uncertain parameter.

Also consider the comparison equation

$$\frac{du}{dt} = g(t, u, \mu), \quad u(t_0) = u_0 \geq 0, \tag{4}$$

where $g \in C[R_+^3, R]$ and $\mu = \mu(\lambda) \geq 0$ is a parameter depending on λ. We assume, for convenience, that the solutions of (3) and (4) exist and unique for $t \geq t_0$. We need the following definition with respect to moving invariant set of the system (3).

Definition 5.3 Let $\rho_0 \leq r_0 \leq r \leq \rho$ depending on λ. Then we say that the set $B = \{x \in R^n : \rho_0 \leq |x| \leq \rho\}$ is *conditionally invariant* with respect to $A = \{x \in R^n : r_0 \leq |x| \leq r\}$ and is *uniformly asymptotically stable (UAS)* relative to (3) if

(i) $r_0 \leq |x_0| \leq r \implies \rho_0 \leq |x(t)| \leq \rho$, $t \geq t_0$;

(ii) given $\epsilon > 0$, and $t_0 \in R_+$

(a) there exists a $\delta = \delta(\epsilon) > 0$ such that $r_0 - \delta < |x_0| < r + \delta \implies \rho_0 - \epsilon < |x(t)| < \rho + \epsilon,\ t \geq t_0$;

(b) there exists a $\delta_0 > 0$ and a $T = T(\epsilon) > 0$ such that $r_0 - \delta_0 < |x_0| < r + \delta_0 \implies \rho_0 - \epsilon < |x(t)| < \rho + \epsilon,\ t > t_0 + T$,

where $x(t) = x(t, t_0, x_0)$ is the solution of (3).

Definition 5.4 Relative to the comparison equation (4), we say that $\Omega = \{u \in R : R_0 \leq u \leq R\}$ is *invariant* and is *UAS* relative to (4) if

(i) $R_0 \leq u_0 \leq R \implies R_0 \leq u(t) \leq R,\ t \geq t_0$;

(ii) given $\epsilon > 0$ and $t_0 \in R_+$,

(a) there exists a $\delta = \delta(\epsilon) > 0$ such that

$$R_0 - \delta < u_0 < R + \delta \implies R_0 - \epsilon < u(t) < R + \epsilon,\quad t \geq t_0;$$

(b) there exists a $\delta_0 > 0$ and a $T = T(\epsilon) > 0$ such that

$$R_0 - \delta_0 < u_0 < R + \delta_0 \implies R_0 - \epsilon < u(t) < R + \epsilon,\quad t \geq t_0 + T;$$

where $u(t) = u(t, t_0, x_0)$ is the solution of (4).

Now we are in a position to state our needed result.

Theorem 5.1 *Assume that*

(A_0) *for each $\lambda \in R^d$, there exist $r = r(\lambda)$, $r_0 = r(\lambda) \geq 0$, $r_0 \leq r$ satisfying $r \to 0$ as $|\lambda| \to 0$ and $r_0 \to \infty$ as $|\lambda| \to \infty$;*

(A_1) *there exists a $V \in C[R_+ \times R^n, R_+]$, $V(t, x)$ is locally Lipschitzian in x for each $t \in R_+$, and for $a, b \in K$, $b(|x|) \leq V(t, x)$ if $|x| \geq r$, $V(t, x) \leq a(|x|)$ if $|x| \leq r_0$;*

(A_2) $D^+V(t, x) \leq g(t, V(t, x), r(\lambda))$ *if* $|x| \geq r$,
$D^+V(t, x) \geq g(t, x, r_0(\lambda))$ *if* $|x| \leq r$;

(A_3) *for each $r_0 \leq r$, there exist $R_0 \leq R$ such that $R = a(r) = b(\rho)$, $R_0 = b(r_0) = a(\rho_0)$, where $\rho_0 \leq r_0 \leq r \leq \rho$ and $R \to 0$ as $r \to 0$ and $R_0 \to \infty$ as $r_0 \to \infty$;*

(A_4) Ω *is invariant and is UAS.*

Then B is conditionally invariant with respect to A and is UAS relative to (3).

As an application of Theorem 5.1, we shall consider the control of uncertain differential system of the form

$$\frac{dx}{dt} = f_0(t, x, \lambda) + B(t, x) F(t, x, u, \lambda),\quad x(t_0) = x_0,\quad t_0 \geq 0, \qquad (5)$$

under the following assumptions:

(A_0) $f_0 \in C[R_+ \times R^n \times \Omega_0, R^n]$, $B \in C[R_+ \times R^n, R^{n \times m}]$ and $F \in C[R_+ \times R^n \times R^m \times \Omega_0, R^n]$, where $\Omega_0 \subset R^d$ is a nonempty set and $u \in R^m$ is the control function;

(A_1) there exist $r_0 = r_0(\lambda) \leq r = r(\lambda)$ such that

$$V'_{f_0}(t,x) \leq -c_1(V(t,x)) \quad \text{if } |x| \geq r,$$
$$V'_f(t,x) \geq -c_2(V(t,x)) \quad \text{if } |x| \leq r_0,$$

where $V \in C'[R_+ \times R^n, R_+]$, $V'_{f_0}(t,x) = V_t(t,x) + V_x(t,x)f_0(t,x,\lambda)$ and $c_1, c_2 \in K$;

(A_2) $b(|x|) \leq V(t,x)$ if $|x| \geq r$, $V(t,x) \leq a(|x|)$ if $|x| \leq r_0$, where $a, b \in K$;

(A_3) for $|x| \geq r$, $u^T F(t,x,u,\lambda) \geq -\beta_1(t,x)|u| + \beta_2(t,x)|u|^2$, where $\beta_1, \beta_2 \in C[R_+ \times R^n, R_+]$, $\beta_1 \leq \beta_2 \tilde{\rho}$, $\beta_1 \leq \kappa$, $\tilde{\rho}, \kappa \in C[R_+ \times R^n, R_+]$;

(A_4) for $|x| \leq r_0$, $u^T F(t,x,u,\lambda) \leq -\gamma_1(t,x)|u| + \gamma_2(t,x)|u|^2$, where $\gamma_1, \gamma_2 \in C[R_+ \times R^n, R_+]$, $\gamma_1 \geq \gamma_2 \tilde{\rho}$, $\gamma_1 \geq \kappa$;

(A_5) $P = [p_\mu \in C[R_+ \times R^n, R^n], \mu > 0]$ is the stabilizing family of controllers satisfying [1]

$$|\alpha|p_\mu = -|p_\mu|\alpha, \quad \text{where } \alpha = B^T V_x^T \text{ and } \eta = \kappa\alpha,$$

and if $\eta > 0$, $|x| \geq r$, $|p_\mu| \geq \tilde{\rho}\left(1 - \frac{r_0}{|\eta|}\right)$.

We are now in a position to prove the following result.

Theorem 5.2 *Assume that the conditions (A_0) to (A_5) hold. Suppose further that $c_2^{-1}(u) \leq c_1^{-1}(u)$. Then the set B is conditionally invariant with respect to A and is UAS relative to (5).*

Proof Let us consider the case $|x| \geq r$. Then we have

$$V'_{f_0}(t,x) \leq -c_1(V(t,x))$$

and

$$\alpha(t,x)F(t,x,p_\mu,\lambda) = -\frac{|\alpha(t,x)|}{|p_\mu(t,x)|} F(t,x,p_\mu,\lambda)p_\mu$$
$$\leq |\alpha(t,x)|[\beta_1(t,x) - \beta_2(t,x)|p_\mu(t,x)|]$$
$$\leq |\alpha(t,x)|\left[\beta_1(t,x) - \beta_2(t,x)\tilde{\rho}(t,x)\left(1 - \frac{r}{|\eta|}\right)\right]$$
$$\leq |\alpha(t,x)|\beta_1(t,x)\frac{r}{|\eta|} \leq r,$$

and consequently,

$$V'_{(5)}(t,x) \leq -c_1(V(t,x)) + r, \quad \text{if } |x| \geq r.$$

Similarly, if $|x| \leq r_0$, we get
$$V'_{f_0}(t,x) \geq -c_2(V(t,x)),$$
and

$$\alpha(t,x)F(t,x,p_\mu,\lambda) = -\frac{|\alpha(t,x)|}{|p_\mu(t,x)|}F(t,x,p_\mu,\lambda)p_\mu$$

$$\geq |\alpha(t,x)|[\gamma_1(t,x) - \gamma_2(t,x)|p_\mu(t,x)|]$$

$$\geq |\alpha(t,x)|\left[\gamma_1(t,x) - \gamma_2(t,x)\tilde{\rho}(t,x)\left(1 - \frac{r_0}{|\eta|}\right)\right]$$

$$\geq |\alpha(t,x)|\gamma_1(t,x)\frac{r_0}{|\eta|} \geq r_0,$$

and, as a result, we have
$$V'_{(5)}(t,x) \geq -c_2(V(t,x)) + r_0, \quad \text{if } |x| \leq r_0.$$
This implies that
$$g(t,u,r) = -c_2(u) + r,$$
$$g(t,u,r_0) = -c_2(u) + r_0,$$

and therefore $u = c_1^{-1}(r) = R$, $u = c_2^{-1}(r_0) = R_0$. Hence, in view of the properties of a, b, c_1, c_2, we can find $\rho_0 \leq r_0 \leq r \leq \rho$ such that $R = a(r) = b(\rho)$, $R_0 = b(r_0) = a(\rho_0)$ and $R_0 \leq R$.

To apply Theorem 5.1, we have to show that the set $\Omega = [u \in R_+ : R_0 \leq u \leq R]$ is invariant and is UAS. In view of the specific nature of the function g, it is not difficult to show this following the proof of Theorem 5.1. See [2, 9, 21, 22, 31] for details.

6 New Meaning of Stability

Lyapunov stability compares the phase-space positions of solutions at exactly simultaneous instants, namely
$$\sup_{t \geq 0} |x(t) - x_0(t)| < \epsilon, \tag{6}$$
where as orbital stability compares at any two unrelated instants, namely,
$$\sup_{t \geq 0} \inf_{s \in R_+} |x(t) - x_0(s)| < \epsilon. \tag{7}$$

Clearly (6) seems too stringent a requirement and (7) seems too loose a requirement. This suggests a middle of the road topology between (7) and (6).

Let $\mathcal{CL}$ be the space of all functions from $\mathbb{R}_+ \to \mathbb{R}_+$ each function $s(t)$ representing a clock and $s(t) = t$ being the perfect clock. Let τ be any topology in $\mathcal{CL}$.

Definition 6.1 The *motion* $x(\cdot, x_0)$ is τ-*stable* if given $\epsilon > 0$ and τ-neighborhood U of the perfect clock, there exists a $\delta > 0$ such that for each y_0 with $|x_0 - y_0| < \delta$, there is a clock $s(\cdot) \in U$ with $|x(t, y_0) - x(s(t), x_0)| < \epsilon$, $t \geq 0$.

As examples, one can consider the topologies

(O) the chaotic topology, where open sets are solely φ and the entire clock space;
(U) the topology defined by the base $U = \big[s(\cdot) \colon \sup_{t \geq 0} |s(t) - s_0(t)| < \epsilon\big]$;
(L) the discrete topology, where every set in $\mathcal{CL}$ is open;
(S) the topology defined by the base

$$S = \big[s(\cdot) \colon |s(0) - s_0(0)| < \epsilon, \sup\{|t_2 - t_1|^{-1}|s(t_2) - s(t_1) - s_0(t_2) + s_0(t_1)| \colon \\ t_1, t_2 \in \mathbb{R}_+, t_1 \neq t_2\} < \epsilon\big].$$

See [28] for further details.

In the investigation of IVPs of differential equations, we have been partial to initial time all along in the sense that only perturb or change the dependent variable or space variable and keep the initial time unchanged. It appears, however, important to vary the starting time as well since it is impossible not to make errors in the starting time. If we do change the initial time for each solution, then we are faced with the problem of comparing any two solutions which differ in the initial starting time. There may be several ways of comparing and to each choice of measuring the difference, we may end up with a different result. In [23], this approach is initiated. Let $x(t, t_0, y_0)$ be the given solution relative which stability is to be considered. Then one can define the stability as follows:

The solution $x(t, t_0, y_0)$ is stable if given $\epsilon > 0$, there exists δ, $\sigma > 0$ such that $|x_0 - y_0| < \delta$, $|t_0 - \tau_0| < \sigma$ implies $|x(t + \tau_0 - t_0, \tau_0, x_0) - x(t, t_0, y_0)| < \epsilon$, $t \geq t_0$. For details see [12, 23]. It would be interesting to absorb this approach in τ-stability and obtain sufficient conditions.

References

[1] Corless, M. and Leitmann, G. (1990). Deterministic control of uncertain systems: a Lyapunov theory approach. *Deterministic Control of Uncertain Systems* (Ed.: A.S.I. Zinober), Chapter 11, Peter Peregrinus, London.
[2] Drici, Z. and Lakshmikantham, V. (1995). Stability of conditionally invariant sets and controlled uncertain systems on time scales. *Math. Probl. in Eng.*, **1**, 1–10.
[3] Hatvani, L. (1985). On partial asymptotic stability and instability III. *Acta Sci. Math*, **49**, 157-167.
[4] Hatvani, L. (1984). On the asymptotic stability of nondecrescent Lyapunov function. *Nonlin. Anal.*, **8**, 67–77.
[5] Koksal, S. and Lakshmikantham, V. (1996). Higher derivatives of Lyapunov functions and cone valued Lyapunov functions. *Nonlin. Anal.*, **26**, 1555–1604.

[6] Ladde, G.S. and Leela, S. (1972). Analysis of invariant sets. *Annali di Mat. Para ed Appl.*, **XCIV**, 283–289.

[7] Lakshmikantham, V. and Leela, S. (1976). On perturbing Lyapunov functions. *Math. Sys. Theory*, **10**, 85–90.

[8] Lakshmikantham, V. and Leela, S. (1977). Cone-valued Lyapunov functions. *Nonlin. Anal.*, **1**, 215–222.

[9] Lakshmikantham, V. and Leela, S. (1995). Controlled uncertain dynamic systems and stability of moving invariant sets. nit Prob. in Nonlin. Anal. in Eng. Sys., **2**, 9–13.

[10] Lakshmikantham, V. and Leela, S. (1969). *Differential and Integral Inequalities*, Vol. I. Academic Press, New York.

[11] Lakshmikantham, V. and Martynyuk, A.A. (1992). Some directions of the developments of Lyapunov direct method in the stability theory. *Prikl. Mekh.*, **28**(3), 3–13 (Russian).

[12] Lakshmikantham, V. and Leela, S. Stability criteria for solutions of differential equations relative to initial time difference. nit J. Nonlin. Diff. Eqns., (to appear).

[13] Lakshmikantham, V., Leela, S. and Martynyuk, A.A. (1989). *Stability Analysis of Nonlinear Systems*. Marcel Dekker, New York.

[14] Lakshmikantham, V., Leela, S. and Martynyuk, A.A. (1990). *Practical Stability of Nonlinear Systems*. World Scientific, Singapore.

[15] Lakshmikantham, V., Leela, S. and Rama Mohana Rao, M. (1991). New directions in the method of vector Lyapunov function. *Nonlin. Anal.*, **16**, 255–262.

[16] Lakshmikantham, V. and Liu, X. (1993). *Stability Analysis in Terms of Two Measures*. World Scientific, Singapore.

[17] Lakshmikantham, V., Leela, S. and Liu, X. (1998). Variational Lyapunov method and stability theory. *Math. Prob. in Eng.*, **3**, 555–571.

[18] Lakshmikantham, V., Matrosov, V.M. and Sivasundaram, S. (1991). *Vector Lyapunov Functions and Stability Analysis*. Kluwer Academic Publications, Amsterdam.

[19] Lakshmikantham, V. and Mohapatra, R. Strict stability of differential systems, *Nonlin. Anal.*, (to appear).

[20] Lakshmikantham, V. and Papageorgiou, N.S. (1994). Cone valued Lyapunov functions and stability theory. *Nonlin. Anal.*, **22**, 381–390.

[21] Lakshmikantham, V. and Sivasundaram, S. (1997). Stability of moving invariant sets and uncertain dynamic systems. *Proc. Conf. Nonlinear Problems in Aerospace*, Daytona Beach, FL.

[22] Lakshmikantham, V. and Vatsala, A.S. (1997). Stability of moving invariant sets. In: *Advances in Nonlinear Dynamics* (Eds.: S. Sivasundaram and A.A. Martynyuk). Gordon and Breach Science Publishers, Amsterdam, Vol. 5, 79–83.

[23] Lakshmikantham, V. and Vatsala, A.S. Differential inequalities with initial time difference. *J. Inequalities and Appl.*, (to appear).

[24] Leitmann, G. (1995). On approach to the control of uncertain dynamic systems. *Appl. Math. Comp.*, **70**, 261–272.

[25] Martynyuk, A.A. (1985). On application of the Lyapunov matrix functions in the theory of stability. *Nonlin. Anal.*, **9**, 1495–1501.

[26] Martynyuk, A.A. (1984) The Lyapunov matrix function. *Nonlin. Anal.*, **8**, 1223–1226.

[27] Abdulin, R.Z., Anapolski, L.Y., Kozlov, R.I., Malikov, A.I., Matrosov, V.M., Voronov, A.A. and Zemljakov, A.S. (1996). *Vector Lyapunov Functions in Stability Theory.* World Federation Publishers Company, Atlanta.

[28] Massera, J.L. (1964). The meaning of stability. *Bol. Fac. Ingen. Agrimens. Montevideo,* **8**, 405–429.

[29] Šiljak, D.D. (1978). *Large Scale Dynamic Systems.* North-Holland, New York.

[30] Šiljak, D.D. (1991). *Decentralized Control of Complex Systems.* Academic Press, New York.

[31] Šiljak, D.D., Ikeda, M. and Ohata, Y. (1991). Parametric stability. *Proc. Universita di Genova-Ohio State University First Conf.*, Birkhäuser, 1–20.

2.6 MATRIX LIAPUNOV FUNCTIONS AND STABILITY ANALYSIS OF DYNAMICAL SYSTEMS*

A.A. MARTYNYUK

Institute of Mechanics of National Academy of Sciences of Ukraine, Kiev, Ukraine

1 Introduction

One can hardly name a branch of natural science or technology in which the problems of stability do not claim the attention of scholars, engineers, and experts who investigate natural phenomena or operate designed machines or systems. If, for a process or a phenomenon, for example, atom oscillations or a supernova explosion, a mathematical model is constructed in the form of a system of differential equations, the investigation of the latter is possible either by a direct (numerical as a rule) integration of the equations or by its analysis by qualitative methods.

The direct Liapunov method based on scalar auxiliary function proves to be a powerful technique of qualitative analysis of the real world phenomena (see [9, 10]).

This paper examines new generalizations of the matrix-valued auxiliary function. Moreover the matrix-valued function is a structure the elements of which compose both scalar and vector Liapunov functions applied in the stability analysis of nonlinear systems (see [4, 5]).

Due to the concept of matrix-valued function developed in the paper, the direct Liapunov method becomes yet more versatile in performing the analysis of nonlinear systems dynamics.

2 Relationship Between the Reference Motion and the Zero Solution

Let $2k$ be the order of the system and y_i, $i = 1, 2, \ldots, 2k$, be its i-th state variable. Using basic physical laws (e.g. the law of the energy conservation and the law of

---

*Advances in Stability Theory (Ed.: A.A. Martynyuk). Stability and Control: Theory, Methods and Applications, Taylor & Francis, London, **13** (2003) 135–151.

the material conservation) we can for a large class of systems get state differential equations in the following scalar form

$$\frac{dy_i}{dt} = Y_i(t, y_1, \ldots, y_{2k}), \quad i = 1, 2, \ldots, 2k, \tag{2.1}$$

or in the equivalent vector form

$$\frac{dy}{dt} = Y(t, y), \tag{2.2}$$

where $y = (y_1, y_2, \ldots, y_{2k})^T \in R^{2k}$ and $Y = (Y_1, Y_2, \ldots, Y_{2k})^T$, $Y\colon \mathcal{T} \times R^{2k} \to R^{2k}$. A motion of (2.2) is denoted by $\eta(t; t_0, y_0)$, $\eta(t_0; t_0, y_0) \equiv y_0$, and the reference motion $\eta_r(t; t_0, y_{r0})$. From the physical point of view the reference motion should be realizable by the system. From the mathematical point of view this means that the reference motion is a solution of (2.2),

$$\frac{d\eta_r(t; t_0, y_{r0})}{dt} \equiv Y[t, \eta_r(t; t_0, y_{r0})]. \tag{2.3}$$

Let the Liapunov transformation of coordinates be used,

$$x = y - y_r, \tag{2.4}$$

where $y_r(t) \equiv \eta_r(t; t_0, y_{r0})$. Let $f\colon \mathcal{T} \times R^{2k} \to R^{2k}$ be defined by

$$f(t, x) = Y[t, y_r(t) + x] - Y[t, y_r]. \tag{2.5}$$

It is evident that

$$f(t, 0) \equiv 0. \tag{2.6}$$

Now (2.2)–(2.5) yield

$$\frac{dx}{dt} = f(t, x). \tag{2.7}$$

In this way, the behavior of perturbed motions related to the reference motion (in total coordinates) is represented by the behavior of the state deviation x with respect to the zero state deviation. The reference motion in the total coordinates y_i is represented by the zero deviation $x = 0$ in state deviation coordinates x_i.

3 Main Results

As already mentioned in the introduction the application of matrix Liapunov functions make it possible to establish easily verified stability conditions for the state $x = 0$ of system (2.7) in terms of the property having a fixed sign of special matrices. The results presented in this section demonstrate the opportunities of the matrix Liapunov functions technique.

Theorem 3.1 *Let the vector-function f in system (2.7) be continuous on $R \times \mathcal{N}$ (on $\mathcal{T}_\tau \times \mathcal{N}$). If there exist*

(1) *an open connected time-invariant neighborhood $\mathcal{G} \subset \mathcal{N}$ of the point $x = 0$;*
(2) *a matrix-valued function $U \in C(R \times \mathcal{N}, R^{m \times m})$ and a vector $y \in R^m$ such that the function $v(t, x, y) = y^T U(t, x) y$ is locally Lipschitzian in x for all $t \in R$ ($t \in \mathcal{T}_\tau$);*
(3) *functions $\psi_{i1}, \psi_{i2}, \psi_{i3} \in K$, $\widetilde{\psi}_{i2} \in CK$, $i = 1, 2, \ldots, m$;*
(4) *$m \times m$ matrices $A_j(y)$, $j = 1, 2, 3$, $\widetilde{A}_2(y)$ such that*

(a)
$$\psi_1^T(\|x\|) A_1(y) \psi_1(\|x\|) \leq v(t, x, y) \leq \widetilde{\psi}_2^T(t, \|x\|) \widetilde{A}_2(y) \widetilde{\psi}_2(t, \|x\|)$$
$$\forall (t, x, y) \in R \times \mathcal{G} \times R^m \quad (\forall (t, x, y) \in \mathcal{T}_\tau \times \mathcal{G} \times R^m);$$

(b)
$$\psi_1^T(\|x\|) A_1(y) \psi_1(\|x\|) \leq v(t, x, y) \leq \psi_2^T(\|x\|) A_2(y) \psi_2(\|x\|)$$
$$\forall (t, x, y) \in R \times \mathcal{G} \times R^m \quad (\forall (t, x, y) \in \mathcal{T}_\tau \times \mathcal{G} \times R^m);$$

(c)
$$D^+ v(t, x, y) \leq \psi_3^T(\|x\|) A_3(y) \psi_3(\|x\|)$$
$$\forall (t, x, y) \in R \times \mathcal{G} \times R^m \quad (\forall (t, x, y) \in \mathcal{T}_\tau \times \mathcal{G} \times R^m).$$

Then, if the matrices $A_1(y)$, $A_2(y)$, $\widetilde{A}_2(y)$, $(y \neq 0) \in R^m$ are positive definite and $A_3(y)$ is negative semi-definite, then

(a) *the state $x = 0$ of system (2.7) is stable (on $\mathcal{T}_\tau$), provided condition (4)(a) is satisfied;*
(b) *the state $x = 0$ of system (2.7) is uniformly stable (on $\mathcal{T}_\tau$), provided condition (4)(b) is satisfied.*

Proof We shall prove assertion (a). Since matrices $A_1(y)$ and $\widetilde{A}_2(y)$ $\forall (y \neq 0) \in R^m$ are positive definite, then $\lambda_m(A_1) > 0$ and $\lambda_M(\widetilde{A}_2) > 0$, where $\lambda_m(\cdot)$ and $\lambda_M(\cdot)$ are minimal and maximal eigenvalues of matrices $A_1(y)$ and $\widetilde{A}_2(y)$ respectively.

Condition (3) of Theorem 3.1 provides the existence of functions $\pi \in K$ and $\rho \in CK$ such that
$$\pi(\|x\|) \leq \psi_1^T(\|x\|) \psi_1(\|x\|)$$

and
$$\rho(t, \|x\|) \geq \widetilde{\psi}_2^T(t, \|x\|) \widetilde{\psi}_2(t, \|x\|).$$

Consequently,
$$\lambda_m(A_1) \pi(\|x\|) \leq v(t, x, y) \quad \forall (t, x, y) \in R \times \mathcal{G} \times R^m$$
$$(\forall (t, x, y) \in \mathcal{T}_\tau \times \mathcal{G} \times R^m) \tag{3.1}$$

and
$$v(t,x,y) \leq \lambda_M(\widetilde{A}_2)\rho(t,\|x\|) \quad \forall\,(t,x,y) \in R \times \mathcal{G} \times R^m$$
$$(\forall\,(t,x,y) \in \mathcal{T}_\tau \times \mathcal{G} \times R^m). \tag{3.2}$$

Since matrix $A_3(y)$ is negative semi-definite, then
$$D^+v(t,x,y) \leq 0 \quad \forall\,(t,x,y \neq 0) \in R \times \mathcal{G} \times R^m$$
$$(\forall\,(t,x,y \neq 0) \in \mathcal{T}_\tau \times \mathcal{G} \times R^m). \tag{3.3}$$

Taking into account (3.1)–(3.3) one can easily see that all conditions of Theorem 5 from [2] are satisfied and the state $x=0$ of system (2.7) is stable (on $\mathcal{T}_\tau$).

The proof of assertion (b) of the Theorem 3.1 is the same, seeing that $\psi_{i2} \in K$.

Theorem 3.2 *Let the vector-function f in system (2.7) be continuous on $R \times R^n$ (on $\mathcal{T}_\tau \times R^n$). If there exist*

(1) *a matrix-valued function $U \in C(R \times R^n, R^{m \times m})$ ($U \in C(\mathcal{T}_\tau \times R^n, R^{m \times m})$) and a vector $y \in R^m$ such that the function $v(t,x,y) = y^{\mathrm{T}} U(t,x) y$ is locally Lipschitzian in x for all $t \in R$ ($t \in \mathcal{T}_\tau$);*

(2) *functions $\varphi_{1i}, \varphi_{2i}, \varphi_{3i} \in KR$, $\widetilde{\varphi}_{2i} \in CKR$, $i = 1,2,\ldots,m$;*

(3) *$m \times m$ matrices $B_j(y)$, $j = 1,2,3$, $\widetilde{B}_2(y)$ such that*

(a) $\varphi_1^{\mathrm{T}}(\|x\|)B_1(y)\varphi_1(\|x\|) \leq v(t,x,y) \leq \widetilde{\varphi}_2^{\mathrm{T}}(t,\|x\|)\widetilde{B}_2(y)\widetilde{\varphi}_2(t,\|x\|)$
$\forall\,(t,x,y) \in R \times R^n \times R^m \quad (\forall\,(t,x,y) \in \mathcal{T}_\tau \times R^n \times R^m);$

(b) $\varphi_1^{\mathrm{T}}(\|x\|)B_1(y)\varphi_1(\|x\|) \leq v(t,x,y) \leq \varphi_2^{\mathrm{T}}(\|x\|)B_2(y)\varphi_2(\|x\|)$
$\forall\,(t,x,y) \in R \times R^n \times R^m \quad (\forall\,(t,x,y) \in \mathcal{T}_\tau \times R^n \times R^m);$

(c) $D^+v(t,x,y) \leq \varphi_3^{\mathrm{T}}(\|x\|)B_3(y)\varphi_3(\|x\|)$
$\forall\,(t,x,y) \in R \times R^n \times R^m \quad (\forall\,(t,x,y) \in \mathcal{T}_\tau \times R^n \times R^m).$

Then, provided that matrices $B_1(y)$, $B_2(y)$ and $\widetilde{B}_2(y)$, $\forall\,(y \neq 0) \in R^m$ are positive definite and matrix $B_3(y)$ is negative definite,

(a) *under condition (3)(a) the state $x = 0$ of system (2.7) is stable in the whole (on $\mathcal{T}_\tau$);*

(b) *under condition (3)(b) the state $x = 0$ of system (2.7) is uniformly stable in the whole (on $\mathcal{T}_\tau$).*

Proof Under conditions (1)–(3)(a) of Theorem 3.2 the function $v(t,x,y)$ is radially unbounded positive definite in the whole (on $\mathcal{T}_\tau$) and weakly decreasing in the whole (on $\mathcal{T}_\tau$). Since the matrix $B_3(y)$, $\forall\,(y \neq 0) \in R^m$ is negative semi-definite, then we have in consequence of condition (3)(c) of Theorem 3.2
$$D^+v(t,x,y) \leq 0 \quad \forall\,(t,x,y \neq 0) \in R \times R^n \times R^m$$
$$(\forall\,(t,x,y \neq 0) \in \mathcal{T}_\tau \times R^n \times R^m).$$

According to Theorem 6 from [2] the state $x = 0$ of system (2.7) is stable in the same manner taking into account conditions (1)–(3)(b) and (3)(c).

Theorem 3.3 *Let the vector-function f in system (2.7) be continuous on $R \times \mathcal{N}$ (on $\mathcal{T}_\tau \times \mathcal{N}$). If there exist*

(1) *an open connected time-invariant neighborhood $\mathcal{G} \subset \mathcal{N}$ of the point $x = 0$;*
(2) *a matrix-valued function $U \in C(R \times \mathcal{N}, R^{m \times m})$ ($U \in C(\mathcal{T}_\tau \times \mathcal{N}, R^{m \times m})$) and a vector $y \in R^m$ such that the function $v(t, x, y) = y^T U(t, x) y$ is locally Lipschitzian in x for all $t \in R$ ($t \in \mathcal{T}_\tau$);*
(3) *functions $\eta_{1i}, \eta_{2i}, \eta_{3i} \in K$, $\widetilde{\eta}_{2i} \in CK$, $i = 1, 2, \ldots, m$;*
(4) *$m \times m$ matrices $C_j(y)$, $j = 1, 2, 3$, $\widetilde{C}_2(y)$ such that*

(a) $$\eta_1^T(\|x\|) C_1(y) \eta_1(\|x\|) \leq v(t, x, y) \leq \widetilde{\eta}_2^T(t, \|x\|) \widetilde{C}_2(y) \widetilde{\eta}_2(t, \|x\|)$$
$$\forall (t, x, y) \in R \times \mathcal{G} \times R^m \quad (\forall (t, x, y) \in \mathcal{T}_\tau \times \mathcal{G} \times R^m);$$

(b) $$\eta_1^T(\|x\|) C_1(y) \eta_1(\|x\|) \leq v(t, x, y) \leq \eta_2^T(\|x\|) C_2(y) \eta_2(\|x\|)$$
$$\forall (t, x, y) \in R \times \mathcal{G} \times R^m \quad (\forall (t, x, y) \in \mathcal{T}_\tau \times \mathcal{G} \times R^m);$$

(c) $$D^* v(t, x, y) \leq \eta_3^T(\|x\|) C_3(y) \eta_3(\|x\|) + m(t, \eta_3(\|x\|))$$
$$\forall (t, x, y) \in R \times \mathcal{G} \times R^m \quad (\forall (t, x, y) \in \mathcal{T}_\tau \times \mathcal{G} \times R^m),$$

where function $m(t, \cdot)$ satisfies the condition

$$\lim \frac{|m(t, \eta_3(\|x\|))|}{\|\eta_3\|} = 0 \quad \text{as} \quad \|\eta_3\| \to 0$$

uniformly in $t \in R$ ($t \in \mathcal{T}_\tau$).

Then, provided the matrices $C_1(y)$, $C_2(y)$, $\widetilde{C}_2(y)$ are positive definite and matrix $C_3(y)$ ($y \neq 0$) $\in R^m$ is negative definite, then

(a) *under condition (4)(a) the state $x = 0$ of the system (2.7) is asymptotically stable (on $\mathcal{T}_\tau$);*
(b) *under condition (4)(b) the state $x = 0$ of the system (2.7) is uniformly asymptotically stable (on $\mathcal{T}_\tau$).*

Proof Following the arguments from the proof of Theorem 3.1 under conditions (1)–(4)(a) the function $v(t, x, y)$ is positive definite on $\mathcal{G}$ (on $\mathcal{T}_\tau \times \mathcal{G}$) and weakly decreasing on $\mathcal{G}$ (on $\mathcal{T}_\tau \times \mathcal{G}$). Consider condition (4)(c). Since $\eta_{3i} \in K$, $i = 1, 2, \ldots, m$ there exists a function $\omega \in K$ such that

$$\omega(\|x\|) \geq \eta_3^T(\|x\|) \eta_3(\|x\|). \tag{3.4}$$

Due to matrix $C_3(y)$ ($y \neq 0$) $\in R^m$ being negative definite all its eigenvalues are negative so that $\lambda_M(C_3) < 0$. Therefore, we get in view of (3.4)

$$D^* v(t, x, y) \leq \lambda_M(C_3) \omega(\|x\|) + m(t, \eta_3(\|x\|))$$
$$\forall (t, x, y \neq 0) \in R \times \mathcal{G} \times R^m \quad (\forall (t, x, y \neq 0) \in \mathcal{T}_\tau \times \mathcal{G} \times R^m). \tag{3.5}$$

Under condition (3.2) for the given neighborhood $\mathcal{G} \subset \mathcal{N}$ of point $x = 0$ a $0 < \mu < 1$ can be taken so that

$$|m(t, \eta(\|x\|))| < -\mu\lambda_M(C_3)\eta_3^T(\|x\|)\eta_3(\|x\|)$$
$$\forall\, (t, x, y \neq 0) \in R \times \mathcal{G} \times R^m \quad (\forall\, (t, x, y \neq 0) \in \mathcal{T}_\tau \times \mathcal{G} \times R^m). \tag{3.6}$$

Together with inequalities (3.5) condition (3.6) yields the estimate

$$D^*v(t, x, y) \leq (1-\mu)\lambda_M(C_3)\omega(\|x\|), \quad \lambda_M(C_3) < 0.$$

Thus, function $D^*v(t, x, y)$ is negative definite on $\mathcal{G}$ (on $\mathcal{T}_\tau \times \mathcal{G}$). Therefore, all conditions of Theorem 7 from [2] are satisfied and the state $x = 0$ of the system (2.7) is asymptotically stable (on $\mathcal{T}_\tau$).

Assertion (b) of Theorem 3.3 is proved in the same manner taking into account that condition (4)(b) ensures function $v(t, x, y)$ decreasing on $\mathcal{G}$ (on $\mathcal{T}_\tau \times \mathcal{G}$).

Theorem 3.4 *Let the vector-function f in system (2.7) be continuous on $R \times R^n$ (on $\mathcal{T}_\tau \times R^n$) and conditions (1)–(3) of Theorem 3.2 are satisfied.*

Then, provided that matrices $B_1(y)$, $B_2(y)$ and $\widetilde{B}_2(y)$ are positive definite and matrix $B_3(y)$ $\forall\,(y \neq 0) \in R^m$ is negative definite,

(a) *under condition (3)(a) of Theorem 3.2 the state $x = 0$ of system (2.7) is asymptotically stable in the whole (on $\mathcal{T}_\tau$);*

(b) *under condition (3)(b) of Theorem 3.2 the state $x = 0$ of system (2.7) is uniformly asymptotically stable in the whole (on $\mathcal{T}_\tau$).*

Proof Under conditions (1)–(3)(a) of Theorem 3.2 the function $v(t, x, y)$ is radially unbounded positive definite in the whole (on $\mathcal{T}_\tau$).

Because matrix $B_3(y)$ $\forall\,(y \neq 0) \in R^m$ is negative definite, proceeding as in the proof of Theorem 3.2 we arrive at the estimate

$$D^*v(t, x, y) \leq \lambda_M(B_3)\varphi_3^T(\|x\|)\varphi_3(\|x\|)$$
$$\forall\, (t, x, y) \in R \times R^n \times R^m \quad (\forall\, (t, x, y) \in \mathcal{T}_\tau \times R^n \times R^m).$$

Since $\varphi_{3i} \in CK$, $i = 1, 2, \ldots, m$, there exist a function $\theta(\|x\|) \in KR$ such that

$$\theta(\|x\|) \geq \varphi_3^T(\|x\|)\varphi_3(\|x\|).$$

Therefore,

$$D^*v(t, x, y) \leq \lambda_M(B_3)\theta(\|x\|), \quad \lambda_M(B_3) < 0$$
$$\forall\, (t, x, y \neq 0) \in R \times R^n \times R^m \quad (\forall\, (t, x, y \neq 0) \in \mathcal{T}_\tau \times R^n \times R^m).$$

Thus, function $D^*v(t, x, y)$ is negative definite in the whole (on $\mathcal{T}_\tau$).

According to Theorem 8 from [2] the state $x = 0$ of system (2.7) is asymptotically stable in the whole (on $\mathcal{T}_\tau$).

The proof of assertion (b) of Theorem 3.4 is similar to the above and takes into account the fact that by conditions (2) and (3) of Theorem 3.2 the function $v(t, x, y)$ is radially unbounded positive definite and decreasing in the whole (on $\mathcal{T}_\tau$).

Theorem 3.5 *Let the vector-function f in system (2.7) be continuous on $R \times \mathcal{N}$ (on $\mathcal{T}_\tau \times \mathcal{N}$). If there exist*

(1) *an open connected time-invariant neighborhood $\mathcal{G} \subset \mathcal{N}$ of the point $x = 0$;*
(2) *a matrix-valued function $U \in C(R \times \mathcal{N}, R^{m \times m})$ and a vector $y \in R^m$ such that the function $v(t, x, y) = y^T U(t, x) y$ is locally Lipschitzian in x for all $t \in R$ ($t \in \mathcal{T}_\tau$);*
(3) *functions σ_{2i}, $\sigma_{3i} \in K$, $i = 1, 2, \ldots, m$, a positive real number Δ_1 and positive integer p, $m \times m$ matrices $F_2(y)$, $F_3(y)$ such that*

(a)
$$\Delta_1 \|x\|^p \leq v(t, x, y) \leq \sigma_2^T(\|x\|) F_2(y) \sigma_2(\|x\|)$$
$$\forall (t, x, y \neq 0) \in R \times \mathcal{G} \times R^m \quad (\forall (t, x, y \neq 0) \in \mathcal{T}_\tau \times \mathcal{G} \times R^m);$$

(b)
$$D^* v(t, x, y) \leq \sigma_3^T(\|x\|) F_3(y) \sigma_3(\|x\|)$$
$$\forall (t, x, y \neq 0) \in R \times \mathcal{G} \times R^m \quad (\forall (t, x, y \neq 0) \in \mathcal{T}_\tau \times \mathcal{G} \times R^m).$$

Then, provided that the matrices $F_2(y)$ ($y \neq 0$) $\in R^m$ are positive definite, the matrix $F_3(y)$ ($y \neq 0$) $\in R^m$ is negative definite and functions σ_{2i}, σ_{3i} are the same magnitude, then the state $x = 0$ of system (2.2) is exponentially stable (on $\mathcal{T}_\tau$).

Proof Under conditions (1) – (4)(a) function $v(t, x, y)$ is positive definite and decreasing (on $\mathcal{T}_\tau$). In fact, we have the estimate

$$v(t, x, y) \leq \lambda_M(F_2) \sigma_2^T(\|x\|) \sigma_2(\|x\|), \quad \lambda(F_2) > 0$$
$$\forall (t, x, y) \in R \times \mathcal{G} \times R^m \quad (\forall (t, x, y \neq 0) \in \mathcal{T}_\tau \times \mathcal{G} \times R^m).$$

Since the functions $\sigma_{3i} \in K$, $i = 1, 2, \ldots, m$, there exists a function $\varkappa \in K$ such that

$$\varkappa(\|x\|) \geq \sigma_2^T(\|x\|) \sigma_2(\|x\|).$$

Therefore

$$\Delta_1 \|x\|^p \leq v(t, x, y) \leq \lambda_M(F_2) \varkappa(\|x\|), \quad \lambda_M(F_2) > 0 \qquad (3.7)$$
$$\forall (t, x, y) \in R \times \mathcal{G} \times R^m \quad (\forall (t, x, y \neq 0) \in \mathcal{T}_\tau \times \mathcal{G} \times R^m).$$

We reduce condition (4)(b) of Theorem 3.5 to the form

$$D^* v(t, x, y) \leq \lambda_M(F_3) \pi(\|x\|), \quad \lambda_M(F_3) < 0 \qquad (3.8)$$
$$\forall (t, x, y) \in R \times \mathcal{G} \times R^m \quad (\forall (t, x, y) \in \mathcal{T}_\tau \times \mathcal{G} \times R^m),$$

where $\pi \in K$ is such that

$$\pi(\|x\|) \geq \sigma_3^T(\|x\|) \sigma_3(\|x\|).$$

Since functions $\varkappa$ and π are of the same magnitude, there exist constants $k_1 > 0$ and $k_2 > 0$ such that

$$k_1 \varkappa(\|x\|) \leq \pi(\|x\|) \leq k_2 \varkappa(\|x\|).$$

We get from inequalities (3.7) and (3.8)

$$D^*v(t,x,y) \leq \lambda v(t,x,y) \qquad (3.9)$$
$$\forall\,(t,x,y \neq 0) \in R \times \mathcal{G} \times R^m \quad (\forall\,(t,x,y \neq 0) \in \mathcal{T}_\tau \times \mathcal{G} \times R^m),$$

where $\lambda = \lambda_M(F_3)\lambda_M^{-1}(F_2)$, $\lambda < 0$.

In view of the estimate from the left in (3.7) we obtain from (3.9)

$$v(t,x,y) \leq v(t_0,x_0,y)\exp(\lambda(t-t_0))$$

and

$$\|\chi(t;t_0 x_0)\| \leq \Delta_1^{-\frac{1}{p}} \lambda_M^{\frac{1}{p}}(F_2) \varkappa^{\frac{1}{p}}(\|x_0\|) \exp\left(\frac{\lambda}{p}(t-t_0)\right). \qquad (3.10)$$

We designate (cf. [3])

$$\alpha = \Delta_1^{-\frac{1}{p}} \lambda_M^{\frac{1}{p}}(F_2), \quad \beta = \frac{\lambda}{p}, \quad \beta < 0.$$

From (3.10) we obtain

$$\|\chi(t;t_0 x_0)\| \leq \alpha \varkappa^{\frac{1}{p}}(\|x_0\|) \exp(\beta(t-t_0)) \qquad \forall\, t \in \mathcal{T}_0, \quad \forall\, t_0 \in \mathcal{T}_i.$$

This proves Theorem 3.5.

Theorem 3.6 *Let the vector-function f in system (2.7) be continuous on $R \times R^n$ (on $\mathcal{T}_\tau \times R^n$). If there exist*

(1) *a matrix-valued function $U \in C(R \times R^n, R^{m \times m})$ ($U \in C(\mathcal{T}_\tau \times R^n, R^{m \times m})$) and a vector $y \in R^m$ such that the function $v(t,x,y) = y^T U(t,x)y$ is locally Lipschitzian in x for all $t \in R$ ($\forall\, t \in \mathcal{T}_\tau$);*

(2) *functions $\nu_{2i}, \nu_{3i} \in KR$, $i = 1,2,\ldots,m$, a positive real number $\Delta_2 > 0$ and a positive integer q;*

(3) *$m \times m$ matrices H_2, H_3 such that*

(a) $$\Delta_2\|x\|^q \leq v(t,x,y) \leq \nu_2^T(\|x\|)H_2(y)\nu_2(\|x\|)$$
$$\forall\,(t,x,y \neq 0) \in R \times R^n \times R^m \quad (\forall\,(t,x,y) \in \mathcal{T}_\tau \times R^n \times R^m);$$

(b) $$D^*v(t,x,y) \leq \nu_3^T(\|x\|)H_3(y)\nu_3(\|x\|)$$
$$\forall\,(t,x,y \neq 0) \in R \times R^n \times R^m \quad (\forall\,(t,x,y \neq 0) \in \mathcal{T}_\tau \times R^n \times R^m).$$

Then, if the matrix $H_2(y)$ $\forall\,(y \neq 0) \in R^m$ is positive definite, the matrix $H_3(y)$ $\forall\,(y \neq 0) \in R^m$ is negative definite and functions ν_{2i}, ν_{3i} are of the same magnitude, the state $x = 0$ of system (2.7) is exponentially stable in the whole (on $\mathcal{T}_\tau$).

Proof of this Theorem is similar to that of Theorem 3.5 taking into account the fact that under conditions of Theorem 3.6 the function $v(t, x, y)$ is radially unbounded (on $\mathcal{T}_\tau$). Inequality (3.10) is replaced by

$$\|\chi(t; t_0 x_0)\| \leq \Delta_2^{-\frac{1}{q}} \lambda_M^{\frac{1}{q}}(H_2) g^{\frac{1}{q}}(\|x_0\|) \exp\left(\frac{\lambda_1}{q}(t - t_0)\right)$$

where $g(\|x\|) \in KR$ and $g(\|x\|) \geq \nu_2^T(\|x\|)\nu_2(\|x\|)$,

$$\lambda_1 = \lambda_M(H_3) k_1 \lambda_M^{-1}(H_2), \qquad k_1 > 0, \quad \lambda_1 < 0.$$

We designate $\beta = \lambda_1 q^{-1}$ and define function $\Phi(\Delta) = \Delta_1^{-\frac{1}{q}} \lambda_M^{\frac{1}{q}}(H_2) g^{\frac{1}{q}}(\Delta)$ whenever $\|x_0\| < \Delta$, $\Delta = +\infty$. Then

$$\|\chi(t; t_0 x_0)\| \leq \Phi(\Delta) \exp\left(\beta(t - t_0)\right), \qquad \beta < 0 \quad \forall\, t \in \mathcal{T}_0, \quad \forall\, t_0 \in \mathcal{T}_i.$$

This proves Theorem 3.6.

Theorem 3.7 *Let the vector-function f in system (2.7) be continuous on $R \times \mathcal{N}$ (on $\mathcal{T}_\tau \times \mathcal{N}$). If there exist*

(1) *an open connected time-invariant neighborhood $\mathcal{G} \subset \mathcal{N}$ of the point $x = 0$;*
(2) *a matrix-valued function $U \in C^1(R \times \mathcal{N}, R^{m \times m})$ ($U \in C^1(\mathcal{T}_\tau \times \mathcal{N}, R^{m \times m})$) and a vector $y \in R^m$;*
(3) *functions $\psi_{1i}, \psi_{2i}, \psi_{3i} \in K$, $i = 1, 2, \ldots, m$, $m \times m$ matrices $A_1(y)$, $A_2(y)$, $G(y)$ and a constant $\Delta > 0$ such that*

(a) $\quad \psi_1^T(\|x\|) A_1(y) \psi_1(\|x\|) \leq v(t, x, y) \leq \psi_2^T(\|x\|) A_2(y) \psi_2(\|x\|)$
$\quad \forall\,(t, x, y) \in R \times \mathcal{G} \times R^m \quad (\forall\,(t, x, y) \in \mathcal{T}_\tau \times \mathcal{G} \times R^m);$

(b) $\quad Dv(t, x, y) \geq \psi_3^T(\|x\|) G(y) \psi_3(\|x\|)$
$\quad \forall\,(t, x, y) \in R \times \mathcal{G} \times R^m \quad (\forall\,(t, x, y) \in \mathcal{T}_\tau \times \mathcal{G} \times R^m);$

(4) *point $x = 0$ belong to $\partial \mathcal{G}$;*
(5) $v(t, x, y) = 0$ *on $\mathcal{T}_0 \times (\partial \mathcal{G} \cap B_\Delta)$, where $B_\Delta = \{x \colon \|x\| < \Delta\}$.*

Then, if matrices $A_1(y)$, $A_2(y)$ and $G(y)$ $\forall\,(y \neq 0) \in R^m$ are positive definite, the state $x = 0$ of system (2.7) is unstable (on $\mathcal{T}_\tau$).

Proof Under conditions (1)–(3)(a) of Theorem 3.7 it is easy to obtain for function $v(t, x, y)$ the estimate

$$\lambda_m(A_1) \gamma(\|x\|) \leq v(t, x, y) \leq \lambda_M(A_2) \zeta(\|x\|)$$
$$\forall\,(t, x, y) \in R \times \mathcal{G} \times R^m \quad (\forall\,(t, x, y) \in \mathcal{T}_\tau \times \mathcal{G} \times R^m). \tag{3.11}$$

Here

$$\gamma \in K \quad \text{and} \quad \gamma(\|x\|) \leq \psi_1^{\mathrm{T}}(\|x\|)\psi_1(\|x\|),$$
$$\zeta \in K \quad \text{and} \quad \zeta(\|x\|) \geq \psi_2^{\mathrm{T}}(\|x\|)\psi_2(\|x\|).$$

Since $\lambda_M(A_1) > 0$, $\lambda_M(A_2) > 0$, then by estimate (3.11) function $v(t,x,y)$ is positive and bounded (on $\mathcal{T}_\tau$). Hence, for every $\delta > 0$ an $x_0 \in \mathcal{G} \cap B_\Delta$ and a $a > 0$ can be found such that $a \geq v(t_0, x_0, y) > 0 \ \forall (y \neq 0) \in R^m$.

Condition (3)(b) of Theorem 3.7 is reduced to the form

$$Dv(t,x,y) \geq \lambda_m(G)\xi(\|x\|), \qquad \lambda_m(G) > 0,$$
$$\forall (t,x,y \neq 0) \in R \times \mathcal{G} \times R^m \quad (\forall (t,x,y) \in \mathcal{T}_\tau \times \mathcal{G} \times R^m). \tag{3.12}$$

Here $\xi \in K$ and $\xi \leq \psi_3^{\mathrm{T}}(\|x\|)\psi_3(\|x\|)$.

In view of (3.11) and (3.12) we have for $\chi(t; t_0, x_0 \in \mathcal{G}$

$$a \geq v(t, \chi(t; t_0, x_0), y) = v(t_0, x_0, y) + \int_{t_0}^{t} Dv(\tau, \chi(\tau; t_0, x_0)y)\,d\tau$$
$$\geq v(t_0, x_0, y) + \lambda_m(G)\xi(\|x_0\|)(t - t_0) \qquad \forall t \in \mathcal{T}_0 \quad (\forall t \in \mathcal{T}_\tau).$$

Hence, it follows that the solution $\chi(t; t_0, x_0)$ must leave neighborhood $\mathcal{G}$ some time later. But because of condition (5) it cannot leave $\mathcal{G}$ through $\partial \mathcal{G} \in B_\Delta$. Consequently, $\chi(t; t_0, x_0)$ leaves the domain B_Δ and the state $x = 0$ of system (2.7) is unstable (on $\mathcal{T}_\tau$).

4 Stability Analysis of Autonomous Large Scale Systems

We consider a large scale systems be decomposed into three subsystems

$$\frac{dx}{dt} = Ax + f(x,y,z),$$
$$\frac{dy}{dt} = By + g(x,y,z), \tag{4.1}$$
$$\frac{dz}{dt} = Cz + h(x,y,z),$$

where $x \in R^{n_1}$, $y \in R^{n_2}$, $z \in R^{n_3}$, $n_1 + n_2 + n_3 = n$; A, B and C are constant matrices of the corresponding dimensions

$$f \in C\left(R^{n_1} \times R^{n_2} \times R^{n_3}, R^{n_1}\right);$$
$$g \in C\left(R^{n_1} \times R^{n_2} \times R^{n_3}, R^{n_2}\right);$$
$$h \in C\left(R^{n_1} \times R^{n_2} \times R^{n_3}, R^{n_3}\right).$$

Moreover, the vector-functions f, g and h vanish for $x = y = z = 0$ and contain variables x, y and z in first power, i.e. the subsystems

$$\frac{dx}{dt} = Ax; \qquad (4.2)$$

$$\frac{dy}{dt} = By; \qquad (4.3)$$

$$\frac{dz}{dt} = Cz; \qquad (4.4)$$

are not complete linear approximation of the system (4.1). Physically speaking this corresponds to the situation when the connections between subsystems (4.2)–(4.4) are carried out by time-invariant linear blocks. For different dynamical properties of subsystems (4.2)–(4.4) sufficient total stability conditions will be established for the state $x = y = z = 0$ of the system (4.1).

The solution algorithm for this problem is based on actual construction of the matrix-valued function

$$U(x, y, z) = [v_{ij}(\cdot)], \qquad v_{ij} = v_{ji} \quad \forall\,(i \neq j) \qquad (4.5)$$

with the elements

$$\begin{aligned}
v_{11}(x) &= x^{\mathrm{T}} P_{11} x, \\
v_{22}(y) &= y^{\mathrm{T}} P_{22} y, \\
v_{33}(z) &= z^{\mathrm{T}} P_{33} z; \\
v_{12}(x, y) &= x^{\mathrm{T}} P_{12} y, \\
v_{13}(x, z) &= x^{\mathrm{T}} P_{13} z, \\
v_{23}(y, z) &= y^{\mathrm{T}} P_{23} z,
\end{aligned} \qquad (4.6)$$

where P_{ii}, $i = 1, 2, 3$, are symmetrical and positive definite matrices, P_{12}, P_{13} and P_{23} are constant matrices. It can be easily verified that for the functions (4.6) there exist estimates

$$\begin{aligned}
v_{11}(x) &\geq \lambda_m(P_{11})\|x\|^2 & &\forall\,(x \neq 0) \in \mathcal{N}_x; \\
v_{22}(y) &\geq \lambda_m(P_{22})\|y\|^2 & &\forall\,(y \neq 0) \in \mathcal{N}_y; \\
v_{33}(z) &\geq \lambda_m(P_{33})\|z\|^2 & &\forall\,(z \neq 0) \in \mathcal{N}_z; \\
v_{12}(x,y) &\geq -\lambda_M^{1/2}\left(P_{12} P_{12}^{\mathrm{T}}\right)\|x\|\|y\| & &\forall\,(x \neq 0,\, y \neq 0) \in \mathcal{N}_x \times \mathcal{N}_y; \\
v_{13}(x,z) &\geq -\lambda_M^{1/2}\left(P_{13} P_{13}^{\mathrm{T}}\right)\|x\|\|z\| & &\forall\,(x \neq 0,\, z \neq 0) \in \mathcal{N}_x \times \mathcal{N}_z; \\
v_{23}(y,z) &\geq -\lambda_M^{1/2}\left(P_{23} P_{23}^{\mathrm{T}}\right)\|y\|\|z\| & &\forall\,(y \neq 0,\, z \neq 0) \in \mathcal{N}_y \times \mathcal{N}_z,
\end{aligned} \qquad (4.7)$$

where $\lambda_m(P_{ii})$ are minimal eigenvalues of matrices P_{ii}, $i = 1,2,3$, $\lambda_M^{1/2}\left(P_{12}P_{12}^{\mathrm{T}}\right)$, $\lambda_M^{1/2}\left(P_{13}P_{13}^{\mathrm{T}}\right)$, $\lambda_M^{1/2}\left(P_{23}P_{23}^{\mathrm{T}}\right)$ are norms of matrices P_{12}, P_{13} and P_{23} respectively.

By means of the function

$$U(x,y,z) = \begin{pmatrix} v_{11}(x) & v_{12}(x,y) & v_{13}(x,z) \\ v_{12}(x,y) & v_{22}(y) & v_{23}(y,z) \\ v_{13}(x,z) & v_{23}(y,z) & v_{33}(z) \end{pmatrix}$$

and the vector $\eta \in R_+^3$, $\eta_i > 0$, $i = 1,2,3$ we introduce the function

$$v(x,y,z,\eta) = \eta^{\mathrm{T}} U(x,y,z)\eta. \tag{4.8}$$

Proposition 4.1 *Let for system (4.1) there exists matrix-valued function (4.5) with elements (4.6) and estimates (4.7). Then for function (4.8) the estimate*

$$v(x,y,z,\eta) \geq u^{\mathrm{T}} H^{\mathrm{T}} P H u$$
$$\forall\, (x \neq 0,\, y \neq 0,\, z \neq 0) \in \mathcal{N}_x \times \mathcal{N}_y \times \mathcal{N}_z \tag{4.9}$$

is satisfied, where $u^{\mathrm{T}} = (\|x\|, \|y\|, \|z\|)$; $H = \mathrm{diag}\,[\eta_1, \eta_2, \eta_3]$,

$$P = \begin{pmatrix} \lambda_m(P_{11}) & -\lambda_M^{1/2}\left(P_{12}P_{12}^{\mathrm{T}}\right) & -\lambda_M^{1/2}\left(P_{13}P_{13}^{\mathrm{T}}\right) \\ -\lambda_M^{1/2}\left(P_{12}P_{12}^{\mathrm{T}}\right) & \lambda_m(P_{22}) & -\lambda_M^{1/2}\left(P_{23}P_{23}^{\mathrm{T}}\right) \\ -\lambda_M^{1/2}\left(P_{13}P_{13}^{\mathrm{T}}\right) & -\lambda_M^{1/2}\left(P_{23}P_{23}^{\mathrm{T}}\right) & \lambda_m(P_{33}) \end{pmatrix}.$$

Together with function (4.8) we shall consider its total derivative

$$Dv(x,y,z,\eta) = \eta^T DU(x,y,z)\eta \tag{4.10}$$

by virtue of system (4.1).

Proposition 4.2 *Let for system (4.1) there exist matrix-valued function (4.5) with elements (4.6). For total derivatives of functions (4.6) by virtue of subsystems (4.2)–(4.4) the following estimates are satisfied*

(1) $(\nabla_x v_{11})^{\mathrm{T}} Ax \leq \rho_{11}\|x\|^2 \quad \forall\, x \in \mathcal{N}_x$;

(2) $(\nabla_x v_{12})^{\mathrm{T}} Ax \leq \rho_{12}\|x\|\|y\| \quad \forall\, (x,y) \in \mathcal{N}_x \times \mathcal{N}_y$;

(3) $(\nabla_x v_{13})^{\mathrm{T}} Ax \leq \rho_{13}\|x\|\|z\| \quad \forall\, (x,z) \in \mathcal{N}_x \times \mathcal{N}_z$;

(4) $(\nabla_y v_{22})^{\mathrm{T}} By \leq \rho_{21}\|y\|^2 \quad \forall\, y \in \mathcal{N}_y$;

(5) $(\nabla_y v_{21})^{\mathrm{T}} By \leq \rho_{22}\|x\|\|y\| \quad \forall\, (x,y) \in \mathcal{N}_x \times \mathcal{N}_y$;

(6) $(\nabla_y v_{23})^{\mathrm{T}} By \leq \rho_{23}\|y\|\|z\| \quad \forall\, (y,z) \in \mathcal{N}_y \times \mathcal{N}_z$;

(7) $(\nabla_z v_{33})^{\mathrm{T}} Cz \leq \rho_{31}\|z\|^2 \quad \forall\, z \in \mathcal{N}_z$;

(8) $(\nabla_z v_{31})^{\mathrm{T}} Cz \leq \rho_{32}\|x\|\|z\| \quad \forall\, (x,z) \in \mathcal{N}_x \times \mathcal{N}_z$;

(9) $(\nabla_z v_{32})^{\mathrm{T}} Cz \leq \rho_{33}\|y\|\|z\| \quad \forall\, (y,z) \in \mathcal{N}_y \times \mathcal{N}_z$,

where $\nabla_u = \partial/\partial u$ and

$$\rho_{11} = \lambda_{\max} \left[P_{11}A + A^{\mathrm T}P_{11} \right],$$
$$\rho_{21} = \lambda_{\max} \left[P_{22}B + B^{\mathrm T}P_{22} \right],$$
$$\rho_{31} = \lambda_{\max} \left[P_{33}C + C^{\mathrm T}P_{33} \right],$$
$$\rho_{12} = \|A^{\mathrm T}P_{12}\|,$$
$$\rho_{13} = \|A^{\mathrm T}P_{13}\|,$$
$$\rho_{22} = \|P_{12}B\|,$$
$$\rho_{23} = \|B^{\mathrm T}P_{23}\|,$$
$$\rho_{32} = \|P_{13}C\|,$$
$$\rho_{33} = \|P_{23}C\|$$

respectively, ρ_{12}, ρ_{13}, ρ_{22}, ρ_{23}, ρ_{32}, ρ_{33} are norms of matrices $A^{\mathrm T}P_{12}$, $A^{\mathrm T}P_{13}$, $P_{12}B$, $B^{\mathrm T}P_{23}$, $P_{13}C$, $P_{23}C$.

Assumption 4.1 There exist constants ρ_{ij}, $i = 1, 2, 3$, $j = 4, 5, \ldots, 12$, such that in open connected neighborhoods $\mathcal{N}_x \subseteq R^{n_1}$, $\mathcal{N}_y \subseteq R^{n_2}$, $\mathcal{N}_z \subseteq R^{n_3}$ or in its product there exist the estimates

(1') $(\nabla_x v_{11})^{\mathrm T} f \leq \rho_{14}\|x\|^2 + \rho_{15}\|x\|\|y\| + \rho_{16}\|x\|\|z\|;$

(2') $(\nabla_x v_{12})^{\mathrm T} f \leq \rho_{17}\|y\|^2 + \rho_{18}\|x\|\|y\| + \rho_{19}\|y\|\|z\|;$

(3') $(\nabla_x v_{13})^{\mathrm T} f \leq \rho_{1.10}\|z\|^2 + \rho_{1.11}\|x\|\|z\| + \rho_{1.12}\|y\|\|z\|;$

(4') $(\nabla_y v_{22})^{\mathrm T} g \leq \rho_{24}\|y\|^2 + \rho_{25}\|x\|\|y\| + \rho_{26}\|y\|\|z\|;$

(5') $(\nabla_y v_{21})^{\mathrm T} g \leq \rho_{27}\|x\|^2 + \rho_{28}\|x\|\|y\| + \rho_{29}\|x\|\|z\|;$

(6') $(\nabla_y v_{23})^{\mathrm T} g \leq \rho_{2.10}\|z\|^2 + \rho_{2.11}\|x\|\|z\| + \rho_{2.12}\|y\|\|z\|;$

(7') $(\nabla_z v_{33})^{\mathrm T} h \leq \rho_{34}\|z\|^2 + \rho_{35}\|x\|\|z\| + \rho_{36}\|y\|\|z\|;$

(8') $(\nabla_z v_{13})^{\mathrm T} h \leq \rho_{37}\|x\|^2 + \rho_{38}\|x\|\|y\| + \rho_{39}\|x\|\|z\|;$

(9') $(\nabla_z v_{23})^{\mathrm T} h \leq \rho_{3.10}\|y\|^2 + \rho_{3.11}\|x\|\|y\| + \rho_{3.12}\|y\|\|z\|.$

Proposition 4.3 If estimates (1)–(9) and (1')–(9') are satisfied, then for all total derivatives of function (4.8) by virtue of system (4.1) the inequality

$$Dv(x, y, z, \eta) \leq u^{\mathrm T} S u \qquad \forall\, (x, y, z) \in \mathcal{N}_x \times \mathcal{N}_y \times \mathcal{N}_z \qquad (4.11)$$

takes place, where

$$S = [\sigma_{ij}], \quad \sigma_{ij} = \sigma_{ji} \ \forall\, (i,j) \in [1,3];$$
$$\sigma_{11} = \eta_1^2(\rho_{11} + \rho_{14}) + 2\eta_1(\eta_2 \rho_{27} + \eta_3 \rho_{37});$$

$$\sigma_{22} = \eta_2^2(\rho_{21} + \rho_{24}) + 2\eta_2(\eta_1\rho_{17} + \eta_3\rho_{3.10});$$
$$\sigma_{33} = \eta_3^2(\rho_{31} + \rho_{34}) + 2\eta_3(\eta_1\rho_{1.10} + \eta_2\rho_{2.10});$$
$$\sigma_{12} = \frac{1}{2}\eta_1^2\rho_{15} + \frac{1}{2}\eta_2^2\rho_{25} + \eta_1\eta_2(\rho_{12} + \rho_{22} + \rho_{18} + \rho_{28})$$
$$+ \eta_3(\eta_1\rho_{38} + \eta_2\rho_{3.11});$$
$$\sigma_{13} = \frac{1}{2}\eta_1^2\rho_{16} + \frac{1}{2}\eta_3^2\rho_{35} + \eta_1\eta_3(\rho_{13} + \rho_{32} + \rho_{1.11} + \rho_{39})$$
$$+ \eta_2(\eta_1\rho_{29} + \eta_3\rho_{2.11});$$
$$\sigma_{23} = \frac{1}{2}\eta_2^2\rho_{26} + \frac{1}{2}\eta_3^2\rho_{36} + \eta_2\eta_3(\rho_{23} + \rho_{33} + \rho_{2.12} + \rho_{3.12})$$
$$+ \eta_1(\eta_2\rho_{19} + \eta_3\rho_{1.12}).$$

Remark 4.1 The dynamical properties of subsystems (4.2)–(4.4) influence only the sign of coefficients ρ_{11}, ρ_{21} and ρ_{31}. The constants ρ_{12}, ρ_{13}, ρ_{22}, ρ_{23}, ρ_{32}, ρ_{33} can always be taken positive and the rest of the constants are independent of matrices A, B and C.

In view of the above remark we introduce the following designations

$$c_{11} = \eta_1^2\rho_{14} + 2\eta_1(\eta_2\rho_{27} + \eta_3\rho_{37});$$
$$c_{22} = \eta_2^2\rho_{24} + 2\eta_2(\eta_1\rho_{17} + \eta_3\rho_{3.10});$$
$$c_{33} = \eta_3^2\rho_{34} + 2\eta_3(\eta_1\rho_{1.10} + \eta_2\rho_{2.10}).$$

Hence we have

$$\sigma_{11} = \eta_1^2\rho_{11} + c_{11}; \quad \sigma_{22} = \eta_2^2\rho_{21} + c_{22}; \quad \sigma_{33} = \eta_3^2\rho_{31} + c_{33}.$$

Proposition 4.4 *The matrix S is negative definite if and only if*

(1) $\eta_1^2\rho_{11} + c_{11} < 0;$

(2) $\eta_1^2\eta_2^2\rho_{11}\rho_{21} + \eta_1^2\rho_{11}c_{22} + \eta_2^2\rho_{21}c_{11} + c_{11}c_{22} - \sigma_{12}^2 > 0;$

(3) $\eta_1^2\rho_{11}\left(\eta_2^2\eta_3^2\rho_{21}\rho_{31} + \eta_2^2\rho_{21}c_{33} + \eta_3^2\rho_{31}c_{22} + c_{22}c_{33} - \sigma_{23}^2\right)$
$+ \eta_2^2\rho_{21}\left(\eta_3^2\rho_{31}c_{11} + c_{11}c_{33} - \sigma_{13}^2\right) + \eta_3^2\rho_{31}\left(c_{11}c_{22} - \sigma_{12}^2\right)$
$+ c_{11}c_{22}c_{33} + 2\sigma_{12}\sigma_{13}\sigma_{23} - c_{11}\sigma_{23}^2 - c_{22}\sigma_{13}^2 - c_{33}\sigma_{12}^2 < 0.$

Remark 4.2 If subsystems (4.2)–(4.4) are nonasymptotically stable, i.e. $\rho_{11} = \rho_{21} = \rho_{31} = 0$, the conditions of Proposition 4.4 become

(1') $c_{11} < 0;$

(2') $c_{11}c_{22} - \sigma_{12}^2 > 0;$

(3') $c_{11}c_{22}c_{33} + 2\sigma_{12}\sigma_{13}\sigma_{23} - c_{11}\sigma_{23}^2 - c_{22}\sigma_{13}^2 - c_{33}\sigma_{12}^2 < 0$

Remark 4.3 If subsystem (4.2) is nonasymptotically stable, subsystem (4.3) is asymptotically stable and (1.7.11) is unstable, i.e. $\rho_{11} = 0$, $\rho_{21} < 0$, $\rho_{31} > 0$, the conditions of Proposition 4.4 become

(1″) $c_{11} < 0$;

(2″) $\eta_2^2 \rho_{21} c_{11} + c_{11} c_{22} - \sigma_{12}^2 > 0$;

(3″) $\eta_2^2 \rho_{21} \left(\eta_3^2 \rho_{31} c_{11} + c_{11} c_{33} - \sigma_{13}^2 \right) + \eta_3^2 \rho_{31} \left(c_{11} c_{22} - \sigma_{12}^2 \right) + c_{11} c_{22} c_{33}$
$+ 2\sigma_{12} \sigma_{13} \sigma_{23} - c_{11} \sigma_{23}^2 - c_{22} \sigma_{13}^2 - c_{33} \sigma_{12}^2 < 0$.

Proposition 4.5 *Matrix S is negative semi-definite iff the inequality signs $<$ and $>$ in Proposition 4.4 are replaced by $\geq$ and $\leq$ correspondingly.*

Function (4.8) and its total derivative (4.10) together with estimates (4.9) and (4.11) allows us to establish sufficient conditions of stability (in the whole) and asymptotic stability (in the whole) for system (4.1).

Theorem 4.1 *Suppose that the system (4.1) be such that*

(1) *in product $\mathcal{N} = \mathcal{N}_x \times \mathcal{N}_y \times \mathcal{N}_z$ there is the matrix-valued function $U : \mathcal{N} \to R^{3 \times 3}$;*

(2) *there exist the vector $\eta \in R_+^3$, $\eta_i > 0$, $i \in [1, 3]$;*

(3) *the matrix P is positive definite;*

(4) *the matrix S is negative semi-definite or equals to zero.*

Then the state $x = y = z = 0$ of the system (4.1) is uniformly stable.

If all estimates mentioned in conditions of Theorem 4.1 are satisfied for $\mathcal{N}_x = R^{n_1}$, $\mathcal{N}_y = R^{n_2}$, $\mathcal{N}_z = R^{n_3}$ and function (4.8) is radially unbounded, the state $x = y = z = 0$ of the system (4.1) is uniformly stable in the whole.

Proof Under all conditions of Theorem 4.1 the conditions of well-known Barbashin-Krasovskii's theorem are satisfied, and hence, the corresponding type of stability of state $x = y = z = 0$ of the system (4.1) takes place (see Theorem 3.2).

Let there exists the domain $\Omega = \{(x, y, z) \in \mathcal{N}, \ 0 \leq v(x, y, z, \eta) < a, \ a \in \overset{\circ}{R}_+\} \subset R^n$ where $Dv(x, y, z, \eta) \leq 0$.

We designate by $\mathcal{M}$ the largest invariant set in Ω where

$$Dv(x, y, z, \eta) = 0.$$

Theorem 4.2 *Suppose that the system (4.1) be such that*

(1) *the conditions (1) – (3) of Theorem 4.1 be satisfied;*

(2) *on the set Ω $Dv(x, y, z, \eta) \leq 0$ i.e. the matrix S is negative semi-definite.*

Then the set $\mathcal{M}$ is attractive relative to the domain Ω, i.e. all motions of system (4.1) starting on set Ω tend to the set $\mathcal{M}$ as $t \to +\infty$.

Proof of this Theorem is similar to that of Theorem 26.1 by Hahn [11].

Theorem 4.3 *Suppose that the system (4.1) is such that*

(1) *the conditions (1) – (3) of Theorem 4.1 are satisfied;*
(2) *the matrix S is negative semi-definite.*

Then the equilibrium state $x = y = z = 0$ of the system (4.1) is uniformly asymptotically stable.

If all estimates mentioned in conditions of Theorem 4.3 are satisfied for $\mathcal{N}_x = R^{n_1}$, $\mathcal{N}_y = R^{n_2}$, $\mathcal{N}_z = R^{n_3}$ and function (4.8) is radially unbounded, the state $x = y = z = 0$ of the system (4.1) is uniformly asymptotically stable in the whole.

The *proof* is similar to that of Theorem 25.2 by Hahn [11].

5 Conclusion

Effective applications of the dynamical systems stability theory is crucially dependent on solving the next three qualitative problems (cf. [2]).

1. How to construct a matrix-valued Liapunov function for a given dynamical system?
2. Which comparison functions mostly relax majorizations of the aggregation procedure, requirements on interactions and stability conditions of the overall large-scale system by assuring simultaneously the order reduction of its aggregation matrix at last to the number of its subsystems?
3. What is the best possible effective estimation of the attraction and/or asymptotic stability domain of an equilibrium state?

The above mentioned results together with [1, 7, 8] enable the progress in solution of all enumerated problems.

The readers can find general results obtained in the direction in the new monograph [6, 12].

References

[1] Djordjević, M.Z. (1983). Stability analysis of large scale systems whose subsystems may be unstable. *Large Scale Systems.*, **5**, 252–262.

[2] Grujić, Lj.T., Martynyuk, A.A. and Ribbens-Pavella, M. (1987). *Large Scale Systems Stability under Structural and Singular Perturbations.* Springer Verlag, Berlin.

[3] He, J.X. and Wang, M.S. (1991). Remarks on exponential stability by comparison functions of the same order of magnitude. *Ann. of Diff. Eqs.*, **7**(4), 409–414.

[4] Martynyuk, A.A. (1986). Liapunov matrix-function and stability of theory. *Proc. IMACS-IFAC Symp.*, June 3–6, IND. Villeneuve d'Ascq, France, 261–265.

[5] Martynyuk, A.A. (1995). Qualitative analysis of nonlinear systems by the method of matrix Lyapunov functions. *Rocky Mountain Journal of Mathematics*, **25**, 397–415.

[6] Martynyuk, A.A. (1998). *Stability by Liapunov's Matrix Functions Method with Applications*. Marcel Dekker, Inc., New York.

[7] Martynyuk, A.A. (1996). Aggregation forms of nonlinear systems. Domains of asymptotic stability. *Prikl. Mekh.*, **32**(4), 3–19 (Russian).

[8] Martynyuk, A.A., Miladzhanov, V.G. and Begmuratov, K.A. (1994). Construction of hierarchical matrix Lyapunov functions. *Journal of Math. Anal. and Appl.*, **185**(1), 129–145.

[9] Rouche, N., Habets, P. and Laloy, M. (1977). *Stability Theory by Liapunov's Direct Method*. Springer-Verlag, New-York.

[10] Zubov, V.I. (1974). *Mathematical Investigation of Automatic Control Systems*. Mashinostroeniye, Leningrad.

[11] Hahn, W. (1967). *Stability of Motion*. Springer-Verlag, Berlin.

[12] Martynyuk, A.A. (2002). *Qualitative Methods in Nonlinear Dynamics. Novel Approaches to Liapunov's Matrix Functions*. Marcel Dekker, Inc., New York.

2.7 STABILITY THEOREMS IN IMPULSIVE FUNCTIONAL DIFFERENTIAL EQUATIONS WITH INFINITE DELAY*

A.A. MARTYNYUK[1], J.H. SHEN[2] and I.P. STAVROULAKIS[3]

[1] *Institute of Mechanics of National Academy of Sciences of Ukraine, Kiev, Ukraine*
[2] *Department of Mathematics, Hunan Normal University, Changsha, China*
[3] *Department of Mathematics, University of Ioannina, Ioannina, Greece*

1 Introduction and Preliminaries

It is now being recognized that the theory of impulsive differential equations is not only richer than the theory of differential equations without impulses but also represents a more natural framework for mathematical modelling of many real world phenomena (cf. [18, 25]). The stability theory of impulsive differential equations goes back to the work of Mil'man and Myshkis [22]. In the last few decades the stability theory of impulsive differential equations marked a rapid development, and most research focuses on impulsive ordinary differential equations. See, for example, [5, 18, 25] and the references cited therein. Now there also exists a well-developed stability theory of functional differential equations (cf. [6–10, 12–17, 26, 32, 35, 37, 38]). However, not so much has been developed in the direction of the stability theory of impulsive functional differential equations. In the few publications dedicated to this subject, earlier works were done by Anokhin [1] and Gopalsamy and Zhang [11]. Recently, stability problems on some linear impulsive delay differential equations are systematically investigated in several papers. See, for example, [2–4, 34, 36]. However, so far stability problems on impulsive functional differential equations in more general form attracted little attention, and the well-known Lyapunov's second method applied to such equations remain neglected unlike in functional differential equations and impulsive ordinary differential equations.

Advances in Stability Theory (Ed.: A.A. Martynyuk). Stability and Control: Theory, Methods and Applications, Taylor & Francis, London, **13** (2003) 153–174.

Recall that during the past 30 years or so, the stability theory of finite and infinite delay functional differential equations based on Lyapunov's direct method has received much attention. See, for example, [6–10, 12–16, 26, 32, 35, 37, 38]. The earliest results on Lyapunov's direct method for such equations tended to be patterned on those for ordinary differential equations with the norm in R^n replaced by the supremum norm in the continuous functions space C (see Krasovskii [17]). Stimulated by the applications of Krasovskii's results, two different directions have taken shape: one is to improve the conditions of Krasovskii's theorems, which is mainly directed toward finding a good formulation for a replacement of the boundedness of vector fields (see Burton [6], Burton and Hatvani [7], Burton and Zhang [8], Busenberg and Cooke [9], Kato [15] and Zhang [37] and the references cited therein); Another is considering Lyapunov function on $R \times R^n$ taking the place of Lyapunov functional on $R \times C$. Such a method was due to Razumikhin [24], which does not need the boundedness of vector fields and is somewhat more convenient in applications. Razumikhin technique including its various variation has also been widely used in the treatment of stability for various functional differential equations (cf. [10, 12–14, 16, 17, 26, 32, 38]). It is well-known that Lyapunov's direct method applied to infinite delay equations is more complicated than to finite delay equations (cf. [8, 10, 12, 14, 15, 26, 38]). On the other hand, Lyapunov's direct method applied to impulsive ordinary differential equations has also attracted growing attention, see Bainov [5] and Samoilenko and Perestyuk [25] and the references cited therein.

This is valid for the investigation of stability properties of impulsive functional differential equations. In [20, 21, 23, 28], by using Lyapunov functionals, the authors studied some systems with finite delay and impulsive Volterra type integro-differential equations. In [30, 31, 33], by employing Lyapunov function and Razumikhin technique, stability problems are discussed for a class of impulsive functional differential equations with finite delay. The results in [30, 31] established the extension of Krasovskii-Razumikhin type theorems and which also imply the persistence of the stability properties of functional differential equations with finite delay under certain impulsive perturbations. The results in [33] make an attempt to achieve a class of impulsive stabilization results which depict the difference between equations with impulses and equations without impulses. An impulsive stabilization result on ordinary differential systems can be found in [19].

In this paper, we consider the system of impulsive functional differential equations with infinite delay of the form

$$x'(t) = F(t, x(\cdot)), \quad t \geq t^*, \quad x \in R^n, \tag{1.1}$$

$$\Delta x = I_k(t, x(t^-)), \quad t = t_k, \quad k \in Z^+. \tag{1.2}$$

We extend a uniform asymptotic stability result of Burton and Zhang [8] (for finite delay equations, see Burton [6]) by employing the Lyapunov functional and examine the persistence of uniform asymptotic stability of (1.1) under the impulsive perturbations (1.2). We also obtain an impulsive stabilization result by employing the Lyapunov function and the Razumikhin technique. It should be noted that it is somewhat more difficult and interesting to derive such an impulsive stabilization result and that, in general, it cannot be achieved by employing an alternative Lyapunov functional.

Let $R = (-\infty, \infty)$, $R^+ = [0, \infty)$. For $x \in R^n$, $|\cdot|$ denotes the Euclidean norm of x. For $t \geq t^* > \alpha \geq -\infty$, $F(t, x(s); \alpha \leq s \leq t)$ or $F(t, x(\cdot))$ is a Volterra type functional (cf. [10]), its values are in R^n and are determined by $t \geq t^*$ and the values of $x(s)$ for $[\alpha, t]$. In the case when $\alpha = -\infty$, the interval $[\alpha, t]$ is understood to be replaced by $(-\infty, t]$. In (1.1), $x'(t)$ denotes the right-hand derivative of x at t. In (1.2), $Z^+ := \{1, 2, \cdots\}$, $\Delta x := x(t) - x(t^-)$, where $x(t^-) = \lim\limits_{s \to t-0} x(s)$. It is assumed that $t^* < t_k < t_{k+1}$ with $t_k \to \infty$ as $k \to \infty$, and $I_k(t, x): [t^*, \infty) \times R^n \to R^n$ is some known functions.

Let $I \subset R$ be any interval. Define $PC(I, R^n) = \{x: I \to R^n,\ x$ is continuous everywhere except at the points $t = t_k \in I$ and $x(t_k^-)$ and $x(t_k^+) = \lim\limits_{t \to t_k + 0} x(t)$ exist with $x(t_k^+) = x(t_k)\}$. For any $t \geq t^*$, $PC([\alpha, t], R^n)$ will be written as $PC(t)$. Define $PCB(t) = \{x \in PC(t): x$ is bounded $\}$. For any $\phi \in PCB(t)$, the norm of ϕ is defined by

$$\|\phi\| = \|\phi\|^{[\alpha, t]} = \sup_{\alpha \leq s \leq t} |\phi(s)|.$$

For given $\sigma \geq t^*$ and $\phi \in PCB(\sigma)$, with (1.1) and (1.2), one associates an initial condition of the form

$$x(t) = \phi(t), \qquad \alpha \leq t \leq \sigma. \tag{1.3}$$

Definition 1.1 A *function* $x(t)$ is called a *solution* corresponding to σ of the initial value problem (1.1)–(1.3) if $x: [\alpha, \beta) \to R^n$ (for some $t^* < \beta \leq \infty$) is continuous for $t \in [\alpha, \beta) \setminus \{t_k, k = 1, 2, \ldots\}$, $x(t_k^+)$ and $x(t_k^-)$ exist and $x(t_k^+) = x(t_k)$, and satisfies (1.1)–(1.3). We denote by $x(t, \sigma, \phi)$ the solution of the initial value problem (1.1)–(1.3).

We suppose that the following conditions (H_1)–(H_4) hold, so that the initial value problem (1.1)–(1.3) exists with unique solution (cf. [29]). The existence results for impulsive functional differential equations can be established based on the considering of piecewise continuous (bounded) initial values functions space $PC(PCB)$ (cf. [10, 27, 29, 33, 34]). We also assume that $F(t, 0) \equiv 0$, $I_k(t, 0) \equiv 0$ so that $x(t) = 0$ is a solution of (1.1) and (1.2), which we call the zero solution.

(H_1) F is continuous on $[t_{k-1}, t_k) \times PC(t)$ for $k = 1, 2, \ldots$, where $t_0 = t^*$. For all $\varphi \in PC(t)$ and $k \in Z^+$, the limit $\lim\limits_{(t, \phi) \to (t_k^-, \varphi)} F(t, \phi) = F(t_k^-, \varphi)$ exists.

(H_2) F is locally Lipschitzian in ϕ in each compact set in $PCB(t)$. More precisely, for every $\gamma \in [t^*, \beta)$ and every compact set $G \subset PCB(t)$ there exists a constant $L = L(\gamma, G)$ such that

$$|F(t, \varphi(\cdot)) - F(t, \psi(\cdot))| \leq L\|\varphi - \psi\|^{[\alpha, t]}$$

whenever $t \in [t^*, \gamma]$ and $\varphi, \psi \in G$.

(H_3) For each $k \in Z^+$, $I_k(t, x) \in C([t^*, \infty) \times R^n, R^n)$.

(H_4) For any $x(t) \in PC([\alpha, \infty), R^n)$, $F(t, x(\cdot)) \in PC([t^*, \infty), R^n)$.

For any $t \geq t^*$ and $\rho > 0$, let

$$PCB_\rho(t) = \{\phi \in PCB(t) : \|\phi\| < \rho\}.$$

Definition 1.2 The *zero solution* of (1.1) and (1.2) is said to be

(S_1) *uniformly stable* (US for short), if for any $\sigma \geq t^*$ and $\varepsilon > 0$, there is a $\delta = \delta(\varepsilon) > 0$ such that $\varphi \in PCB_\delta(\sigma)$ implies $|x(t, \sigma, \varphi)| \leq \varepsilon$ for $t \geq \sigma$.

(S_2) *uniformly asymptotically stable* (UAS), if it is US, and there exists a $\delta > 0$ such that for any $\varepsilon > 0$ there is a $T = T(\varepsilon) > 0$ such that $\sigma \geq t^*$ and $\varphi \in PCB_\delta(\sigma)$ imply $|x(t, \sigma, \varphi)| \leq \varepsilon$ for $t \geq \sigma + T$.

Let $S(H) = \{x \in R^n : |x| < H\}$. We denote the following Lyapunov like functions and functionals.

Definition 1.3 A *function* $V(t, x) : [t^*, \infty) \times S(H) \to R^+$ belongs to class ν_0 if

(A_1) V is continuous on each of the sets $[t_{k-1}, t_k) \times S(H)$ and for all $x \in S(H)$ and $k \in Z^+$, the limit $\lim_{(t,y) \to (t_k^-, x)} V(t, y) = V(t_k^-, x)$ exists.

(A_2) V is locally Lipschitzian in x and $V(t, 0) \equiv 0$.

Definition 1.4 A *functional* $V(t, \phi) : [t^*, \infty) \times PCB(t) \to R^+$ belongs to class $\nu_0(\cdot)$ if

(B_1) V is continuous on each of the sets $[t_{k-1}, t_k) \times PCB(t)$ and for all $\varphi \in PCB(t)$ and $k \in Z^+$, the limit $\lim_{(t,\phi) \to (t_k^-, \varphi)} V(t, \phi) = V(t_k^-, \varphi)$ exists.

(B_2) V is locally Lipschitzian in ϕ and $V(t, 0) \equiv 0$.

Definition 1.5 A *functional* $V(t, \phi)$ belongs to class $\nu_0^*(\cdot)$ if $V \in \nu_0(\cdot)$ and for any $x \in PC([\alpha, \infty), R^n)$, $V(t, x(\cdot))$ is continuous for $t \geq t^*$.

Remark 1.1 The class ν_0 is an analogue of Lyapunov functions as introduced in [5]. The class $\nu_0(\cdot)$ is an analogue of Lyapunov functionals. It is noted that the class $\nu_0^*(\cdot)$ will play an important role in the application of Lyapunov functional method to impulsive functional differential equations. Since it is difficult for one to depict the impulsive perturbations in (1.2) by using the functionals in the class $\nu_0(\cdot)$, one has to introduce the class $\nu_0^*(\cdot)$ so that it is possible to use the functions in the class ν_0 to depict the impulsive perturbations. A function class which is similar to $\nu_0^*(\cdot)$ was introduced in [28]. It should be pointed out that such a class $\nu_0^*(\cdot)$ is common in applications.

Let $V \in \nu_0$, for any $(t,x) \in [t_{k-1}, t_k) \times S(H)$, the right hand derivative $V'(t,x)$ along the solution $x(t)$ of (1.1) and (1.2) is defined by

$$V'(t, x(t)) = \limsup_{h \to 0+} \{V(t+h, x(t+h)) - V(t, x(t))\}/h.$$

Let $V \in \nu_0(\cdot)$, for any $(t,\phi) \in [t_{k-1}, t_k) \times PCB(t)$, the right hand derivative $V'(t,\phi)$ along the solution $x(t)$ of (1.1) and (1.2) is defined by

$$V'(t, x(\cdot)) = \limsup_{h \to 0+} \{V(t+h, x(\cdot)) - V(t, x(\cdot))\}/h.$$

We say a *function* $W: [0, \infty) \to [0, \infty)$ *belongs to class* $\Re$ if W is continuous and strictly increasing and satisfies $W(0) = 0$. We say a *function* $\eta: [t^*, \infty) \to [0, \infty)$ *belongs to class PIM* if η is measurable function such that for any $l > 0$ and every $\varepsilon > 0$, there exist $T \geq t^*$ and $\delta > 0$ such that $[t \geq T, Q \subset [t-l, t]$, a measurable set, and $\mu(Q) \geq \varepsilon]$ imply

$$\int_Q \eta(t)\, dt \geq \delta,$$

where $\mu(Q)$ is the measure of the set Q.

Throughout this paper, we assume that there exists $H_1 \in (0, H)$ such that $x \in S(H_1)$ implies that $x + I_k(t_k, x) \in S(H)$ for $k \in Z^+$.

2 Main Results

Our first result employs a Lyapunov functional which belongs to class $\nu_0(\cdot)$ and is of the form $V(t, \phi(\cdot)) = V_1(t, \phi(t)) + V_2(t, \phi(\cdot))$, where $V_1 \in \nu_0$ and $V_2 \in \nu_0^*(\cdot)$. This result extends a result by Burton and Zhang [8] and examines the persistence of uniform asymptotic stability of (1.1) under the impulsive perturbations (1.2). The proof of this result requires the following lemma.

Lemma 2.1 *Let $\{x_n\}$ be a sequence of Lebesgue integrable functions, $x_n: [0, 1] \to [0, 1]$. Let $\eta(t) \in PIM$ and $g: [0, \infty) \to [0, \infty)$ be continuous, $g(0) = 0$, $g(r) > 0$ for $r > 0$, and let g be nondecreasing. If there exists $\alpha > 0$ such that $\int_0^1 x_n(t)\, dt \geq \alpha$ for all n, then there exists $\beta > 0$ such that $\int_0^1 \eta(t)g(x_n(t))\, dt \geq \beta$ for all n.*

Proof Let $A_k = \{t \in [0,1]: x_k(t) > \alpha/2\}$, A_k^c be the complement of A_k on $[0,1]$, and let μ_k be the measure of A_k. Then $\mu_k \geq \alpha/2$. To see this, if $\mu_k < \alpha/2$, then

$$\alpha \leq \int_0^1 x_k(t)\,dt = \int_{A_k} x_k(t)\,dt + \int_{A_k^c} x_k(t)\,dt < \frac{\alpha}{2} + \frac{\alpha}{2},$$

a contradiction. (The first $\alpha/2$ follows from $\mu_k < \alpha/2$ and $x_k(t) \leq 1$. The second $\alpha/2$ follows from $A_k^c \leq 1$ and $x_k(t) \leq \alpha/2$). Hence, for any k, we have

$$\int_0^1 \eta(t)g(x_k(t))\,dt \geq \int_{A_k} \eta(t)g(x_k(t))\,dt \geq \int_{A_k} \eta(t)g(\alpha/2)\,dt$$

$$= g(\alpha/2) \int_{A_k} \eta(t)\,dt \geq g(\alpha/2)\delta := \beta,$$

where $\delta = \delta(\alpha) > 0$ is some constant. This completes the proof.

Remark 2.1 We note that Lemma 2.1 extends the Lemma in [6] where the sequence $\{x_n\}$ was assumed to be continuous functions having continuous derivatives.

Theorem 2.1 *Let $V_1(t,x) \in \nu_0$, $V_2(t,\phi) \in \nu_0^*(\cdot)$, $W_i \in \Re$ $(i = 1,\ldots,5)$ and $\eta \in PIM$. Let the function $\Phi \in C(R^+, R^+)$ be bounded and satisfy $\Phi \in L^1(R^+)$. Assume that the following conditions hold:*

(i) $W_1(|\phi(t)|) \leq V(t,\phi(\cdot)) \leq W_2(|\phi(t)|) + W_3\left(\int_\alpha^t \Phi(t-s)W_4(|\phi(s)|)\,ds\right)$, *where*
$V(t,\phi(\cdot)) = V_1(t,\phi(t)) + V_2(t,\phi(\cdot)) \in \nu_0(\cdot)$;

(ii) *for $k \in Z^+$ and $x \in S(H_1)$,*

$$V_1(t_k, x + I_k(t_k^-, x)) \leq \psi_k(V_1(t_k^-, x)),$$

where $\psi_k(s) \in C(R^+, R^+)$, $\psi_k(s) \geq s$ for $s \geq 0$, $\psi_k(s_1 + s_2) \geq \psi_k(s_1) + \psi_k(s_2)$ for $s_1, s_2 \geq 0$, $\psi_k(s)/s$ are nondecreasing for $s > 0$, and there exists a constant $M \geq 1$ such that for $a > 0$,

$$\sum_{k=1}^\infty \frac{\psi_k(a) - a}{a} < \infty, \qquad \frac{\psi_k(\psi_{k-1}(\cdots(\psi_1(a))\cdots))}{a} \leq M, \quad k \in Z^+;$$

(iii) *if $x(t) = x(t,\sigma,\varphi)$ is a solution of (1.1) and (1.2) with $\sigma \geq t^*$ and $\varphi \in PCB(\sigma)$, then*

$$V'(t, x(\cdot)) \leq -\eta(t)W_5(|x(t)|).$$

Then the zero solution of (1.1) and (1.2) is uniformly asymptotically stable.

Proof Let $\varepsilon > 0$ ($\varepsilon < H_1$) be given and choose a positive number $\delta < \varepsilon$ such that $MW_2(\delta) \leq W_1(\varepsilon)/2$ and $MW_3(JW_4(\delta)) \leq W_1(\varepsilon)/2$, where $J = \int_0^\infty \Phi(u)\,du$. Let $\sigma \geq t^*$, $\varphi \in PCB_\delta(\sigma)$ and $x(t) = x(t,\sigma,\varphi)$. Set $V_1(t) = V_1(t,x(t))$, $V_2(t) = V_2(t,x(\cdot))$ and $V(t) = V_1(t) + V_2(t)$. Let $\sigma \in [t_{m-1}, t_m)$ for some $m \in Z^+$, where $t_0 = t^*$. Then for $\alpha \leq t \leq \sigma$ we have

$$W_1(|x(t)|) \leq V(t) \leq W_2(\delta) + W_3\left(W_4(\delta)\int_\alpha^t \Phi(t-s)\,ds\right)$$

$$= W_2(\delta) + W_3\left(W_4(\delta)\int_0^{t-\alpha} \Phi(u)\,du\right)$$

$$\leq W_2(\delta) + W_3\left(W_4(\delta)\int_0^\infty \Phi(u)\,du\right)$$

$$= W_2(\delta) + W_3(JW_4(\delta)) \leq M^{-1}W_1(\varepsilon).$$

From (iii) we have $V'(t) \leq 0$ for $\sigma \leq t < t_m$, which implies $V(t) \leq V(\sigma) \leq M^{-1}W_1(\varepsilon)$ for $\sigma \leq t < t_m$. Thus, by condition (ii) we have

$$V(t_m) = V_1(t_m) + V_2(t_m) = V_1(t_m, x(t_m^-) + I_m(t_m, x(t_m^-))) + V_2(t_m)$$
$$\leq \psi_m(V_1(t_m^-, x(t_m^-))) + \psi_m(V_2(t_m))$$
$$\leq \psi_m(V_1(t_m^-) + V_2(t_m^-)) = \psi_m(V(t_m^-))$$
$$\leq \psi_m(V(\sigma)) \leq \psi_m(M^{-1}W_1(\varepsilon)) \leq W_1(\varepsilon).$$

Similarly,

$$V(t) \leq V(t_m) \leq \psi_m(V(\sigma)), \quad t_m \leq t < t_{m+1},$$
$$V(t_{m+1}) \leq \psi_{m+1}(\psi_m(V(\sigma))) \leq \psi_{m+1}(\psi_m(M^{-1}W_1(\varepsilon))) \leq W_1(\varepsilon).$$

By induction, one can prove in general that for $i = 0, 1, \ldots,$

$$V(t) \leq \psi_{m+i}(\cdots(\psi_m(V(\sigma)))\cdots) \leq W_1(\varepsilon), \quad t_{m+i} \leq t < t_{m+i+1},$$
$$V(t_{m+i+1}) \leq \psi_{m+i+1}(\cdots(\psi_m(V(\sigma)))\cdots) \leq W_1(\varepsilon).$$

Thus
$$W_1(|x(t)|) \leq V(t) \leq W_1(\varepsilon), \quad t \geq \sigma,$$

or $|x(t)| \leq \varepsilon$ for $t \geq \sigma$. This proves the uniform stability (US).

Next we will prove UAS. For $\varepsilon = \varepsilon_1 \leq \min\{H_1, W_4^{-1}(1)\}$ find δ of uniform stability. Then $V(t) \leq W_1(\varepsilon_1)$ for $t \geq \sigma$ and $|x(t)| \leq \varepsilon_1$ for $t \geq \alpha$. Let $\varepsilon > 0 (\varepsilon < \varepsilon_1)$ be given. We must find $T > 0$ such that $[\sigma \geq t^*, \varphi \in PCB_\delta(\sigma), t \geq \sigma + T]$ imply that $|x(t, \sigma, \varphi)| \leq \varepsilon$.

For this $\varepsilon > 0$, find $\theta > 0$ such that

$$W_2[W_4^{-1}(\theta/K)] + W_3(2\theta) \leq M^{-1} W_1(\varepsilon), \tag{2.1}$$

where we let $\Phi(t) \leq K$ for $t \geq 0$. Now, find $r > 1$ with

$$W_4(\varepsilon_1) \int_r^\infty \Phi(u)\, du < \theta. \tag{2.2}$$

For $t \geq \sigma + r$, we have

$$W_1(|x(t)|) \leq V(t, x(\cdot)) \leq W_2(|x(t)|) + W_3\left(\int_\alpha^t \Phi(t-s) W_4(|x(s)|)\, ds\right)$$

$$\leq W_2(|x(t)|) + W_3\left(\int_\alpha^{t-r} \Phi(t-s) W_4(\varepsilon_1)\, ds + \int_{t-r}^t \Phi(t-s) W_4(|x(s)|)\, ds\right)$$

$$\leq W_2(|x(t)|) + W_3\left(W_4(\varepsilon_1) \int_r^{t-\alpha} \Phi(u)\, du + \int_{t-r}^t K W_4(|x(s)|)\, ds\right)$$

$$\leq W_2(|x(t)|) + W_3\left(W_4(\varepsilon_1) \int_r^\infty \Phi(u)\, du + K \int_{t-r}^t W_4(|x(s)|)\, ds\right)$$

$$\leq W_2(|x(t)|) + W_3\left(\theta + K \int_{t-r}^t W_4(|x(s)|)\, ds\right).$$

We will find a s_1 such that

$$|x(s_1)| < W_4^{-1}(\theta/Kr) := \varepsilon_2, \tag{2.3}$$

$$\int_{s_1-r}^{s_1} W_4(|x(s)|)\, ds < \frac{\theta}{K}. \tag{2.4}$$

For any $r_1 \geq \sigma$, from $V'(t, x(\cdot)) \leq -\eta(t) W_5(|x(t)|)$, we have for $t > r_1$,

$$V(t, x(\cdot)) \leq V(r_1, x(\cdot)) - \int_{r_1}^{t} \eta(s) W_5(|x(s)|) \, ds + \sum_{r_1 < t_i \leq t} [V(t_i) - V(t_i^-)]$$

$$\leq W_1(\varepsilon_1) + \sum_{r_1 < t_i \leq t} [V_1(t_i) - V_1(t_i^-)] - \int_{r_1}^{t} \eta(s) W_5(|x(s)|) \, ds$$

$$\leq W_1(\varepsilon_1) + \sum_{r_1 < t_i \leq t} V_1(t_i^-) \left(\frac{\psi_i(V_1(t_i^-))}{V_1(t_i^-)} - 1 \right) - \int_{r_1}^{t} \eta(s) W_5(|x(s)|) \, ds$$

$$\leq W_1(\varepsilon_1) + \sum_{i=1}^{\infty} W_1(\varepsilon_1) \left(\frac{\psi_i(W_1(\varepsilon_1))}{W_1(\varepsilon_1)} - 1 \right) - \int_{r_1}^{t} \eta(s) W_5(|x(s)|) \, ds$$

$$= W_1(\varepsilon_1)[1 + G(W_1(\varepsilon_1))] - \int_{r_1}^{t} \eta(s) W_5(|x(s)|) \, ds,$$

where we let

$$G(W_1(\varepsilon_1)) = \sum_{i=1}^{\infty} [\psi_i(W_1(\varepsilon_1)) - W_1(\varepsilon_1)]/W_1(\varepsilon_1).$$

Thus, there exists a $T_1 \geq r$ such that $|x(t)| \geq W_4^{-1}(\theta/Kr) = \varepsilon_2$ fails for some value of t on every interval of length T_1. Hence, there exists $\{s_n\} \to \infty$ as $n \to \infty$ such that $|x(s_n, \sigma, \varphi)| < \varepsilon_2$. In particular, we choose

$$s_n \in [\sigma + (n-1)T_1, \sigma + nT_1], \quad n = 2, 3, \ldots.$$

The length of the intervals is independent of σ and φ.

Now, we claim that there exists some s_j such that

$$\int_{s_j - r}^{s_j} K W_4(|x(s)|) \, ds < \theta.$$

Otherwise, for all s_n,

$$\int_{s_n - r}^{s_n} K W_4(|x(s)|) \, ds \geq \theta, \quad n = 2, 3, \ldots. \tag{2.5}$$

Set $x_n(t) = W_4(|x(-rt + s_n)|)$ (then $0 \leq x_n(t) \leq 1$) so that by Lemma 2.1 we have

$$\int_{s_n - r}^{s_n} \eta(s) W_5(|x(s)|) \, ds \geq \beta$$

for some $\beta = \beta(r, K, \theta) > 0$. Then, for $t > s_{2n}$ we have

$$V'(t, x(\cdot)) \leq -\eta(t) W_5(|x(t)|)$$

so that

$$V(t, x(\cdot)) \leq V(\sigma, \varphi) - \int_\sigma^t \eta(s) W_5(|x(s)|)\, ds + \sum_{\sigma < t_i \leq t} [V(t_i) - V(t_i^-)]$$

$$\leq W_1(\varepsilon_1)[1 + G(W_1(\varepsilon_1))] - \sum_{i=1}^n \int_{s_{2i}-r}^{s_{2i}} \eta(s) W_5(|x(s)|)\, ds$$

$$\leq W_1(\varepsilon_1)[1 + G(W_1(\varepsilon_1))] - n\beta < 0$$

if

$$n > W_1(\varepsilon_1)[1 + G(W_1(\varepsilon_1))]/\beta. \tag{2.6}$$

Hence, if (2.6) holds, then s_n fails to exist with (2.5) holding. We choose n as the smallest integer such that (2.6) holds. Then, we have

$$|x(s_n, \sigma, \varphi)| < \varepsilon_2,$$

and

$$\int_{s_n-r}^{s_n} K W_4(|x(s)|)\, ds < \theta.$$

This s_n may be seen as s_1 in (2.3) and (2.4).

Set $q = \min\{k \in Z^+ : t_k > s_n\}$. Then, we have for $s_n \leq t < t_q$,

$$W_1(|x(t)|) \leq V(t, x(\cdot)) \leq V(s_n, x(\cdot))$$

$$\leq W_2(|x(s_n)|) + W_3\left(\theta + K \int_{s_n-r}^{s_n} W_4(|x(s)|)\, ds\right)$$

$$< W_2(W_4^{-1}(\theta/Kr)) + W_3(2\theta) < M^{-1} W_1(\varepsilon).$$

By condition (ii) and similar arguments as in the proof of uniform stability, we can prove that for $i = 0, 1, \ldots,$

$$V(t) \leq \psi_{q+i}(\psi_{q+i-1}(\cdots(\psi_q(M^{-1} W_1(\varepsilon)))\cdots)), \quad t_{q+i} \leq t < t_{q+i+1}.$$

Thus, by condition (ii) we obtain that

$$W_1(|x(t)|) \leq V(t, x(\cdot)) \leq W_1(\varepsilon), \quad t \geq s_n.$$

Now, let $T = 2nT_1$. Then $t \geq \sigma + T$ implies $|x(t,\sigma,\varphi)| \leq \varepsilon$. The proof for UAS is complete.

Our second result employs a Lyapunov function which belongs to class ν_0 together with certain Razumikhin techniques. This result shows that certain impulsive perturbations may make an unstable system uniformly stable and even uniformly asymptotically stable. It should be pointed out that, in general, such a result cannot be achieved by employing Lyapunov functionals.

Theorem 2.2 *Suppose that there are functions $V \in \nu_0$, $W_1, W_2 \in \Re$, $q(s)$, $H(s), P(s) \in C(R^+, R^+)$ such that q is nonincreasing with $q(s) > 0$ for $s > 0$, $H(0) = 0$, $H(s) > 0$ for $s > 0$, $P(s)$ is strictly increasing with $P(0) = 0$, $P(s) > s$ for $s > 0$. Assume that the following conditions hold:*

(i) $W_1(|x|) \leq V(t,x) \leq W_2(|x|)$, $(t,x) \in [t^*, \infty) \times S(H_1)$;

(ii) *for any solution $x(t)$ of (1.1) and (1.2), $V(s, x(s)) \leq P(V(t, x(t)))$ for $\max\{\alpha, t - q(V(t, x(t)))\} \leq s \leq t$, implies that*

$$V'(t, x(t)) \leq g(t) H(V(t, x(t))),$$

where $g: [t^*, \infty) \to R^+$, locally integrable;

(iii) *for $k \in Z^+$ and $x \in S(H_1)$,*

$$V(t_k, x + I_k(t_k, x)) \leq \psi_k(V(t_k^-, x)),$$

where $\psi_k \in C(R^+, R^+)$ with $\psi_k(s) \leq P^{-1}(s)$ for $s \geq 0$ and $k \in Z^+$, where P^{-1} denotes the inverse of the function P;

(iv) *there exist constants $\lambda_1 > 0$, $\lambda_2 > 0$ and $A > 0$, such that for all $k \in Z^+$ and $\mu > 0$,*

$$\lambda_1 \leq t_{k+1} - t_k \leq \lambda_2, \quad \text{and}$$

$$\int_\mu^{P(\mu)} \frac{du}{H(u)} - \int_{t_k}^{t_{k+1}} g(s)\,ds \geq A.$$

Then the zero solution of (1.1) and (1.2) is UAS.

Proof We first prove US. For given $\varepsilon > 0$ ($\varepsilon \leq H_1$), we may choose a $\delta = \delta(\varepsilon) > 0$ such that $P(W_2(\delta)) \leq W_1(\varepsilon)$. For $\sigma \geq t^*$, $\varphi \in PCB_\delta(\sigma)$, let $x(t) = x(t, \sigma, \varphi)$ be the solution of (1.1) and (1.2) and let $V(t) = V(t, x(t))$. Let $\sigma \in [t_{m-1}, t_m)$ for some $m \in Z^+$, where $t_0 = t^*$. Then we have for $\alpha \leq t \leq \sigma$,

$$W_1(|x(t)|) \leq V(t) \leq W_2(|x(t)|) \leq W_2(\delta) < P(W_2(\delta)).$$

We claim that

$$V(t) \leq P(W_2(\delta)), \quad \sigma \leq t < t_m. \tag{2.7}$$

Indeed, if (2.7) does not hold, then there exists a $s_1 \in (\sigma, t_m)$ such that

$$V(s_1) > P(W_2(\delta)) > W_2(\delta) \geq V(\sigma).$$

This implies that there exists a $s_2 \in (\sigma, s_1)$ such that

$$V(s_2) = P(W_2(\delta)), \quad V(t) \leq P(W_2(\delta)), \quad \sigma \leq t \leq s_2,$$

and also there exists a $s_3 \in [\sigma, s_2)$ such that

$$V(s_3) = W_2(\delta), \quad V(t) \geq W_2(\delta), \quad s_3 \leq t \leq s_2.$$

Thus, for $s_3 \leq t \leq s_2$ we have

$$V(s) \leq P(W_2(\delta)) \leq P(V(t)), \quad \max\{\alpha, t - q(V(t))\} \leq s \leq t.$$

In view of condition (ii) we have

$$V'(t) \leq g(t)H(V(t)), \quad s_3 \leq t \leq s_2,$$

and so

$$\int_{V(s_3)}^{V(s_2)} \frac{du}{H(u)} \leq \int_{s_3}^{s_2} g(s)\,ds \leq \int_{t_{m-1}}^{t_m} g(s)\,ds.$$

On the other hand, let $\mu = W_2(\delta)$ in condition (iv) we have

$$\int_{V(s_3)}^{V(s_2)} \frac{du}{H(u)} = \int_{W_2(\delta)}^{P(W_2(\delta))} \frac{du}{H(u)} \geq \int_{t_{m-1}}^{t_m} g(s)\,ds + A > \int_{V(s_3)}^{V(s_2)} \frac{du}{H(u)}.$$

This is a contradiction and so (2.7) holds. From (2.7) and condition (iii) we have

$$V(t_m) \leq \psi_m(V(t_m^-, x(t_m^-))) = \psi_m(V(t_m^-)) \leq P^{-1}(V(t_m^-)) \leq W_2(\delta). \tag{2.8}$$

Similarly, we have

$$V(t) \leq P(W_2(\delta)), \quad t_m \leq t < t_{m+1}, \quad V(t_{m+1}) \leq W_2(\delta).$$

By induction, we can prove that

$$V(t) \leq P(W_2(\delta)), \quad t_{m+i} \leq t < t_{m+i+1},$$
$$V(t_{m+i+1}) \leq W_2(\delta), \quad i = 0, 1, \ldots.$$

Thus, we have

$$W_1(|x(t)|) \leq V(t) \leq P(W_2(\delta)) \leq W_1(\varepsilon), \quad t \geq \sigma,$$

which shows that the zero solution of (1.1) and (1.2) is US.

Next, we will prove UAS. For $\varepsilon = H_1$ find δ of uniform stability such that $P(W_2(\delta)) = W_1(H_1)$ and $\sigma \geq t^*$, $\varphi \in PCB_\delta(\sigma)$ imply

$$V(t, x(t)) \leq P(W_2(\delta)), \quad |x(t)| \leq H_1, \quad t \geq \sigma,$$

where $x(t) = x(t, \sigma, \varphi)$.

Now let $\varepsilon > 0$ ($\varepsilon < H_1$) be given. We will prove that there exists a $T = T(\varepsilon) > 0$ such that $\varphi \in PCB_\delta(\sigma)$ implies that $|x(t)| \leq \varepsilon$ for $t \geq \sigma + T$. Since $A > 0$, we can find a smallest positive integer N such that

$$P(W_2(\delta)) \leq P^{-1}(W_1(\varepsilon)) + ANH(P^{-1}(W_1(\varepsilon))). \tag{2.9}$$

This N is independent of σ and φ.

Let $\gamma = \gamma(\varepsilon) = q(P^{-1}(P^{-1}(W_1(\varepsilon))))$. Let $\sigma \in [t_{m-1}, t_m)$ for some $m \in Z^+$ and let

$$m_i = \min\{k \in Z^+ : t_k - t_{m_{i-1}} \geq \gamma\}, \quad i = 1, 2, \ldots, N,$$

where we let $m_0 = m$. We consider all the intervals $J_i := [t_{m_{i-1}}, t_{m_i}]$, $i = 1, 2, \ldots, N$. In view of the piecewise continuous properties of $V(t)$ on J_i and

$$V(t_k) = V(t_k, x(t_k^-) + I_k(t_k, x(t_k^-)))$$
$$\leq \psi_k(V(t_k^-)) \leq P^{-1}(V(t_k^-)) \leq V(t_k^-),$$

we see that $\sup\{V(t) : t \in J_i\} := L_i$ exists and satisfies either $L_i = V(t_{m_{i-1}})$ or $L_i = V(r_i^-)$ for some $r_i \in (t_{m_{i-1}}, t_{m_i}]$. Without loss of the generality, we assume that $L_i = V(r_i^-)$, $i = 1, 2, \ldots, N$. For the case when $L_j = V(t_{m_{j-1}})$ for some $j \in \{1, 2, \ldots, N\}$, the proof is similar and is omitted. Let $m^* = m^*(\varepsilon)$ be the smallest positive integer such that $m^*\lambda_1 \geq \gamma$. It is easy to see that $\sum_{i=1}^{N}(t_{m_i} - t_{m_{i-1}}) \leq N(m^*\lambda_1 + \lambda_2)$. Let $T = T(\varepsilon) = N(m^*\lambda_1 + \lambda_2) + \lambda_2$. We will prove that

$$|x(t)| \leq \varepsilon, \quad t \geq \sigma + T.$$

To this end, we first prove that if

$$V(r_i^-) \leq P^{-1}(W_1(\varepsilon)) \quad \text{for some} \quad i \in \{1, 2, \ldots, N\}, \tag{2.10}$$

then

$$V(t) \leq W_1(\varepsilon), \quad t \geq t_{m_N}. \tag{2.11}$$

From (2.10) we have
$$V(t) \leq P^{-1}(W_1(\varepsilon)) < W_1(\varepsilon), \quad t_{m_i-1} \leq t \leq t_{m_i}. \tag{2.12}$$

We claim that
$$V(t) \leq W_1(\varepsilon), \quad t_{m_i} \leq t < t_{m_i+1}. \tag{2.13}$$

If (2.13) does not hold, then there exists a $s_1 \in (t_{m_i}, t_{m_i+1})$ such that
$$V(s_1) > W_1(\varepsilon) > P^{-1}(W_1(\varepsilon)) \geq V(t_{m_i}).$$

This implies that there exists a $s_2 \in (t_{m_i}, s_1)$ such that
$$V(s_2) = W_1(\varepsilon), \quad V(t) \leq W_1(\varepsilon), \quad t_{m_i} \leq t \leq s_2 \tag{2.14}$$

and also there exists a $s_3 \in [t_{m_i}, s_2)$ such that
$$V(s_3) = P^{-1}(W_1(\varepsilon)), \quad V(t) \geq P^{-1}(W_1(\varepsilon)), \quad s_3 \leq t \leq s_2. \tag{2.15}$$

From (2.12), (2.14) and (2.15) we have for $s_3 \leq t \leq s_2$,
$$V(s) \leq W_1(\varepsilon) \leq P(V(t)), \quad t - \gamma \leq s \leq t,$$

and so by the definition of γ we have
$$V(s) \leq P(V(t)), \quad \max\{\alpha, t - q(V(t))\} \leq s \leq t.$$

By condition (ii) we have
$$V'(t) \leq g(t) H(V(t)), \quad s_3 \leq t \leq s_2.$$

This yields
$$\int_{V(s_3)}^{V(s_2)} \frac{du}{H(u)} \leq \int_{s_3}^{s_2} g(s)\, ds < \int_{t_{m_i}}^{t_{m_i+1}} g(s)\, ds + A$$
$$\leq \int_{P^{-1}(W_1(\varepsilon))}^{W_1(\varepsilon)} \frac{du}{H(u)} = \int_{V(s_3)}^{V(s_2)} \frac{du}{H(u)}.$$

This is a contradiction and so (2.13) holds. From (2.13) and condition (iii) we have
$$V(t_{m_i+1}) \leq \psi_{m_i+1}(V(t_{m_i+1}^-)) \leq P^{-1}(V(t_{m_i+1}^-)) \leq P^{-1}(W_1(\varepsilon)).$$

By induction, we can prove in general that

$$V(t) \leq W_1(\varepsilon), \quad t_{m_i+k} \leq t < t_{m_i+k+1},$$
$$V(t_{m_i+k+1}) \leq P^{-1}(W_1(\varepsilon)), \quad k = 0, 1, 2, \ldots. \tag{2.16}$$

This shows that if (2.10) holds for some $i \in \{1, 2, \ldots, N\}$ then (2.11) holds.

Next we will prove that there exists a $i \in \{1, 2, \ldots, N\}$ such that (2.10) holds. Otherwise, assume that for all $i = 1, 2, \ldots, N$, $V(r_i^-) > P^{-1}(W_1(\varepsilon))$. In the following we will prove that this assumption leads to a contradiction. To this end, we show that

$$V(r_i^-) \leq V(r_0^-) - iAH(P^{-1}(W_1(\varepsilon))), \quad i = 0, 1, \ldots, N, \tag{2.17}_i$$

where $V(r_0^-) := P(W_2(\delta))$. Clearly $(2.17)_0$ holds. Now suppose $(2.17)_j$ holds for some j $(0 \leq j < N)$. We must prove that $(2.17)_{j+1}$ holds. We first claim that

$$V(r_{j+1}^-) \leq V(r_j^-). \tag{2.18}$$

In fact, since $V(t) \leq V(r_j^-)$ for $t_{m_j-1} \leq t \leq t_{m_j}$, it follows that

$$V(t_{m_j}) \leq \psi_{m_j}(V(t_{m_j}^-)) \leq P^{-1}(V(t_{m_j}^-)) \leq P^{-1}(V(r_j^-)) < V(r_j^-).$$

By similar arguments to that of the proof of (2.16), we can obtain

$$V(t) \leq V(r_j^-), \quad t_{m_j+k} \leq t < t_{m_j+k+1},$$
$$V(t_{m_j+k+1}) \leq P^{-1}(V(r_j^-)), \quad k = 0, 1, \ldots,$$

and so (2.18) holds. Next, we consider two possible cases:

Case 1. $P^{-1}(W_1(\varepsilon)) < V(r_{j+1}^-) \leq P^{-1}(V(r_j^-))$.

In this case, by condition (iv) we have

$$\int_{V(r_{j+1}^-)}^{P(V(r_{j+1}^-))} \frac{du}{H(u)} \geq A,$$

and so

$$V(r_{j+1}^-) \leq P(V(r_{j+1}^-)) - AH(P^{-1}(W_1(\varepsilon)))$$
$$\leq V(r_j^-) - AH(P^{-1}(W_1(\varepsilon))) \leq V(r_0^-) - (j+1)AH(P^{-1}(W_1(\varepsilon))).$$

Case 2. $P^{-1}(V(r_j^-)) < V(r_{j+1}^-) \leq V(r_j^-)$.

Let $r_{j+1} \in (t_{m_j+k}, t_{m_j+k+1}]$ for some $k \in Z^+ \cup \{0\}$. Then for $k = 0$, we have $V(t_{m_j+k}) = V(t_{m_j}) \leq \psi_{m_j}(V(t_{m_j}^-)) \leq P^{-1}(V(r_j^-))$, and for $k \neq 0$ we also have $V(t_{m_j+k}) \leq \psi_{m_j+k}(V(t_{m_j+k}^-)) \leq P^{-1}(V(r_{j+1}^-)) \leq P^{-1}(V(r_j^-))$. Therefore, there exists a $r^* \in [t_{m_j+k}, r_{j+1})$ such that

$$V(r^*) = P^{-1}(V(r_j^-)), \tag{2.19}$$

and

$$P^{-1}(V(r_j^-)) \leq V(t) \leq P(W_2(\delta)), \quad r^* \leq t < r_{j+1}. \tag{2.20}$$

From (2.18) and (2.20) we have for $r^* \leq t < r_{j+1}$,

$$V(s) \leq V(r_j^-) \leq P(V(t)), \quad t - \gamma \leq s \leq t,$$

and so for $r^* \leq t < r_{j+1}$

$$V(s) \leq P(V(t)), \quad \max\{\alpha, t - q(V(t))\} \leq s \leq t.$$

By condition (ii) we have

$$V'(t) \leq g(t)H(V(t)), \quad r^* \leq t < r_{j+1},$$

which implies

$$\int_{V(r^*)}^{V(r_{j+1}^-)} \frac{du}{H(u)} \leq \int_{r^*}^{r_{j+1}} g(s)\,ds \leq \int_{t_{m_j+k}}^{t_{m_j+k+1}} g(s)\,ds.$$

This, together with (2.19) and condition (iv), yields

$$\int_{P^{-1}(V(r_j^-))}^{V(r_{j+1}^-)} \frac{du}{H(u)} \leq -A + \int_{P^{-1}(V(r_j^-))}^{V(r_j^-)} \frac{du}{H(u)}.$$

Accordingly, we obtain

$$\int_{V(r_{j+1}^-)}^{V(r_j^-)} \frac{du}{H(u)} \geq A.$$

This leads to

$$V(r_{j+1}^-) \leq V(r_j^-) - AH(P^{-1}(W_1(\varepsilon))) \leq V(r_0^-) - (j+1)AH(P^{-1}(W_1(\varepsilon))).$$

By combining the cases 1 and 2, we may conclude that $(2.17)_{j+1}$ holds. By induction, we see that $(2.17)_i$ hold for all $i = 0, 1, \ldots, N$. Therefore, we obtain

$$V(r_N^-) \leq V(r_0^-) - NAH(P^{-1}(W_1(\varepsilon)))$$
$$= P(W_2(\delta)) - NAH(P^{-1}(W_1(\varepsilon))) \leq P^{-1}(W_1(\varepsilon)). \quad \text{(by (2.9))}$$

This contradicts the assumption that $V(r_i^-) > P^{-1}(W_1(\varepsilon))$ for all $i = 1, 2, \ldots, N$. Thus, there is a $i \in \{1, 2, \ldots, N\}$ such that (2.10) holds and so (2.11) holds. Since $\sigma + T = \sigma + N(m^*\lambda_1 + \lambda_2) + \lambda_2 \geq t_{m_N}$, it follows that

$$W_1(|x(t)|) \leq V(t) \leq W_1(\varepsilon), \quad t \geq \sigma + T.$$

The proof is now complete.

Remark 2.2 In Theorem 2.2 if $\psi_k(s) = \psi(s)$, $k = 1, 2, \ldots$, where $\psi(s) \in C(R^+, R^+)$, strictly increasing with $\psi(s) < s$ for $s > 0$, then we may let the function $P(s) = \psi^{-1}(s)$ without structuring another P-function.

From the proof of uniform stability part in Theorem 2.2, we see that in the following weaker conditions uniform stability is achieved.

Theorem 2.3 *Let functions $V(t, x)$, W_1, W_2, P, ψ_k, g and H as in Theorem 2.2. Assume that conditions (i) and (ii) in Theorem 2.2 hold and that*

(iii) *for any solution $x(t)$ of (1.1) and (1.2), $V(s, x(s)) \leq P(V(t, x(t)))$ for $\alpha \leq s \leq t$ implies that*

$$V'(t, x(t)) \leq g(t)H(V(t, x(t)));$$

(iv) *for any $\mu > 0$ and $k \in Z^+$*

$$\int_\mu^{P(\mu)} \frac{du}{H(u)} > \int_{t_k}^{t_{k+1}} g(s)\, ds.$$

Then the zero solution of (1.1) and (1.2) is US.

3 Examples

In this section, two illustrative examples are given. In particular, we describe by Example 3.2 the significance of impulsive stabilization.

Example 3.1 Consider the equation

$$x'(t) = -A(t)x(t) + k(t) \int_{-\infty}^{t} C(t-s)x(s)\,ds, \quad t \geq 0, \qquad (3.1)$$

$$\Delta x = b_k x(t^-), \quad t = t_k, \quad k \in Z^+, \qquad (3.2)$$

where $A(t) \in C(R^+, R^+)$, $k, C \in C(R^+, R)$, $\{b_k\}$ is a sequence of numbers. Assume that the following conditions are satisfied:

(i) $\int_0^\infty |C(u)|\,du < \infty$, $\Phi(t) := \int_t^\infty |k(u)||C(u)|\,du \in L^1[0,\infty)$ and $\Phi(t) \leq K$ for some constant $K > 0$;

(ii) there exists constants $\gamma > 0$, $\beta > 0$ such that $\gamma|k(t)| \leq |k(t-s)|$ for $t \geq 0$, $-\infty < s \leq t$, $\beta \times \gamma \geq 1$, and $\eta(t) := A(t) - \beta \int_t^\infty |k(u-t)||C(u-t)|\,du \in PIM$;

(iii) $\sum_{k=1}^\infty |b_k| < \infty$.

Then the zero solution of (3.1) and (3.2) is UAS.

In fact, let $V(t, x(\cdot)) = V_1(t, x(t)) + V_2(t, x(\cdot))$, where

$$V_1(t,x) = |x|, \quad V_2(t,x(\cdot)) = \beta \int_{-\infty}^{t} \int_{t}^{\infty} |k(u-s)||C(u-s)|\,du\,|x(s)|\,ds.$$

Let $\psi_k(s) = (1 + |b_k|)s$ and $M = \prod_{k=1}^\infty (1 + |b_k|)$. It is not difficult to see that conditions (i) and (ii) of Theorem 2.1 are satisfied. If $x(t)$ is any solution of (3.1) and (3.2), then

$$V'(t,x(\cdot)) \leq -A(t)|x(t)| + |k(t)| \int_{-\infty}^{t} |C(t-s)||x(s)|\,ds$$

$$+ \beta \int_{t}^{\infty} |k(u-t)||C(u-t)|\,du\,|x(t)| - \beta \int_{-\infty}^{t} |k(t-s)||C(t-s)||x(s)|\,ds$$

$$\leq -\left(A(t) - \beta \int_{t}^{\infty} |k(u-t)||C(u-t)|\,du\right)|x(t)| = -\eta(t)|x(t)|.$$

Thus, condition (iii) of Theorem 2.1 is satisfied. By Theorem 2.1, the zero solution of (3.1) and (3.2) is UAS.

Example 3.2 Consider the equation

$$x'(t) = A(t)x(t) + B(t)x(t-\tau) + \int_{-\infty}^{0} C(t,u,x(t+u))\,du, \quad t \geq 0, \quad (3.3)$$

$$\Delta x = I_k(t, x(t^-)), \quad t = t_k, \quad k \in Z^+, \quad (3.4)$$

where $\tau > 0$, $A(t), B(t) \in C(R^+, R)$, $A(t) \leq a$, $|B(t)| \leq b$, $C(t,u,v)$ is continuous on $R^+ \times (-\infty, 0] \times R$, $I_k(t,x) \in C([R^+ \times R, R)$, and $|x + I_k(t_k, x)| \leq c|x|$, $k = 1, 2, \ldots$, where c is a constant. Let the following conditions be satisfied:

(i) $|C(t,u,v)| \leq m(u)|v|$, $t \geq 0$, and

$$\int_{-\infty}^{0} m(u)\,du \leq M;$$

(ii) $0 < c < 1$, and there exist constants $\lambda_2 \geq \lambda_1 > 0$ such that

$$\lambda_1 \leq t_{k+1} - t_k \leq \lambda_2 < -\frac{\ln c}{a + bc^{-1} + Mc^{-1}}, \quad k \in Z^+.$$

Then the zero solution of (3.3) and (3.4) is UAS.
In fact, let $V(t,x) = V(x) = \frac{1}{2}x^2$, $\psi_k(s) = \psi(s) = c^2 s$, $H(s) = s$. Then

$$V(x + I_k(t_k, x)) = \frac{1}{2}[x + I_k(t_k, x)]^2 \leq \frac{1}{2}c^2 x^2 = \psi(V(x)).$$

From (ii) we can choose a positive constant A such that

$$t_{k+1} - t_k \leq -\frac{2\ln c + A}{2(a + bc^{-1} + Mc^{-1} + A)}, \quad k \in Z^+. \quad (3.5)$$

From (i) we see that there exists a continuous function $q: (0, \infty) \to (0, \infty)$, $q(s) \geq \tau$ for $s > 0$, q is nonincreasing, such that

$$\int_{-\infty}^{-q(s)} m(u)\,du \leq A\sqrt{2s}.$$

By Theorem 2.3, we can easily obtain the uniform stability. Thus, without loss of generality, we may assume that $\|x(t)\|^{(-\infty, t]} \leq 1$. Let $P(s) = \psi^{-1}(s)$. Then $P(s) > s$ for $s > 0$. If

$$V(s, x(s)) \leq P(V(t, x(t))), \quad \max\{-\infty, t - q(V(t, x(t)))\} \leq s \leq t,$$

then we have

$$V'(t,x(t)) \leq ax^2(t) + |B(t)||x(t)||x(t-\tau)| + |x(t)|\int_{-\infty}^{t} m(v-t)|x(v)|\,dv$$

$$\leq ax^2(t) + bc^{-1}x^2(t) + |x(t)|\int_{-\infty}^{t-q(V(t,x(t)))} m(v-t)|x(v)|\,dv$$

$$+ |x(t)|\int_{t-q(V(t,x(t)))}^{t} m(v-t)|x(v)|\,dv$$

$$\leq (a+bc^{-1})x^2(t) + |x(t)|\int_{-\infty}^{-q(V(t,x(t)))} m(u)\,du + c^{-1}x^2(t)\int_{-\infty}^{0} m(u)\,du$$

$$\leq (a+bc^{-1}+Mc^{-1})x^2(t) + A|x(t)|\sqrt{2V(t,x(t))}$$

$$= (a+bc^{-1}+Mc^{-1}+A)x^2(t)$$

$$= g(t)H(V(t,x(t))),$$

where $g(t) = 2(a+bc^{-1}+Mc^{-1}+A)$. From (3.5) we see that for all $\mu > 0$ and $k \in Z^+$

$$\int_{\mu}^{P(\mu)} - \int_{t_k}^{t_{k+1}} g(s)\,ds = \int_{\mu}^{c^{-2}\mu} \frac{du}{u} - 2\int_{t_k}^{t_{k+1}} (a+bc^{-1}+Mc^{-1}+A)\,dt$$

$$= -2\ln c - 2(a+bc^{-1}+Mc^{-1}+A)(t_{k+1}-t_k) \geq A.$$

Thus, we may conclude from Theorem 2.2 that the zero solution of (3.3) and (3.4) is UAS.

Remark 3.1 We note that in the above Example 3.2 the uniform asymptotic stability may be caused by the impulsive perturbations. To see this, we note the fact that under those conditions on equation (3.3) given in this example, it is possible that the zero solution of (3.3) is unstable (because we may allow $A(t) > 0$); another fact is that impulsive equations (3.3) and (3.4) reduce to equation (3.3) without impulses if and only if all $I_k(t,x) \equiv 0$, however, in this case condition $0 < c < 1$ in this example will not be satisfied.

Acknowledgement

This work was done while the second author was visiting the Department of Mathematics, University of Ioannina, in the framework of a Post-doctoral scholarship offered by IKY (State Scholarships Foundation), Athens, Greece.

References

[1] Anokhin, A.V. (1995). On linear impulsive systems for functional differential equations. *Soviet Math. Dokl.*, **33**, 220–223.

[2] Anokhin, A.V., Berezansky, L. and Braverman, E. (1995). Exponential stability of linear delay impulsive differential equations. *J. Math. Anal. Appl.*, **193**, 923–941.

[3] Bainov, D.D., Covachev, V. and Stamova, I. (1994). Stability under persistent disturbances of impulsive differential-difference equations of neutral type. *J. Math. Anal. Appl.*, **187**, 790–808.

[4] Bainov, D.D., Kulev, G. and Stamova, I. (1995). Global stability of the solutions of impulsive differential difference equations. *SUT J. of Math.*, **31**, 55–71.

[5] Bainov, D.D. and Simeonov, P.S. (1989). *Systems with Impulsive Effect: Stability Theory and Applications*. Chichester, Ellis Horwood.

[6] Burton, T.A. (1978). Uniform asymptotic stability in functional differential equations. *Proc. Amer. Math. Soc.*, **68**, 195–199.

[7] Burton, T.A. and Hatvani, L. (1989). Stability theorems for nonautonomous functional differential equations by Liapunov functionals. *Tohoku Math. J.*, **41**, 65–104.

[8] Burton, T.A. and Zhang Shunian. (1986). Unified boundedness, periodicity and stability in ordinary and functional differential equations. *Ann. Mat. Pura. Appl.*, **145**, 129–158.

[9] Busenberg, S.V. and Cooke, K.L. (1984). Stability conditions for linear non-autonomous delay differential equations. *Q. Appl. Math.*, **42**, 295–306.

[10] Driver, R.D. (1962). Existence and stability of solutions of a delay-differential system. *Arch. Rat. Mech. Anal.*, **10**, 401–426.

[11] Gopalsamy, K. and Zhang, B.G. (1989). On delay differential equations with impulses. *J. Math. Anal. Appl.*, **139**, 110–122.

[12] Grimmer, R. and Seifert, G. (1975). Stability properties of Volterra integro-differential equations. *J. Differential Equations*, **19**, 142–166.

[13] Haddock, J.R. and Terjeki, J. (1983). Liapunov-Razumikhin functions and an invariance principle for functional differential equations. *J. Differential Equations.*, **48**, 95–122.

[14] Kato, J. (1973). On Liapunov-Razumikhin type theorems for functional differential equations. *Funkc. Ekvacioj*, **16**, 225–239.

[15] Kato, J. (1978). Stability problems in functional differential equations with infinite delay. *Funkc. Ekvacioj*, **21**, 63–80.

[16] Kolmanovskii, V.B. and Nosov, V.R. (1986). *Stability of Functional Differential Equations*. Academic Press, New York.

[17] Krasovskii, N.N. (1963). *Stability of Motion*. Stanford Univ. Press, Stanford.

[18] Lakshmikantham, V., Bainov, D.D. and Simeonov, P.S. (1989). *Theory of Impulsive Differential Equations*. World Scientific, Singapore.

[19] Liu, X. (1993). Impulsive stabilization of nonlinear systems. *IMA J. of Math. Control Information*, **10**, 11–19.
[20] Martynyuk, A.A. and Rizaev, A. (2000). On application of matrix-valued functionals for stability analysis of systems with delay. *Dopovidi Nats. Akad. Nauk of Ukraine*, **2**, 15–19.
[21] Martynyuk, A.A. and Sun Zhen qi (1998). A Lyapunov's matrix-valued functional and stability of systems with delay. *Doklady Akad. Nauk*, **359**(2), 165–167 (Russian).
[22] Mil'man, V.D. and Myshkis, A.D. (1960). On the stability of motion in the presence of impulses. *Siberian Math. J.*, **1**, 233–237.
[23] Rama Mohana Rao, M. and Srivastava, S.K. (1992). Stability of Volterra integro-differential equations with impulsive effect. *J. Math. Anal. Appl.*, **163**, 47–59.
[24] Razumikhin, B.S. (1960). An application of Liapunov's method to a problem on the stability of systems with lag. *Autom. Remote Cont.*, **21**, 740–748.
[25] Samoilenko, A.M. and Perestyuk, N.A. (1995). *Impulsive Differential Equations*. World Scientific, Singapore.
[26] Seifert, G. (1974). Liapunov-Razumikhin conditions for asymptotic stability in functional differential equations of Volterra type. *J. Differential Equations*, **16**, 45–52.
[27] Shen, J.H. (1996). The existence of nonoscillatory solutions of delay differential equations with impulses. *Applied Math. and Comput.*, **77**, 153–165.
[28] Shen, J.H. (1996). On some asymptotic stability results of impulsive integro differential equations. *Chin. Math. Ann.*, **17A**, 759–765.
[29] Shen, J.H. (1996). Existence and uniqueness of solutions for a class of infinite delay functional differential equations with applications to impulsive differential equations. *Huaihua Teach. Coll.*, **15**, 45–51 (Chinese).
[30] Shen, J.H. (1999). Razumikhin techniques in impulsive functional differential equations. *Nonlinear Analysis*, **36**, 119–130.
[31] Shen, J.H. and Yan, J. (1998). Razumikhin type stability theorems for impulsive functional differential equations. *Nonlinear Analysis*, **33**, 519–531.
[32] Taniguchi, T. (1995). Asymptotic behavior theorems for non-autonomous functional differential equations via Lyapunov-Razumikhin method. *J. Math. Anal. Appl.*, **189**, 715–730.
[33] Yan, J. and Shen, J.H. (1999). Impulsive stabilization of functional differential equations by Lyapunov-Razumikhin functions. *Nonlinear Analysis*, **37**, 245–255.
[34] Yan, J. and Zhao, A. (1998). Oscillation and stability of linear impulsive delay differential equations. *J. Math. Anal. Appl.*, **227**, 187–194.
[35] Yoshizawa, T. (1966). *Stability Theory by Liapunov's Second Method*. Math. Soc. Japan, Tokyo.
[36] Yu, J.S. and Zhang, B.G. (1996). Stability theorems for delay differential equations with impulses. *J. Math. Anal. Appl.*, **199**, 162–175.
[37] Zhang, B. (1996). A stability theorem in functional differential equations. *Differential and Integral Eqns.*, **9**, 199–208.
[38] Zhang, S. (1989). Razumikhin techniques in infinite delay equations. *Sci. Sinica*, **A32**, 38–51.

2.8 THE ASYMPTOTIC BEHAVIOUR OF SOLUTIONS TO STOCHASTIC FUNCTIONAL DIFFERENTIAL EQUATIONS WITH FINITE DELAYS BY THE LYAPUNOV-RAZUMIKHIN METHOD*

T. TANIGUCHI

Department of Mathematics, Kurume University, Miimachi, Kurume, Fukuoka 830, Japan

Dedicated to Professor F. Uchida on his 60th birthday.

1 Introduction

In this paper we consider the moment asymptotic behaviors of solutions to a stochastic functional differential equation with finite delay $r > 0$ by the Lyapunov-Razumikhin method:

$$dX(t) = f(t, X_t)dt + g(t, X_t)d\beta(t), \quad t \geq t_0 \geq 0, \quad (1)$$
$$X_{t_0} = \varphi,$$

where $f \colon [-r, \infty) \times C \to R^d$ and $g \colon [-r, \infty) \times C \to R^{d \times n}$ are continuous functions, $\beta(t)$ is the n-dimensional standard Brownian motion and $\varphi \colon [-r, 0] \times \Omega \to R^d$ is a continuous process with $\varphi(s) \colon \Omega \to R^d$ a F_{t_0}-measurable function for all $s \in [-r, 0]$ and $E\|\varphi\|_C^2 < \infty$. See below for the definition of C and $\|\cdot\|_C$. We obtain Theorems 3.4–3.6 as the main theorems. Regarding application we discuss the almost sure asymptotic behaviors of solutions to (1). See Theorems 4.2 and 4.3.

The stability theorems of solutions to a deterministic functional differential equation with finite delay have been studied by many authors using the Lyapunov functionals and it has been shown that finite delays cause many interesting phenomena

---
*Advances in Stability Theory (Ed.: A.A. Martynyuk). Stability and Control: Theory, Methods and Applications, Taylor & Francis, London, **13** (2003) 175–188.

(see [1] and [2]). On the other hand, what about the solutions to the stochastic FDE with finite delay (e.g. Mackey and Nechaeva [7])? Unfortunately, however, it seems that there are not many papers on the asymptotic stability of solutions by the Lyapunov functionals, except very special cases, for the case of the stochastic FDE with finite delay. Moreover, it would seem in general, very difficult to construct Lyapunov functionals that satisfy the conditions in the asymptotic stability theorems of the solutions to a stochastic FDE with finite delay (see Kolmanovskii and Myshkis [4], and Kolmanovskii and Nosov [5]). In this paper we consider the stochastic version of the asymptotic behavior theorems [14, 15] of solutions to the functional differential equation with finite delay and we generalize the theorems of [16]. Then, since the theorems presented are based on the Lyapunov-Razumikhin method which is given by two Lyapunov functions, we can avoid the difficulty of the constructions of the Lyapunov functionals.

The contents of this paper are as follows. In Section 2 we cover preliminaries. In Section 3 we consider the convergence of $EW(t, X(t))$ as $t \to \infty$. In Section 4 we discuss the almost sure Lyapunov exponent. In Section 5 we present examples.

2 Preliminaries

Let R^d be the d-dimensional Euclidian space and let R^+ be a set of all nonnegative real numbers. For a continuous function $x: [-r, \infty) \to R^d$ we define the continuous function $x_t: [-r, 0] \to R^d$ by setting $x_t(s) = x(t+s)$, $s \in [-r, 0]$. Let $C = C([-r, 0], R^d)$ be the space of all continuous functions $\psi: [-r, 0] \to R^d$ with norm $\|\psi\|_C = \sup\{|\psi(s)|: s \in [-r, 0]\} < \infty$, where $|\cdot|$ is the Euclidian norm.

Let $(\Omega, F, \{F_t\}_{t \geq 0}, P)$ be the complete probability space with a right continuous increasing family $\{F_t\}_{t \geq 0}$ of sub-σ-algebra of F, where each F_t contains all the P-null sets and let $MC(t)$ be the space of all F_t-measurable C-valued functions $\varphi: \Omega \to C$ with seminorm $E\|\varphi\|_C^2 = E \sup_{-r \leq s \leq 0} |\varphi(\omega)(s)|^2 < \infty$. For a continuous F_t-adapted stochastic process $X(t): \Omega \to R^d$, $t \geq t_0 - r$ we have the continuous F_t-adapted C-valued stochastic process $X_t: \Omega \to C$, $t \geq t_0 \geq 0$ by setting $X_t(\omega)(s) = X(t+s)(\omega)$, $s \in [-r, 0]$, $t \geq t_0$ [10].

In this paper we assume that the functions f and g satisfy enough additional smoothness conditions to ensure the unique, global solution of (1). A continuous function $V: [-r, \infty) \times R^d \to R^+$ is said to belong to the space $C^{(1,2)}([-r, \infty) \times R^d, R^+)$ if $\partial V(t, x)/\partial t$, $V_x(t, x) \equiv (\partial V(t, x)/\partial x_1, \ldots, \partial V(t, x)/\partial x_d)$, and $V_{xx}(t, x) \equiv [\partial^2 V(t, x)/(\partial x_i \, \partial x_j)]_{i,j=1}^d$ exist and are continuous on $[-r, \infty) \times R^d$. Let $LV(t, X(t)) = V_t(t, X(t)) + V_x(t, X(t))f(t, X_t) + (1/2) \operatorname{tr}\{V_{xx}(t, X(t))g(t, X_t) g(t, X_t)^T\}$ to equation (1).

Throughout this paper let $X(t) = X(t; t_0, \varphi)$ denote the unique, global solution of (1) for any initial value (t_0, φ), $\varphi \in MC(t_0)$, $t_0 \geq 0$ and we always assume that $E|LV(t, X(t))|$ is locally bounded in $t \in [t_0, \infty)$.

Definition 2.1 A continuous *function* $p\colon [0,\infty) \to R^+$ is said to be *integrally positive* if $\int_J p(s)\,ds = \infty$ holds on every set $J = \bigcup_{m=1}^{\infty} (a_m, b_m)$ such that $0 \le a_1$, $a_m < b_m \le a_{m+1}$, $b_m - a_m \ge \delta > 0$ $(m = 1, 2, 3, \ldots)$ for some fixed $\delta > 0$.

3 The Convergence of $EV(t, X(t))$

In this section we discuss the convergence of $EV(t, X(t))$ as $t \to \infty$ and we consider under what conditions $EV(t, X(t))$ approaches zero as $t \to \infty$. In this paper we assume that any solution to (1) exists globally and is unique. Thus, from the proof of Theorem 3.2 [16] we have the following theorem.

Theorem 3.1 *Suppose that there exists a continuous function* $V \in C^{(1,2)}$ $([-r, \infty) \times R^d, R^+)$, *and the following condition is satisfied*:

(a) *there exists an integrable function* $\gamma\colon [-r, \infty) \to R^+$ *such that* $ELV(t, X(t)) \le \gamma(t)$, $t \ge t_0$, *whenever* $EV(t, X(t)) = \sup_{-r \le s \le 0} EV(t+s, X(t+s))$, $t \ge t_0$.

Then, for each solution $X(t) = X(t; t_0, \varphi)$ *of (1)*

$$\sup_{-r \le s \le 0} EV(u+s, X(u+s)) + \int_{u-r}^{\infty} \gamma(\tau)\,d\tau \ge \sup_{-r \le s \le 0} EV(t+s, X(t+s))$$

at any time $t > u \ge t_0$.

Throughout this paper let $\|V_t\|_D := \sup_{-r \le s \le 0} EV(t+s, X(t+s))$. Then, we have

Theorem 3.2 *Suppose that all the conditions of Theorem 3.1 are satisfied. Then* $\lim_{t \to \infty} \|V_t\|_D$ *exists*.

Proof By Theorem 3.1 we have the set $\{\|V_t\|_D, t \ge t_0\}$ is bounded. Now suppose that $\lim_{t \to \infty} \|V_t\|_D$ does not exist. Then there exist two sequences $\{u_n\}$, $\{t_n\}$ and real numbers $\alpha \ge 0$, $\beta \ge 0$ such that $u_n, t_n \to \infty$ and $\|V_{u_n}\|_D \to \alpha$ and $\|V_{t_n}\|_D \to \beta$ as $n \to \infty$ with $0 \le \alpha < \beta < \infty$. Without loss of generality we may assume that $u_n < t_n < u_{n+1}$ $(n = 1, 2, 3, \ldots)$.

Then, by Theorem 3.1

$$\|V_{u_n}\|_D + \int_{u_n - r}^{\infty} \gamma(\tau)\,d\tau \ge \|V_{t_n}\|_D.$$

Since $\gamma \in L^1(0, \infty)$, there exists an $N > 0$ such that $\int_{u_n - r}^{\infty} \gamma(\tau)\,d\tau < (\beta - \alpha)/2$ for all integers $n \ge N$. Thus, letting $n \to \infty$, we have $(\beta - \alpha)/2 \ge \beta - \alpha > 0$, which is a contradiction. Therefore, the proof of the theorem is complete.

Theorem 3.3 *Suppose that all the conditions of Theorem 3.1 are satisfied. Furthermore, suppose that there exist an integrable function $\psi\colon [0,\infty) \to R^+$ and for every $\varepsilon > 0$ a continuous function $h = h(\varepsilon)\colon [0,\infty) \to R^+$ with $h(u) > u$ for $u > 0$ and $h(0) = 0$, a $K = K(\varepsilon) > 0$ and a time $T = T(\varepsilon) > t_0$ such that for any solution $X(t) = X(t;t_0,\varphi)$ of (1)*

$$ELV(t, X(t)) \leq K\psi(t), \quad t \geq T,$$

whenever $\|V_t\|_D \leq 2\varepsilon$, $EV(t, X(t)) \geq \varepsilon$ *and* $\|V_t\|_D \leq h(EV(t, X(t)))$, $t \geq T$. *Then*, $\lim\limits_{t\to\infty} EV(t, X(t))$ *exists.*

Proof By Theorem 3.2, there exists an $\alpha \geq 0$ such that $\lim\limits_{t\to\infty} \|V_t\|_D = \alpha$. Thus we get a sequence $\{t_n\}$ with $t_n \to \infty$ as $n \to \infty$ such that $EV(t_n, X(t_n)) \to \alpha$ as $n \to \infty$. If $\alpha = 0$, then the proof of the theorem is complete.

Next, assume $\alpha > 0$. Now, suppose that the conclusion of the theorem does not hold. Then, there exists a sequence $\{\tau_n\}$ with $\tau_n \to \infty$ as $n \to \infty$ such that $\{EV(\tau_n, X(\tau_n))\}$ converges as $n \to \infty$ to some $\alpha_1 \geq 0$ with $\alpha_1 \neq \alpha$. Here, without loss of generality, we may moreover assume $\tau_n < t_n < \tau_{n+1}$ ($n = 1, 2, 3, \ldots$). Then, by Theorem 3.1,

$$\|V_{t_n}\|_D + \int_{t_n-r}^{\infty} \gamma(\tau)\,d\tau \geq EV(\tau_{n+1}, X(\tau_{n+1})).$$

Thus we have $\alpha_1 < \alpha$. Hence, let $\lambda > 0$ be a real number such that $0 < \lambda < \alpha < 2\lambda$ and let $\delta(\lambda) = \min\{|h(u) - u| \colon \lambda \leq u \leq 2\lambda\}$, where $h = h(\lambda)$. Without loss of generality we may suppose $\lambda < \alpha_1 < \alpha$ and $0 < \alpha - \alpha_1 < \delta(\lambda)/3$. Then, we can choose a sufficiently large natural integer N and two new sequences $\{\tau'_n\}$ and $\{t'_n\}$ with $\tau'_n < t'_n < \tau'_{n+1}$ and $\tau'_n \to \infty$ as $n \to \infty$ such that $EV(\tau'_n, X(\tau'_n)) = (\alpha + 2\alpha_1)/3$, $EV(t'_n, X(t'_n)) = (2\alpha + \alpha_1)/3$ and $(\alpha + 2\alpha_1)/3 < EV(t, X(t)) < (2\alpha + \alpha_1)/3$ for all $t \in (\tau'_n, t'_n)$ with $n \geq N$. Furthermore, we may assume that $|\|V_t\|_D - \alpha| < \delta(\lambda)/3$ and $\lambda \leq \|V_t\|_D \leq 2\lambda$ for all $t > \tau'_N$, without loss of generality. Thus we obtain that

$$\|V_t\|_D - EV(t, X(t)) \leq |\|V_t\|_D - \alpha| + |\alpha - EV(\tau'_n, X(\tau'_n))|$$
$$+ |EV(\tau'_n, X(\tau'_n)) - EV(t, X(t))| < \frac{\delta(\lambda)}{3} + \frac{\delta(\lambda)}{3} + \frac{\delta(\lambda)}{3}$$
$$\leq h(EV(t, X(t))) - EV(t, X(t))$$

for all $t \in (\tau'_n, t'_n)$ with $n \geq N$, which implies that $\|V_t\|_D \leq h(EV(t, X(t)))$ for all $t \in (\tau'_n, t'_n)$ with $n \geq N$. Thus, by the condition of the theorem we have that

$$\frac{(\alpha - \alpha_1)}{3} = EV(t'_n, X(t'_n)) - EV(\tau'_n, X(\tau'_n)) \leq \int_{\tau'_n}^{t'_n} K\psi(\tau)\, d\tau, \quad n \geq N,$$

which yields a contradiction letting $n \to \infty$ because $\psi \in L^1(0, \infty)$. Therefore, the proof of the theorem is complete.

Theorem 3.4 *Let $c > 0$ and $\xi \colon [0, \infty) \to R^+$ be a positive real number and an integrable function, respectively. Suppose that there exist continuous functions V, $W \in C^{(1,2)}([-r, \infty) \times R^d, R^+)$ and V satisfies condition (a) of Theorem 3.1. And assume that for every $\varepsilon > 0$, there exists an $H = H(\varepsilon) > 0$, a continuous function $h = h(\varepsilon) \colon [0, \infty) \to R^+$ with $h(u) > u$ for $u > 0$, $h(0) = 0$, and a time $T = T(\varepsilon) > t_0$ such that*

(a) *there exists a nondecreasing, integrable function $m(t) > 0$ such that $ELW(t, X(t)) \leq m(t)$ for all $t \geq T$ whenever $\|V_t\|_D \leq 2\varepsilon$, $EV(t, X(t)) \geq \varepsilon$ and $\|V_t\|_D \leq h(EV(t, X(t)))$ for all $t \geq T$;*

(b) *for any sequence $\{\tau_k\}$ with $\tau_k \to \infty$ as $k \to \infty$, $\sum_{k=1}^{\infty} \frac{1}{m(\tau_k)} = \infty$;*

(c) *$ELV(t, X(t)) \leq -HcEW(t, X(t)) + \xi(t)$ for all $t \geq T$ whenever $\|V_t\|_D \leq 2\varepsilon$, $EV(t, X(t)) \geq \varepsilon$ and $\|V_t\|_D \leq h(EV(t, X(t)))$ for all $t \geq T$.*

Furthermore, assume that if $EW(t, X(t))$ does not approach zero as $t \to \infty$, then $EV(t, X(t))$ also does not approach zero as $t \to \infty$. Then $EW(t, X(t))$ approaches zero as $t \to \infty$ for any solution $X(t)$ of (1).

Proof Suppose that there exists a solution $X(t)$ of (1) such that $EW(t, X(t))$ does not approach zero as $t \to \infty$. Then, by Theorem 3.3 and the hypothesis there exists a $\beta > 0$ such that $EV(t, X(t))$ approaches β as $t \to \infty$. Because of $\beta > 0$, we can choose a $\lambda > 0$ with $\lambda < \beta < 2\lambda$. Let $h = h(\lambda)$ and let $\sigma(\lambda) = \min\{h(u) - u \colon \lambda \leq u \leq 2\lambda\}$. Then we can take a $T_0 = T(\lambda) > t_0$ such that $\|V_t\|_D < \beta + \sigma(\lambda)/2$, $\|V_t\|_D < 2\lambda$, $EV(t, X(t)) \geq \lambda$ and $|EV(t, X(t)) - \beta| < \sigma(\lambda)/2$ for all $t \geq T_0$. Thus, for all $t \geq T_0$, $\|V_t\|_D < \beta + \sigma(\lambda)/2 < \sigma(\lambda) + EV(t, X(t)) \leq h(EV(t, X(t)))$ since $\lambda \leq EV(t, X(t)) \leq 2\lambda$. First, assume that there exist two sequences $\{t'_j\}$, $\{t_j\}$ with t'_j, $t_j \to \infty$ as $j \to \infty$, $t_j < t'_j < t_{j+1}$ ($j = 1, 2, 3, \ldots$) and some $\delta > 0$ such that $EW(t_j, X(t_j)) = \delta/2$, $EW(t'_j, X(t'_j)) = \delta$ and $\delta/2 < EW(t, X(t)) < \delta$ for all $t \in (t_j, t'_j)$. Then, by condition (a) of Theorem 3.4 we have that

$$\frac{\delta}{2} = EW(t'_j, X(t'_j)) - EW(t_j, X(t_j)) = \int_{t_j}^{t'_j} ELW(s, X(s))\, ds \leq m(t'_j)(t'_j - t_j).$$

Thus, we have that $\delta/[2m(t'_j)] \leq (t'_j - t_j)$. Moreover, by condition (c) of Theorem 3.4

$$EV(t'_k, X(t'_k)) \leq EW(t_1, X(t_1)) + \sum_{j=1}^{k} \int_{t_j}^{t'_j} -\frac{Hc\delta}{2} ds + \int_0^\infty \xi(s) ds$$

$$\leq EV(t_1, X(t_1)) - \sum_{j=1}^{k} \frac{Hc\delta^2}{4m(t'_j)} + \int_0^\infty \xi(s) ds,$$

which yields a contradiction letting $k \to \infty$ since $\int_0^\infty \xi(\tau) d\tau < \infty$.

Next, suppose that there exists a time $T_1 > t_0$ and $\delta_0 > 0$ such that $EW(t, X(t)) \geq \delta_0$ for all $t \geq T_1$. Then, let $T_2 = \max\{T_1, T(\lambda)\}$ and

$$EV(t, X(t)) = EV(T_2, X(T_2)) + E \int_{T_2}^{t} LV(s, X(s)) ds$$

$$\leq EV(T_2, X(T_2)) - Hc\delta_0(t - T_2) + \int_{T_2}^{t} \xi(s) ds, \quad t \geq T_2,$$

which yields a contradiction as $t \to \infty$ because $\xi \in L^1(0, \infty)$. Therefore the proof of the theorem is complete.

Theorem 3.5 *Let $p: [0, \infty) \to R^+$ and $\xi: [0, \infty) \to R^+$ be an integrally positive function and an integrable function, respectively. Suppose that there exist continuous functions $V, W \in C^{(1,2)}([-r, \infty) \times R^d, R^+)$ and V satisfies condition (a) of Theorem 3.1. And assume that for every $\varepsilon > 0$, there exist an $H = H(\varepsilon) > 0$, a continuous function $h = h(\varepsilon): [0, \infty) \to R^+$ with $h(u) > u$ for $u > 0$, $h(0) = 0$ and a time $T = T(\varepsilon) > t_0$ such that*

(a) *there exists a positive real number $M > 0$ such that $ELW(t, X(t)) \leq M$ for all $t \geq T$ whenever $\|V_t\|_D \leq 2\varepsilon$, $EV(t, X(t)) \geq \varepsilon$ and $\|V_t\|_D \leq h(EV(t, X(t))$ for all $t \geq T$;*
(b) *$ELV(t, X(t)) \leq -Hp(t)EW(t, X(t)) + \xi(t)$ for all $t \geq T$ whenever $\|V_t\|_D \leq 2\varepsilon$, $EV(t, X(t)) \geq \varepsilon$ and $\|V_t\|_D \leq h(EV(t, X(t)))$ for all $t \geq T$.*

Furthermore, assume that if $EW(t, X(t))$ does not approach zero as $t \to \infty$, then $EV(t, X(t))$ also does not approach zero as $t \to \infty$. Then $EW(t, X(t))$ approaches zero as $t \to \infty$ for any solution $X(t)$ of (1).

Proof The proof of the theorem is carried out by the same method as in Theorem 3.4, noting that the function $p: [0, \infty) \to R^+$ is integrally positive.

Example 3.1 Let r be a positive constant. Consider the equation

$$dX(t) = -[t^2 \sin^2 t]X(t-r)\,dt + \xi(t)\,d\beta(t), \quad t \geq t_0,$$

where $\beta(t)$ is the one-dimensional standard Brownian motion and $\xi \colon [0,\infty) \to R^+$ is an integrable function. Then, since $t^2 \sin^2 t$, $t \geq 0$ is integrally positive, we have that $E|X(t)|^2 \to 0$ as $t \to \infty$, by Theorem 3.5.

Theorem 3.6 *Suppose that all the conditions of Theorem 3.3 are satisfied and there exists a closed set Q in R^d with $0 \in Q$. Furthermore suppose that the following conditions are satisfied:*

(a) *for any ε_1, $\varepsilon > 0$ there exist a $\delta = \delta(\varepsilon_1,\varepsilon)$, a $T = T(\varepsilon_1,\varepsilon) > t_0$ and $\xi \in L^1(0,\infty)$ with $\xi(t) \geq 0$ such that at any time $t \geq T$ satisfying $\varepsilon_1/2 < d(Q, EX(t)) < \varepsilon_1$,*

$$ELV(t,X(t)) \leq -\delta E|f(t,X_t)| + \xi(t), \quad t \geq T,$$

whenever $\|V_t\|_D \leq 2\varepsilon$, $EV(t,X(t)) \geq \varepsilon$ and $\|V_t\|_D \leq h(EV(t,X(t)))$, $t \geq T$;

(b) *if $EV(t,X(t)) \to 0$ as $t \to \infty$, then $EX(t) \to 0$ as $t \to \infty$;*

(c) *if there exist a $\gamma > 0$ and for every $\varepsilon > 0$ there exists a $T^* = T(\varepsilon) > t_0$ such that $\mathrm{dis}(Q, EX(t)) \geq \gamma$ for all $t \geq T^*$ and*

$$\int_{t_0}^{\infty} ELV(\tau, X(\tau))\,d\tau = -\infty$$

whenever $\|V_t\|_D \leq 2\varepsilon$, $EV(t,X(t)) \geq \varepsilon$ and $\|V_t\|_D \leq h(EV(t,X(t)))$ for all $t \geq T^*$.

Then, $EX(t)$ approaches the set Q as $t \to \infty$, where $X(t) = X(t;t_0,\varphi)$ is any solution of (1).

Proof By Theorem 3.3 we have a constant $\beta \geq 0$ such that $EV(t,X(t)) \to \beta$ as $t \to \infty$. If $\beta = 0$, then the proof is complete.

Suppose $\beta > 0$.

If there exists a solution $X(t)$ of (1) such that $EX(t)$ does not approach the set Q as $t \to \infty$, then we have two cases as follows:

Case 1. There exist two sequences $\{t_j\}$, $\{t'_j\}$ with $t_j, t'_j \to \infty$ as $j \to \infty$, $t_j < t'_j < t_{j+1}$ $(j = 1,2,3,\ldots)$ and some $\gamma_1 > 0$ such that

$$\mathrm{dis}(Q, EX(t_j)) = \gamma_1, \quad \mathrm{dis}(Q, EX(t'_j)) = \frac{\gamma_1}{2},$$

$$\frac{\gamma_1}{2} < \mathrm{dis}(Q, EX(t)) < \gamma_1$$

for all $t \in (t_j, t'_j)$. Then, because of $\beta > 0$, we can take a $\lambda > 0$ such that $\lambda < \beta < 2\lambda$. Let $\sigma = \sigma(\lambda) = \min\{|h(u) - u|: \lambda \leq u \leq 2\lambda\}$, where $h = h(\lambda)$. From the proof of Theorem 3.4 we have a time $T_0 = T(\lambda)$ such that $\|V_t\|_D \leq 2\lambda$, $EV(t, X(t)) \geq \lambda$ and $h(EV(t, X(t))) \geq \|V_t\|_D$ for all $t \geq T_0$. Thus, by condition (a) of Theorem 3.6 we have a $\delta = \delta(\gamma_1, \lambda)$ and a $T_1 = T(\gamma_1, \lambda) > T_0$ such that

$$ELV(t, X(t)) \leq -\delta E|f(t, X_t)| + \xi(t), \quad t \geq T_1$$

at any time $t \geq T_1$ satisfying $\gamma_1/2 < \text{dis}(Q, EX(t)) < \gamma_1$. Hence without loss of generality, we can suppose that $t_i > T_1$ $(i = 1, 2, 3, \ldots)$. Thus, by using the condition of Theorem 3.3, and noting that $\gamma/2 = |EX(t'_j) - EX(t_j)| \leq \int_{t_j}^{t'_j} E|f(\tau, X_\tau)| d\tau$,

$$EV(t'_k, X(t'_k)) \leq EV(t_1, X(t_1)) + E\int_{t_1}^{t'_k} LV(\tau, X(\tau)) d\tau$$

$$\leq EV(t_1, X(t_1)) + \sum_{j=1}^{k} \int_{t_j}^{t'_j} [-\delta E|f(\tau, X_\tau)| + \xi(\tau)] d\tau + \int_{t_1}^{\infty} K\psi(\tau) d\tau$$

$$\leq EV(t_1, X(t_1)) - \delta\gamma_1 \frac{k}{2} + \int_{t_1}^{\infty} [K\psi(\tau) + \xi(\tau)] d\tau,$$

where $K = K(\lambda)$, which yields a contradiction letting $k \to \infty$ since $\psi, \xi \in L^1(0, \infty)$, $\psi \geq 0$ and $\xi \geq 0$.

Case 2. There exist some $\varepsilon > 0$ and some time $T_2 > T(\lambda)$ such that $EX(t) \in U(Q, \varepsilon)^c$ for all $t \geq T_2$, where $U(Q, \varepsilon) = \{x \in R^d: \text{dis}(Q, x) \leq \varepsilon\}$. Then by condition (c) of Theorem 3.6 we have that

$$EV(t, X(t)) \leq EV(T_2, X(T_2)) + \int_{T_2}^{t} ELV(\tau, X(\tau)) d\tau,$$

which yields a contradiction as $t \to \infty$. Therefore, the proof of the theorem is complete.

4 The Almost Sure Lyapunov Exponent

In this section we discuss almost sure asymptotic behavior of solutions of the stochastic functional differential equation with finite delays (1). First we prove

Theorem 4.1 *Suppose that there exists a continuous function $V \in C^{(1,2)}([-r,0] \times R^d, R^+)$ and the following conditions are satisfied:*

(a) *there exist positive real numbers $c, \lambda > 0$ and a continuous function $\theta: [0,\infty) \to R^+$ such that $0 < \lambda < c$ and $e^{\lambda t}\theta(t)$ is an integrable function, and*

$$ELV(t, X(t)) \leq -cEV(t, X(t)) + \theta(t), \quad t \geq t_0,$$

whenever $e^{\lambda r}EV(t, X(t)) \geq \sup_{-r \leq s \leq 0} EV(t+s, X(t+s))$, $t \geq t_0$;

(b) *there exist a real number $q > 1$ and a continuous function $\xi: [0,\infty) \to R^+$ such that $e^{\lambda t}\xi(t)$ is an integrable function, and for every $\varepsilon > 0$ there exists a $T = T(\varepsilon) > \max\{t_0, r\}$ such that*

$$ELV(t, X(t)) \leq -cEV(t, X(t)) + \xi(t), \quad t \geq T(\varepsilon),$$

whenever $\|V_t\|_D \leq 2\varepsilon$, $EV(t, X(t)) \geq \varepsilon e^{-\lambda t}$ and $\|V_t\|_D \leq qe^{\lambda r}EV(t, X(t))$, $t \geq T(\varepsilon)$.

Then, $e^{\lambda t}EV(t, X(t)) \to 0$ as $t \to \infty$, that is, there exists a $B = B(t_0, \varphi) < \infty$ such that $EV(t, X(t)) \leq BE\|\varphi\|_C^2 e^{-\lambda(t-t_0)}$ for all $t \geq t_0$.

Proof Let $U(t, x) = e^{\lambda t}V(t, x)$, $x \in R^d$, $t \geq -r$ with $0 < \lambda < c$. If $EU(t, X(t)) = \|U_t\|_D$, $t \geq t_0$ holds, then $e^{\lambda r}EV(t, X(t)) \geq \|V_t\|_D$, $t \geq t_0$. Therefore, by condition (a) of Theorem 4.1, we obtain that

$$ELU(t, X(t)) \leq e^{\lambda t}(\lambda EV(t, X(t)) - cEV(t, X(t)) + \theta(t)) \leq e^{\lambda t}\theta(t), \quad t \geq t_0.$$

Therefore U satisfies condition (a) of Theorem 3.1. Next, for every $\varepsilon > 0$, whenever $\|U_t\|_D \leq 2\varepsilon$, $EU(t, X(t)) \geq \varepsilon$ and $\|U_t\|_D \leq qEU(t, X(t))$ for all $t \geq T(\varepsilon)$, we get that $\|V_t\|_D \leq 2\varepsilon$, $EV(t, X(t)) \geq \varepsilon e^{-\lambda t}$ and $\|V_t\|_D \leq qe^{\lambda r}EV(t, X(t))$ for all $t \geq T(\varepsilon)$. Thus, by condition (b) of Theorem 4.1 we obtain for all $t \geq T(\varepsilon)$

$$ELU(t, X(t)) \leq -(c-\lambda)EU(t, X(t)) + e^{\lambda t}\xi(t).$$

Thus, since all the conditions of Theorem 3.4 are satisfied, the proof of the theorem is complete.

Next we present two theorems on almost sure Lyapunov exponent of solutions to equation (1).

Theorem 4.2 *Suppose that all the conditions of Theorem 4.1 and the following conditions are satisfied:*

(a) *there exists a $c_1 > 0$ such that $c_1|x|^2 \leq V(t, x)$ for all $x \in R^d$ and $t \geq t_0$;*

(b) *there exists a positive real number $Q > 0$ and a positive continuous function $\zeta: [0,\infty) \to R^+$ with $0 \leq \zeta(t) \leq B_1 e^{-\lambda t}$, $t \geq t_0$ ($B_1 > 0$ is a constant) such that for any solution $X(t) = X(t; t_0, \varphi)$ of (1)*

$$E|f(t, X_t)|^2 + E|g(t, X_t)|^2 \leq Q\|X_t\|_D^2 + \zeta(t), \quad t \geq t_0.$$

Then, there exists a positive real number $A_1 = A_1(t_0, \varphi) > 0$ such that
$$E\|X_t\|_C^2 \leq A_1 e^{-\lambda(t-t_0)}, \quad t \geq t_0 + r$$
and $\limsup\limits_{t\to\infty}(1/t)\log|X(t;t_0,\varphi)| \leq -\lambda/2$, almost surely.

Proof First of all, by simple calculations we have
$$E\|X_t\|_C^2 \leq 3E|X(t-r)|^2 + 3(1+r)\int_{t-r}^{t}[Q\|X_u\|_D^2 + \zeta(u)]\,du.$$

Therefore, by Theorem 4.1
$$E\|X_t\|_C^2 \leq \frac{3B}{c_1} E\|\varphi\|_C^2 \, e^{\lambda(r+t_0)} e^{-\lambda t}$$
$$+ 3(1+r)\left(\frac{QB}{c_1} e^{\lambda(r+t_0)} E\|\varphi\|_C^2 + B_1\right)\frac{1}{\lambda} e^{\lambda r} e^{-\lambda t}.$$

Thus we have the first desired conclusion. Next, let $\varepsilon_k > 0$. Then, by the condition and the Chebyshev inequality we have a positive real number $B_2 = B_2(t_0, \varphi) > 0$ such that
$$P\left(\sup_{k \leq t \leq k+1} |X(t)|^2 > \varepsilon_k\right) \leq \frac{B_2}{\varepsilon_k} e^{-\lambda(k-t_0)}.$$

Set $\varepsilon_k = B_2 e^{-(\lambda-\rho)(k-t_0)}$ for each positive integer $k > t_0 + r$, where ρ is any real number with $0 < \rho < \lambda$. Thus, by the Borel-Cantelli lemma we have the second desired conclusion.

Theorem 4.3 *Suppose that all the conditions of Theorem 4.1, condition (a) of Theorem 4.2 and the following condition are satisfied.*

(a) *There exist real numbers $Q > 0$ and $\mu \geq 0$ with $0 \leq \mu < \lambda < c$ and a positive continuous function $\zeta\colon [0, \infty) \to R^+$ with $0 \leq \zeta(t) \leq B_1 e^{-\lambda t}$, $t \geq 0$ ($B_1 > 0$ is a constant) such that for any solution $X(t)$ of (1)*
$$E|f(t, X_t)|^2 + E|g(t, X_t)|^2 \leq Q e^{\mu t}\|X_t\|_D^2 + \zeta(t), \quad t \geq t_0.$$

Then, $\limsup\limits_{t\to\infty}(1/t)\log|X(t;t_0,\varphi)| \leq -(\lambda-\mu)/2$, *almost surely.*

5 Example

Let $r > 0$ be a real number and let $r_1, r_2\colon [0, \infty) \to R^+$ be continuous functions with $0 \leq r_1(t), r_2(t) \leq r < \infty$. For a matrix $G = (g_{ij})$, set
$$\|G\|^2 := \sum_{i,j=1}^{m} g_{ij}^2.$$

Example 5.1 Let A, B and C be $d \times d$-matrices. Suppose that all the eigenvalues of A have negative real parts.

Then it is well-known that there exists a positive definite and symmetric matrix D such that $A^T D + DA = -I$ (I = identity matrix). Let α_i ($i = 1, 2, 3, \ldots, d$) be the eigenvalues of D. Since D is positive definite and symmetric, $\beta_1 := \min\{\alpha_i\} > 0$ $\beta_2 := \max\{\alpha_i\} > 0$ and it holds that $\beta_1 |x|^2 \leq x^T D x \leq \beta_2 |x|^2$ for all $x \in R^d$. Now consider the stochastic delay differential equation:

$$dX(t) = [AX(t) + BX(t - r_1(t)) + \theta(t)] \, dt + [CX(t - r_2(t)) + \xi(t)] \, d\beta(t), \quad (2)$$

where $\theta(t)$, $\xi(t)$ are d-dimensional continuous vector functions and $\beta(t)$ is the one-dimensional standard Brownian motion. Let $\delta_1 > 0$ be the solution of the equation $1 = \delta + 2e^{(\delta/\beta_2)r} \|D\|(\beta_2/\beta_1)(\|B\| + \|C\|^2) + \|D\|$ and assume that $e^{(\delta_1/\delta_2)t}|\theta(t)|^2$ and $e^{(\delta_2/\beta_2)t}|\xi(t)|^2$ are integrable, bounded functions. Then,

$$\limsup_{t \to \infty} \frac{1}{t} \log |X(t)| \leq -\frac{\delta_1}{2\beta_2},$$

almost surely.

Proof We prove only the case where $\theta(t) \geq 0$ for all $t \geq t_0$. Let $\delta > 0$ be a real number such that $0 < \delta < \delta_1$. Then, we can choose a $q = q(\delta) > 1$ such that $0 < \delta < 1 - 2qe^{(\delta/\beta_2)r}\|D\|(\beta_2/\beta_1)(\|B\| + \|C\|^2) - \|D\|$. Let $c = [1 - 2qe^{\lambda r}\|D\|(\beta_2/\beta_1)(\|B\| + \|C\|^2) - \|D\|]/\beta_2$, where $\lambda := \delta/\beta_2$.

Let $V(t, x) = V(x) = x^T D x$ for $x \in R^d$. Thus, whenever $\sup_{-r \leq s \leq 0} EV(X(t + s)) \leq qe^{\lambda r} EV(X(t))$, we have that $\beta_1 E|X(t + s)|^2 \leq \beta_2 qe^{\lambda r} E|X(t)|^2$ for all $s \in [-r, 0]$. Therefore,

$$ELV(X(t)) = E\{X(t)^T A^T D X(t) + X(t)^T D A X(t)$$
$$+ X(t - r_1(t))^T B^T D X(t) + X(t)^T D B X(t - r_1(t)) + 2\theta(t)^T D X(t)$$
$$+ \mathrm{tr}\,\{D(CX(t - r_2(t)) + \xi(t))(CX(t - r_2(t)) + \xi(t))^T\}\}$$

$$\leq -E|X(t)|^2 + 2\sqrt{q \frac{\beta_2}{\beta_1}} e^{\lambda r/2} \|D\| \|B\| E|X(t)|^2$$

$$+ 2\|D\| \left[qe^{\lambda r} \|C\|^2 \frac{\beta_2}{\beta_1} E|X(t)|^2 + |\xi(t)|^2 \right] + 2\|D\| |\theta(t)| E|X(t)|$$

$$\leq -\left[1 - 2qe^{\lambda r} \|D\| \frac{\beta_2}{\beta_1} (\|B\| + \|C\|^2) - \|D\| \right] E|X(t)|^2$$
$$+ \|D\|(|\theta(t)|^2 + 2|\xi(t)|^2)$$
$$\leq -cEV(X(t)) + \|D\|(|\theta(t)|^2 + 2|\xi(t)|^2),$$

which implies that condition (b) of Theorem 4.1 is satisfied. Condition (a) of Theorem 4.1 holds also since $ELV(X(t)) \leq -cEV(X(t)) + 2\|D\|(|\theta(t)|^2 + |\xi(t)|^2)$ whenever $e^{\lambda r}EV(X(t)) \geq \sup_{-r \leq s \leq 0} EV(X(t+s))$. Therefore by Theorem 4.2 we have that $\limsup_{t\to\infty}(1/t)\log|X(t)| \leq -\delta/(2\beta_2)$, almost surely. Thus, letting $\delta \to \delta_1$, the proof of the example is complete.

Example 5.2 Let $a, b\colon [0,\infty) \to [0,\infty)$ be continuous functions such that $\alpha_1 > a(t) \geq \alpha \geq b(t)^2 \geq 0$ for all $t \geq 0$ with constants $\alpha_1, \alpha > 0$. Let $0 < r < 1$ and let $\delta_3 > 0$ be the solution of the equation $1 = \delta + r^2 e^{2\alpha\delta r}$. Consider the stochastic delay differential equation

$$dX(t) = -a(t)X(t)dt + \left[b(t)\int_{-r_1(t)}^{0} X(t+u)\,du + \xi(t)\right]d\beta(t), \quad t \geq 0, \quad (3)$$

where $\xi\colon [0,\infty) \to (-\infty,\infty)$ is continuous and $|\xi(t)|^2 e^{2\alpha\delta_3 t}$ is an integrable, bounded function. Then, for any solution $X(t)$ of (3),

$$\limsup_{t\to\infty} \frac{1}{t}\log|X(t)| \leq -\alpha\delta_3,$$

almost surely.

Proof Let $\delta > 0$ be a real number such that $0 < \delta < \delta_3$. Then, we can choose a real number $q = q(\delta) > 1$ such that $1 - qr^2 e^{2\alpha\delta r} > \delta > 0$.

Let $V(x) = x^2/2$, $x \in R^1$. Let $c = 2\alpha(1 - qr^2 e^{2\alpha\delta r})$. Then, $c > 2\alpha\delta$. And whenever $EV(X(t+s)) \leq qe^{2\alpha\delta r}EV(X(t))$ for all $s \in [-r,0]$,

$$ELV(X(t)) = -a(t)EX(t)^2 + qr^2 e^{2\alpha\delta r}b(t)^2 EX(t)^2 + \xi(t)^2$$
$$\leq -2\alpha(1 - qr^2 e^{2\alpha\delta r})EV(X(t)) + \xi(t)^2 = -cEV(X(t)) + \xi(t)^2.$$

Thus condition (b) of Theorem 4.1 is satisfied. Since condition (a) of Theorem 4.1 also holds, by Theorem 4.2 we obtain that $\limsup_{t\to\infty}(1/t)\log|X(t)| \leq -\alpha\delta$, almost surely. Thus, letting $\lambda \to \lambda_3$, the proof of the example is complete.

Example 5.3 Let $\tau\colon [-r,0] \to R^+$ be a nondecreasing, continuous function, and let the matrix A be the same as in Example 5.1 and the function $\xi\colon [0,\infty) \to R^d$ be continuous. Consider the stochastic functional differential equation

$$dX(t) = AX(t)\,dt + [a_2(t,X_t) + \xi(t)]\,d\beta(t), \quad t \geq 0, \quad (4)$$

where $a_2\colon [0,\infty) \times C([-r,0], R^d) \to R^d$ is a continuous function with the global Lipschitz condition in $\psi \in C$ and condition

$$|a_2(t,\psi)|^2 \leq \int_{-r}^{0} |\psi(s)|^2 \, d\tau(s).$$

Set $\tau_0 := \tau(0) - \tau(-r)$. Let $\delta_4 > 0$ be the solution of the equation $1 = \delta + 2\tau_0(\beta_2/\beta_1)\|D\|e^{(\delta/\beta_2)r}$, where β_1 and β_2 are the same as in Example 5.1. Suppose that $|\xi(t)|^2 e^{(\delta_4/\beta_2)}$ is an integrable, bounded function. Then, for any solution $X(t)$ of (4)

$$\limsup_{t\to\infty} \frac{1}{t} \log|X(t)| \leq -\frac{\delta_4}{2\beta_2},$$

almost surely.

Proof Let $V(x) = x^T D x$, $x \in R^d$, where the matrix D is the same as in Example 5.1. Let $\delta > 0$ be a real number such that $0 < \delta < \delta_4$. Then we can choose a $q = q(\delta) > 1$ such that $0 < \delta < 1 - 2q\|D\|\tau_0(\beta_2/\beta_1)e^{(\delta/\beta_2)r}$. Let $c = (1/\beta_2)(1 - 2q\|D\|\tau_0(\beta_2/\beta_1)e^{(\delta/\beta_2)r})$. Then, whenever $EV(X(t+s)) \leq qe^{(\delta/\beta_2)r}EV(X(t))$ for all $s \in [-r, 0]$, we get that $\beta_1 E|X(t+s)|^2 \leq \beta_2 q e^{(\delta/\beta_2)r} E|X(t)|^2$ for all $s \in [-r, 0]$. Thus,

$$\begin{aligned} ELV(X(t)) &\leq -E|X(t)|^2 + 2E\|D\|(|a_2(t,X_t)|^2 + |\xi(t)|^2) \\ &\leq -\left(1 - 2q\|D\|\tau_0\left(\beta_2/\beta_1\right)e^{(\delta/\beta_2)r}\right)E|X(t)|^2 + 2\|D\|\|\xi(t)|^2 \\ &\leq -cEV(X(t)) + 2\|D\|\|\xi(t)|^2. \end{aligned}$$

Therefore, by Theorem 4.2 we have that $\limsup_{t\to\infty}(1/t)\log|X(t)| \leq -\delta/(2\beta_2)$, almost surely. Thus, letting $\delta \to \delta_4$, the proof is complete.

References

[1] Gopalsamy, K. (1992). *Stability and Oscillations in Delay Differential Equations of Population Dynamics*. Kluwer Academic Publishers, Dordrecht.

[2] Hale, J. (1977). *Theory of Functional Differential Equations*. Springer-Verlag, New York.

[3] Hasminskii, R.Z. (1980). *Stochastic Stability of Differential Equations*. Sijthoff and Noordhoff, Maryland.

[4] Kolmankvskii, V. and Myshkis, A. (1992). *Applied Theory of Functional Differential Equations*. Kluwer Academic Publishers, Dordrecht.

[5] Kolmanovskii, V. and Nosov, V. (1986). *Stability of Functional Equations*. Academic Press, London.

[6] Kushner, H.J. (1968). On the stability of processes defined by stochastic difference-differential equations. *J. Diff. Eqns*, **4**, 424–443.

[7] Mackey, M.C. and Nechaeva, I.G. (1994). Noise and stability in differential delay equations. *J. of Dynamics and Diff. Equations*, **6**(3), 395–426.

[8] Marcus, C.M. and Westervelt, R.M. (1989). Stability of analog neural networks with delay. *Phys. Rev. A*, **39**, 347–359.

[9] Mizel, V.J. and Trutzer, V. (1984). Stochastic hereditary equations: existence and asymptotic stability. *J. Integral Eqns*, **7**, 1–72.

[10] Mohammed, S.-E.A. (1984). *Stochastic Functional Differential Equations.* Pitman Research Notes in Math. Vol. 99, Longman, UK.
[11] Rodkina, A.E. (1984). On existence and uniqueness of solution of stochastic differential equations with heredity. *Stochastics*, **12**, 197–200.
[12] Scheutzow, M. (1984). Qualitative behaviour of stochastic delay equations with a bounded memory. *Stochastics*, **12**, 41–80.
[13] Taniguchi, T. (1992). Successive approximations to solution of stochastic differential equations, *J. Diff. Eqns*, **96**, 152–169.
[14] Taniguchi, T. (1995). Asymptotic behavior theorems for non-autonomous functional differential equations via Lyapunov-Razumikhin method. *J. Math. Anal. Applic.*, **189**, 715–730.
[15] Taniguchi, T. (1996). Asymptotic behavior of solutions of functional differential equations with finite delays. *J. Math. Anal. Applic.*, **199**, 776–786.
[16] Taniguchi, T. (1996). Moment asymptotic behavior and almost sure Lyapunov exponent of stochastic functional differential equations with finite delays. *Stochastics and Stochastic Reports*, **58**, 191–208.

2.9 A NON-STANDARD APPROACH TO THE STUDY OF THE DYNAMIC SYSTEM STABILITY*

V.A. VUJIČIĆ

Mathematical Institute SANU, Beograd, Yugoslavia

0 Introduction

An important difference between the study of the dynamic system stability in mathematics and the dynamic systems in mechanics is pointed out here. Regarding the fact that all the differential equations of the mechanical system motion as well as the differential equations of the disturbed motion can be reduced to the general form of the dynamic system's differential equations it is argued and proved that such generalized equations are neither invariant with respect to punctual transformations, while they are being mapped from a set of rectilinear coordinates into a set of curvilinear ones. This is simple to prove by comparing the differential equations of motion of the same object with respect to the rectilinear and curvilinear coordinate system in the Euclidean space as well as respective equations upon the tangent and cotangent manifolds. The thesis is set forth that the mechanical system stability cannot be reliably estimated unless the corresponding differential equations of motion, especially the disturbance equations, exactly describe the observed motion. The general invariant criterion of the balanced state stability and the mechanical system motion is proved. Lagrange's theorem about the system's balanced state stability is generalized while its application to one characteristic instance of the rheonomic system is shown.

---

*Advances in Stability Theory (Ed.: A.A. Martynyuk). Stability and Control: Theory, Methods and Applications, Taylor & Francis, London, **13** (2003) 189–200.

1 On Covariant Differential Equations of the Mechanical System Motion

The concept of the dynamic system stability is not unanimously determined. The words "stability", "dynamics" and "systems", taken separately, have a general meaning, while the motion stability theory asks for an exact unambiguous determination. At the mathematical base of the "dynamic system" concept there are the following differential equations

$$\frac{dx}{dt} = f(x,t), \quad x \in R^n. \tag{1.1}$$

In mechanics the notion "dynamics" implies that part of mechanics in which forces and their mutual relations are studied. All the differential equations of the mechanical system motion can be formally reduced to form (1.1), but such equations are not invariant with respect to various coordinate systems; thus, they do not speak about the same attributes of the objects' motion, let alone about their stability. This is very clearly shown in the case of a free material point's motion of constant mass m acted upon by force Y. With respect to the orthonormal coordinate system $y := (y^1, y^2, y^3) \in E^3$ the differential equations of motion

$$\dot{y} := \frac{dy}{dt} = \frac{p}{m}, \tag{1.2}$$

$$\dot{p} = Y(y, p, t) \tag{1.3}$$

are in accordance with the basic theorem of mechanics or Newton's law stating that impulse derivative p is equal to force Y. However, if the same motion is described by means of any curvilinear coordinates $x = (x^1, x^2, x^3)$ equations (1.3) obtain a different form:

$$\dot{x}^i = a^{ij}(x) p_j, \tag{1.4}$$

$$\dot{p}_i = -\frac{\partial a^{kj}}{\partial x^i} p_k p_j + X_i, \tag{1.5}$$

where X are forces Y expressed by means of coordinates x, that is $X = Y \frac{\partial y}{\partial x}$. It becomes obvious that $\dot{p} = X$, as is the case in equation (1.3). Equations (1.3) and (1.5) are invariant with respect to transformation $X = Y \frac{\partial y}{\partial x}$ only by means of covariant derivative

$$\frac{Dp_i}{dt} := \left(\frac{dp_i}{dt} - \Gamma^j_{ik}(x) p_j \frac{dx^k}{dt} \right). \tag{1.6}$$

A similar, but much more general statement can be derived for the mechanical system motion from N material points with holonomic scleronomic constraints. Lagrange's equations of motion of the first kind

$$m_i \frac{d\dot{y}^i}{dt} = Y_i + \sum_{\mu=1}^{k} \lambda_\mu \frac{\partial f_\mu}{\partial y^i}$$

$$f_\mu(y^1, \ldots, y^{3N}) = 0, \qquad (1.7)$$

$$m_{3i-2} \equiv m_{3i-1} \equiv m_{3i}$$

are not equivalent to Hamilton's equations [9]

$$\dot{p}_i = -\frac{\partial H}{\partial q^i} + Q_i^*, \quad \dot{q}^i = \frac{\partial H}{\partial p_i}, \qquad (1.8)$$

$$q \in M^{3N-k} =: M, \quad (p; q) \in T^*M,$$

where $q = (q^1, \ldots, q^n)^T$ are Lagrange's independent generalized coordinates; T^*M is cotangent configuration manifold. By means of these standard Hamilton's equations the balanced state stability of the system balance is most often discussed.

In further study of the motion stability of one and the same mechanical system described by equations (1.8) we will point to two non-equivalent system of the disturbance differential equations. Disturbed equations of motion of Syng [5]

$$\frac{D^2 \xi^i}{dt^2} + R^i_{jkl} \dot{q}^j \xi^k \dot{q}^l = \nabla_l Q^i \xi^l \qquad (1.9)$$

are not equivalent to covariant differential equations of disturbance [7].

$$\frac{D\eta_i}{dt} = \psi_i(t, \eta, \xi) \qquad (1.10)$$

$$\frac{D\xi^j}{dt} = a^{ij} \eta_i, \qquad (1.11)$$

where are $\xi = \delta q$ and $\eta = \delta p$.

It is logical that the conclusions concerning stability or instability of the null solutions of differential equations (1.9) and equations (1.10) will not be identical.

All the forces present in mechanics are formulated by means of real vector functions with accuracy up to a certain constant. But those constants, involving the gravitational one, do not have only one single numerical value. Instead, they have values of a limited set of real numbers whose value depends on the nature of an object as well as on the medium that its moves through. The influence of these parameters upon the motion stability has a more essential importance than that of the initial conditions of phase variables. Due to this it is of primary importance to know both the origin and the invariant form of the differential equations of motion as well as the way of determining disturbances. On the basis of such covariant equations it is also possible to prove a general invariant criterion of stability.

In order to do this, let's observe N material points of mass m_ν, $(\nu = 1, \ldots, N)$ whose radius vectors are r_ν, $\nu = 1, \ldots N$. Let material points be connected by k independent constraints

$$f_\mu(y_\nu^1, y_\nu^2, y_\nu^3, \tau(t)) = f_\mu(y^0, y^1, \ldots, y^{3N}) = 0, \quad y^0 = \tau(t), \qquad (1.12)$$

where the notations are introduced

$$y_\nu^1 =: y^{3\nu-2}, \quad y_\nu^2 =: y^{3\nu-1}, \quad y_\nu^3 =: y^{3\nu}, \qquad y^0 = \tau(t). \qquad (1.13)$$

Mechanics reliably starts from the D'Alembert's principle of dynamic balance and the Newton-Laplace's principle of determinacy that the known differential equations of motion result from:

$$m_\nu \frac{d v_\nu}{dt} = F_\nu + R_\nu, \qquad (1.14)$$

$$f_\nu(\vec{r}_1, \ldots, \vec{r}_N, \tau) = 0, \quad \tau = \tau(t),$$

where are $R_\nu = \sum_{\mu=1}^{k} \lambda_\mu \operatorname{grad}_\nu f_\mu$ main vectors of the smooth constraints reactions forces. The configurational manifold is generated by matrix relation

$$\left| \frac{\partial f_\mu}{\partial y} \right| \neq 0. \qquad (1.15)$$

On the basis of the implicit function theorem in the domain allowed for by relations (1.15) the holonomic constraint equations can be written in parametric form [8,9]

$$r_\nu = r_\nu(q^0, q^1, \ldots, q^n), \quad n = 3N - k,$$
$$q := (q^0, \ldots, q^n) \in M^{n+1}; \quad q^0 = \tau(t). \qquad (1.16)$$

Scalar multiplication of equations (1.14) by coordinate vectors $\dfrac{\partial r_\nu}{\partial q^\alpha}$ and adding with respect to index ν these equations projections upon the coordinate directions of manifold M in the covariant form are obtained

$$a_{\alpha\beta} \frac{D \dot{q}^\beta}{dt} = Q_\alpha, \quad (\alpha, \beta = 0, 1, \ldots, n), \qquad (1.17)$$

or

$$a_{i\beta} \frac{D\dot{q}^\beta}{dt} = Q_i, \quad (i=1,\ldots,n), \tag{1.17a}$$

$$a_{0\beta} \frac{D\dot{q}^\beta}{dt} = Q_0, \tag{1.17b}$$

or

$$\dot{q}^\alpha = a^{\alpha\beta}(q)p_\alpha, \tag{1.18}$$

$$\frac{Dp_\alpha}{dt} = Q_\alpha, \tag{1.19}$$

where $a_{\alpha\beta}(q)$ are coordinates of inertia tensor [9], $a_{\alpha\beta} = a_{\beta\alpha}(q^0, q^1, \ldots, q^n)$ are generalized impulses, whereas Q denotes generalized forces that may also explicitly depend on time only if forces F in (1.14) are time functions.

2 Invariant Criterion of the Balanced Position Stability

The concept of the system's equilibrium state implies rest of the observed bodies in particular position $q^\alpha = q_0^\alpha$ and all the generalized velocities are equal to zero so that. The equilibrium state equations, consequently, spring from equations (1.19), that is,

$$Q_\alpha(q,\dot{q})|_{\dot{q}=0} = 0 \tag{2.1}$$

so that the solutions of equations (2.1) determine the equilibrium state of the material system.

Definition 2.1 The equilibrium state of the mechanical system implies a set of solutions q of equations (2.1) and $\dot{q}(t) = 0$. The equilibrium position of the mechanical system implies a position $q^\alpha = q_0^\alpha$ on the coordinate manifolds whose coordinates satisfy equations (2.1).

Definition 2.2 If at any randomly given number $A > 0$, regardless of how small it is not, such a real number λ can be chosen for which all the initial disturbances are constrained by the relation

$$\delta_{\alpha\beta}q^\alpha(t_0)q^\beta(t_0) + \delta^{\alpha\beta}p_\alpha(t_0)p_\beta(t_0) \leq \lambda, \tag{2.2}$$

and for every $t \geq t_0$ the inequality is satisfied

$$\delta_{\alpha\beta}q^\alpha q^\beta + \delta^{\alpha\beta}p_\alpha p_\beta < A, \tag{2.3}$$

the undisturbed equilibrium state $q^\alpha = q_0^\alpha$, $\dot{q}^\alpha = 0$ is stable; otherwise, it is unstable. As in the previous proposition, $\delta_{\alpha\beta}$ and $\delta^{\alpha\beta}$ are Kronecker's symbols.

Stability Criterion *If for the differential equations of motion of the scleronomic system (1.17) the positively definite function $W(t, q^1, \ldots q^n)$ could be found, such that it is*

$$\frac{\partial W}{\partial t} + \left(Q_i + \frac{\partial W}{\partial q^i}\right)\dot{q}^i \leq 0 \quad (i = 1, \ldots, n), \tag{2.4}$$

the equilibrium state $q = q_0$, $\dot{q} = 0$ is stable.

Proof With the conjunction that there is function W, the function

$$V = \frac{1}{2} a_{ij}(q^1, \ldots, q^n)\dot{q}^i \dot{q}^j + W(q^1, \ldots, q^n, t) \tag{2.5}$$

is positively definite since kinetic energy

$$E_k = \frac{1}{2} a_{ij} \dot{q}^i \dot{q}^j$$

is, by its definition, positively definite. The derivative with respect to time of function (2.5) is since it is

$$\dot{V} = a_{ij} \frac{D\dot{q}^i}{dt} \dot{q}^j + \frac{\partial W}{\partial t} + \frac{\partial W}{\partial q^i} \dot{q}^i,$$

while $\dot{V} = \dfrac{dV}{dt} = \dfrac{DV}{dt}$.

If equations (1.17a), $q^0 \equiv 0$ are kept in mind, the previous derivative is reduced to the form

$$\frac{\partial W}{\partial t} + Q_i \dot{q}^i + \frac{\partial W}{\partial q^i} a^{ij} p_j = \frac{\partial W}{\partial t} + \left(Q_i + \frac{\partial W}{\partial q^i}\right)\dot{q}^i, \tag{2.6}$$

and the criterion is proved by this. If the system is autonomous, function W should be looked for only depending upon the coordinates, so that condition (2.6) is reduced to

$$\left(Q_i + \frac{\partial W}{\partial q^i}\right)\dot{q}^i \leq 0, \tag{2.7}$$

The previous theorem is also valid for mechanical systems with rheonomic constraints. Condition (2.7) changes only if indices $i, j = 1, \ldots, n$ take on values $\alpha, \beta = 0, 1, \ldots, n$. Therefore, three additional addends are obtained:

$$\left(Q_\alpha + \frac{\partial W}{\partial q^\alpha}\right)\dot{q}^\alpha = \left(Q_i + \frac{\partial W}{\partial q^i}\right)\dot{q}^i + \left(Q_0 + \frac{\partial W}{\partial q^0}\right)\dot{q}^0 \leq 0 \tag{2.8}$$

The proof is identical to the previous one, except for the fact that the indices in equations (1.17) remain in the range $0, 1, \ldots, n$.

3 Generalization of Lagrange's Theorem

For the mechanical system whose motion is described by Lagrange's equations of the second kind

$$\frac{d}{dt}\frac{\partial E_k}{\partial \dot{q}^i} - \frac{\partial E_k}{\partial q^i} = -\frac{\partial E_p}{\partial q^i}, \quad i=1,\ldots,n. \tag{3.1}$$

Lagrange's theorem states that the mechanical conservative system balance position is stable in some domain A if potential energy E_p has isolated minimum there. On the basis of the criterion (2.8) Lagrange's theorem can be generalized for non-conservative systems as well. Indeed, if generalized forces consist of conservative, $-\frac{\partial E_p}{\partial q}$, and non-conservative Q^* forces, that is,

$$Q_\alpha = -\frac{\partial E_p}{\partial q^\alpha} + Q^*_\alpha, \tag{3.2}$$

Lagrange's theorem can be formulated in a more general way.

Theorem 3.1 *The position of balance in domain A of a non-conservative system is stable if potential energy E_p has isolated minimum in the balanced position, while power $Q^*_\alpha \dot{q}^\alpha$ of non-conservative forces Q^* is not positive.*

Proof 1 Starting from the fact that $E_p(q^0, q^1, \ldots, q^n)$ is a positively definite function, for W can be chosen function E_p. Substituting (3.2) in relation (2.8), it follows

$$\left(Q_\alpha + \frac{\partial W}{\partial q^\alpha}\right)\dot{q}^\alpha \leq 0, \tag{3.3}$$

which proves the above-stated generalized theorem.

Proof 2 In the case taken into consideration, the differential equations of motion (3.1) can be written in the form

$$\frac{D}{dt}\frac{\partial E_k}{\partial \dot{q}^\alpha} = -\frac{\partial E_p}{\partial q^\alpha} + Q^*_\alpha. \tag{3.4}$$

Since $V = E_k + E_p$ a positive definite function, the functions derivative along the solution of equations (3.4) is

$$Q^*_\alpha \dot{q}^\alpha \leq 0 \tag{3.5}$$

which is in accordance with the first proof.

Rheonomic Systems. The motion stability of the dynamic system theory,s rheonomic system (1.1) comprises the non-autonomous system concept. In this area there is an even greater difference between the mathematical dynamic systems and the corresponding systems in mechanics; thus it can be said that this area of

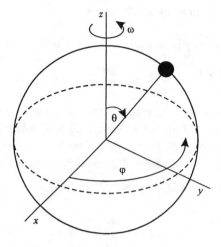

Figure 3.1. Bead on a rotating circle.

stability in mechanics is still open. As a proof of this statement, it is enough to stress that some authors describe one and the same system by means of n independent Lagrange's generalized coordinates $q \in M^n$, while others do it by means of $n+1$ coordinates $(q, q^0) \in M^{n+1}$. This further assumes that some authors observe kinetic energy as a homogeneous square positively-definitive form of the generalized energy, while others regard it as a sum of three forms $T = \sigma_{s=1}^{3} T_s$, where s is homogeneity degree. Hence it further follows that the basic matrix tensors of manifold M^n and M^{n+1}, those that are relevant for the motion stability, are different. In this approach, the question of stability with respect to the rheonomic coordinate q^0 and corresponding generalized impulse p_0 is not clear enough. For this reason we are here quoting an example from the current and very important book [2] and [4].

Example 3.1 Considered the motion of a bead along a vertical circle of radius r which rotates with angular velocity ω around the vertical axis passing through the center O of the circle. The manifold M^{n+1} is the sphere.

Let x, y and z be cartesian coordinates in E^3 with origin O and vertical axis z. Let φ be the angle of the plane of the circle with the plane xOz. By hypothesis, $\varphi = \omega t$; ω is constant. For constraint $f_1 = r - \text{const} = 0$ and condition $\varphi = \omega t =: q^0$, the point position at M^{n+1} and E^3 is mutually mapped by the relations $y^1 = r \sin q \cos q^0$, $y^2 = r \sin q \sin q^0$, $y^3 = r \cos q$. Kinetic energy is

$$E_k = \frac{m}{2}\left(\dot{y}_1^2 + \dot{y}_2^2 + \dot{y}_3^2\right) = \frac{mr^2}{2}\left(\dot{q}^2 + (\dot{q}^0)^2 \sin^2 q\right) \geq 0.$$

The generalized force Q, by its definition, is

$$Q = -mg\frac{\partial z}{\partial q} = mg\sin q,$$

while force Q_0 is determined by means of equation (1.17b), that is,

$$\frac{D}{dt}\frac{\partial E_k}{\partial \dot{q}^0} = \frac{D}{dt}(mr^2\dot{q}^0\sin^2 q) = Q_0$$

From balance equation $Q = 0$ and $Q_0 = 0$ balance positions $q = 0, \varphi$ are obtained.

Stability is investigated by means of criteria (2.4). For W let's choose, upon interval $[0, \pi]$, the function

$$W = mgr(1 - \cos q)$$

The stable balance positions satisfy the relations

$$\left(Q + \frac{\partial W}{\partial q}\right)\dot{q} + \left(Q_0 + \frac{\partial W}{\partial q^0}\right)\dot{q}^0 = (2mgr\sin q + mr^2(\dot{q}^0)^2\sin 2q)\dot{q} \leq 0,$$

that is, $mr(g + r\omega^2\cos q))\dot{q}\sin q \leq 0$. Since $\sin q$ is positive upon interval $[0, \pi]$, it should be that

$$(g + r\omega^2\cos q)\dot{q} \leq 0.$$

Therefore, balance state $(\dot{q} = 0, q = \pi)$ is stable if $g > r\omega_2$.

The same results is achieved by the generalized Lagrange's theorem. Indeed, the potential energy is

$$V := E_p = -\int_s(Qdq + Q_0dq^0) = -\int_s\left(mrg\sin qdq + \frac{Dp_0}{dt}dq^0\right)$$

$$= mgr(\cos q - 1) - \int_s a_{00}(q)\dot{q}^0 D\dot{q}^0,$$

since

$$Dp_0 = D(a_{00}\dot{q}^0) = a_{00}D\dot{q}^0 = mr^2\sin^2 qD\dot{q}^0.$$

The condition

$$\frac{d^2V}{dq^2} = -mr(g\cos q + r\omega^2\cos 2q),$$

determines the stable in positions $q = \pi$.

4 Invariant Criterion of Motion Stability

The concept of the "invariant criterion" implies general measurement standard in all the coordinate systems for estimating stability of some undisturbed mechanical system's motion. As such, it comprises stability of the equilibrium position and state, stability of stationary motions and, in general, of motion of mechanical systems whose disturbance equations are of coordinate shape (1.10).

Theorem 4.1 *If for the differential equations of disturbance (1.10) there is such a positively definitive function W of disturbance $\xi^0, \xi^1, \ldots, \xi^n$ and time t that the expression is*

$$\frac{\partial W}{\partial t} + a^{\alpha\beta}\left(\Psi_\alpha + \frac{\partial W}{\partial \xi^\alpha}\right)\eta_\beta \leq 0 \qquad (4.1)$$

smaller or equal to zero, the undisturbed state of the mechanical system's motion is stable.

Proof As can be seen from equation

$$\Psi_\gamma := \sum_{\nu=1}^{N}\left(F_\nu^* - F_\nu\right)\cdot\frac{\partial r_\nu}{\partial q^\gamma} = \Psi_\gamma(\xi,\eta,t)$$

functions Ψ_α for undisturbed motion $\xi^\alpha = 0$, $\eta_\alpha = 0$ are equal to zero, that is, $\Psi_\alpha(0,0,t) = 0$. The function

$$V = \frac{1}{2}a^{\alpha\beta}\eta_\alpha\eta_\beta + W(\xi,t) \qquad (4.2)$$

is positively definite, since it is $\overset{a}{\alpha\beta}$ positively definite matrix of the functions upon M, while W is a positively definite function of disturbance ξ^α and t. As a scalar invariant, V is a tensor of zero order. That is why ordinary derivative $\dfrac{dV}{dt}$ is equal to the natural derivative

$$\frac{DV}{dt} = a^{\alpha\beta}\frac{D\eta_\alpha}{dt}\eta_\beta + \frac{\partial W}{\partial \xi^\alpha}\frac{D\xi^\alpha}{dt} + \frac{\partial W}{\partial t} \qquad (4.3)$$

which necessarily has to be smaller or identical to zero. By substitution of the natural derivatives from equations (1.10) in (4.3) it is obtained that

$$\frac{DV}{dt} = \frac{\partial W}{\partial t} + a^{\alpha\beta}\Psi_\alpha\eta_\beta + \frac{\partial W}{\partial \xi^\alpha}a^{\alpha\beta}\eta_\beta$$

and this, along with the criterion requirement, is reduced to

$$\frac{\partial W}{\partial t} + a^{\alpha\beta}\left(\Psi_\alpha + \frac{\partial W}{\partial \xi^\alpha}\right)\eta_\beta \leq 0. \qquad (4.4)$$

Therefore, the stability criterion is proved. If functions Ψ_α are explicitly independent of t, then function W should also be looked for only in its dependence on disturbances ξ^α, that is, $W = W(\xi^0,\xi^1,\ldots\xi^n)$, so that expressions (4.4) is reduced to

$$a^{\alpha\beta}\left(\Psi_\alpha + \frac{\partial W}{\partial \xi^\alpha}\right)\eta_\beta \leq 0. \qquad (4.5)$$

If the mechanical system's constraints do not depend on time, q^0, ξ^0, η_0 and Ψ_0 vanish, so that expression (4.4), is reduced to

$$\frac{\partial W}{\partial t} + a^{ij}\left(\Psi_i + \frac{\partial W}{\partial \xi^i}\right)\eta_j \leq 0, \tag{4.6}$$

while expression (4.5) is reduced to

$$a^{ij}\left(\Psi_i + \frac{\partial W}{\partial \xi^i}\right)\eta_j \leq 0. \tag{4.7}$$

All the expressions of the previously given criterion for the equilibrium state stability appear as consequences of expression (4.1) if ξ and η are regarded as disturbances of motion or if q and p are regarded as disturbances of equilibrium state $q = q_0$; $p = 0$.

5 Comments and References

The expression "the stability of the mechanical system motion" is used in the sense of the classical analytical mechanics of Lagrange and Hamilton [2, 9]. The attribute "nonstandard" points to our approach to the theory about body motion is different from standard analytical dynamics, especially with the rheonomic constraint systems [8, 9]. It also stresses that the "dynamic systems" (1.1) are different in the sense of invariance and determination of the differential equations of the system motion (1.19) as well as from the differential equations of disturbance (1.10). If the differential equations of motion of the material points or bodies are not well composed, neither they nor their disturbance equations are the ones to base reliable conclusions about undisturbed motions or undisturbed balance state upon. The general invariant criterion about the balance stability and the mechanical system motion on the basis of the Liapunov's theory is derived. However, function W is introduced in the criterion expression, namely, the function dependent only upon the position, that is, of twice less variables than Liapunov's function V. On the basis of this criterion the Lagrange-Dirichlet's theorem is generalized. The example are given in order to state more clearly the differences between our approach and the known standard approaches from the mechanical system stability [2].

References

[1] Angelitch, T.P. (1968). Tensorkalkul nebst Anwendungen, *Die Grundlehren der Mathematishen Wissenschaften*, Band 141, *Mathematische Hilfsmittel des Ingenieurs*, Teil III, Springer-Verlag, 167–230.

[2] Arnold, V.I. (1981). *Mathematical Methods of Classical Mechanics*. Springer-Verlag, Berlin.
[3] Martynyuk, A.A., Lakshmikantham, V. and Leela, S. (1989). *Stability of Motion: Method of Integral Inequalities*. Naukova Dumka, Kiev (Russian).
[4] Rubinovskij, V.N. and Samsonov, V.A. (1988). *Stability of Stationary Motions*. Nauka, Moscow (Russian).
[5] Syng, J.L. (1936). *Tensorial Methods in Dynamics*. Toronto.
[6] Vujičić, V.A. (1968). Über die stabilitat der stationären bevegungen. *ZAMM*, **48**, 291–293 (Germany).
[7] Vujičić, V.A. (1971). Covariant equations of disturbed motion of mechanical systems. *Tensor*, **22**, 41–47.
[8] Vujičić, V.A. (1987). The modification of analytical dynamics systems. *Tensor*, **46**, 418–431.
[9] Vujičić, V.A. and Martynyuk A.A. (1991). *Some Problems of Mechanics of Nonautonomous Systems*. Mat. Inst. SANU Beograd and Inst. Mekh. AN Ukraine, Kiev (Russian).
[10] Vujičić, V.A. (1995). Energy exchange theorems in systems with time-dependent constraints. *Teor. Prim. Mehanika*, **21**, 105–121.

Part 3
STABILITY OF SOLUTIONS TO PERIODIC DIFFERENTIAL SYSTEMS

3.1 A SURVEY OF STARZHINSKII'S WORKS ON STABILITY OF PERIODIC MOTIONS AND NONLINEAR OSCILLATIONS*

Yu.A. MITROPOL'SKII[1], A.A. MARTYNYUK[2] and V.I. ZHUKOVSKII[3]

[1] *Institute of Mathematics of National Academy of Sciences of Ukraine, Kiev, Ukraine*
[2] *Institute of Mechanics of National Academy of Sciences of Ukraine, Kiev, Ukraine*
[3] *The Russian Correspondence Institute of Textile and Light Industry, Moscow, Russia*

1 V.M.Starzhinskii's life

Viacheslav Mikhailovich Starzhinskii died on December, 1993 at the age of 76. He was a world-wide known specialist in the field of theory of stability and nonlinear oscillations. He was also a leader in the scientific community, unselfish with his time and always concerned with the general welfare of his colleagues.

Viacheslav Mikhailovich was born into teacher's family on March 10, 1918 in Lemeshevichi village of the Pinsk region of Belorussia. His gift for mathematics was evident early in his childhood.

In 1941 he graduated from the Mechanical and Mathematical Department of Moscow University. During the Second World War he worked for several military research institutions. In 1945 Starzhinskii entered the post-graduate course of the Scientific Research Institute of Mechanics of Moscow University and in 1948 he brilliantly defended his master's thesis entitled "Some problems of theory of servosystems". After that he joined the All-Union (now Russian) Correspondence Institute of Textile and Light Industry (Moscow) where he held the professorship of theoretical mechanics till the end of his life. In 1957 V.M.Starzhinskii defended his doctor's thesis entitled "Some problems of stability of periodic motions". This work was published in complete form in the Proceedings of the American Mathematical Society.

Advances in Stability Theory (Ed.: A.A. Martynyuk). Stability and Control: Theory, Methods and Applications, Taylor & Francis, London, **13** (2003) 201–215.

Professor V.M.Starzhinskii was an Honoured Scientist of the Russian Federative Republic and Member of the editorial boards of several publishers.

* * *

V.M.Starzhinskii published more than 150 works (including 27 monographs and textbooks). His works cover the following fields:

1. The second Lyapunov method: first, second, third and fourth order equations;

2. The stability of periodic motions: estimations of characteristic constants in the second and n-th order systems; the theory of parametric resonance Maté and Hill equations;

3. Oscillations of substantially nonlinear systems, combination of the Lyapunov and Poincaré methods, oscillating chains, energy jump, damped oscillating systems, computation of normal modes; normal modes for third, fourth and sixth order systems;

4. Application of parametric resonance theory to acoustic and electromagnetic waveguides;

5. Dynamics of a solid body: dimensionless form of the Euler-Poisson equations, oscillations of a heavy body with a fixed point, exclusive cases of Kovalevskaya gyroscope motion, QP-procedure for Kovalevskaya's case.

6. Applied problems: calculation of thread tension, elastic shaft, dynamical stability of rods, problem of three bodies, torsion oscillations of crank-shafts, pendulum on spring, thread mechanics, servosystems, cyclical accelerators.

In the present review we consider some problems of stability of periodic motions, the mathematical theory of parametric resonance, the theory of vibration of substantially nonlinear systems, the use of the theory of normal modes of analytical autonomous systems of ordinary differential equations, and the problems of application. Within the above-mentioned areas a number of unsolved problems are formulated.

2 Stability of Periodic Motions

This problem is the subject of many profound investigations. Here we point out, following [9, 24, 25], the sufficient conditions of asymptotic stability of the trivial solution of linear differential equations with constant coefficients. These conditions are nonimprovable within the framework of the second method of Lyapunov [24'] and for V-functions which are quadratic forms with constant coefficients. For example, for the equation

$$\ddot{x} + s(t)\dot{x} + p(t)x = 0 \qquad (2.1)$$

with real piecewise continuous bounded coefficients satisfying the inequalities

$$0 < l \leq s(t) \leq L; \quad 0 < m \leq p(t) \leq M \quad (t_0 \leq t < \infty), \qquad (2.2)$$

these conditions become

$$L < \frac{M + 2\sqrt{Mm} + 5m}{\sqrt{M} - \sqrt{m}}; \qquad t > \sqrt{M} - \sqrt{m}. \qquad (2.3)$$

In particular, for $s(t) \equiv \alpha$ only the second inequality (2.3) remains, and becomes $\alpha > \sqrt{M} - \sqrt{m}$. The conditions (2.3) remain valid also for the nonlinear equation (2.1) with the functions $s = s(t, x, \dot{x})$ and $p = p(t, x, \dot{x})$, satisfying the inequalities (2.2) for all values of t, x, $\dot{x}$. Similar conditions (see [4, 9]) are obtained for equations of the third and fourth orders.

The Lyapunov method of estimating the characteristic constant is extended in [31] to systems of linear differential equations with periodic coefficients. To begin with, let the equations of the first approximation of the perturbed motion of a dynamical system be given in the form

$$\frac{dx_1}{dt} = p_{11}(t)x_1 + p_{12}(t)x_2; \qquad \frac{dx_2}{dt} = p_{21}(t)x_1 + p_{22}(t)x_2, \qquad (2.4)$$

where p_{ij} ($i, j = 1, 2$) are real, piecewise continuous periodic functions of t with a period T.

In [2, 3, 5–7, 9] and [C] (Section VII.3) it is shown that an orthogonal transformation with T-periodic coefficients can reduce any system (2.4) to a form where the functions p_{12} and p_{21} are of constant sign. This is in fact assumed below. We denote

$$\int_0^T p_{11}(t)\, dt = \alpha, \qquad \int_0^T p_{22}(t)\, dt = \beta$$

and without loss of generality, put $\alpha \geq \beta$. For $\alpha + \beta > 0$ the nonperturbed motion is unstable in view of the Liouville expression. In what follows we shall assume that $\alpha + \beta \leq 0$. Taking into account $\alpha - \beta \geq 0$, we write the estimate for α in the form

$$\beta \leq \alpha \leq -\beta \quad (\beta \leq 0).$$

The following Theorems (see [9] and [C], Section VII.3) hold.

Let in the system (2.4) the functions p_{12} and p_{21} be of constant sign (i.e. with preservation of the sign they can become zero, but so that their product is not identical to zero on $[0, T]$) and furthermore, of the same sign. If, besides, the average value of one of the functions p_{11} or p_{22} is nonnegative, then the nonperturbed motion is unstable.

Corollary 2.1 *If in the equation*

$$\ddot{x} + s(t)\dot{x} + p(t)x = 0, \quad (s(t+T) \equiv s(t), \quad p(t+T) \equiv p(t)) \qquad (2.5)$$

the periodic coefficient $p(t)$ is nonpositive for all values of t $(0 \leq t \leq T)$, then the trivial solution of (2.5) is unstable regardless of the behaviour of the second periodic coefficient $s(t)$.

In the case when the conditions of the theorem are not satisfied a number of criteria defining the regions of stability and instability of the nonperturbed motion in a parameter space are presented in [8, 9, 14 and [C], Section VII.3]. Similar criteria are given in [9] and [C] (Section VII.3) for critical cases [16'], but only for the stability and instability of the trivial solution of the system (2.4). There yet another way of investigating the stability of the trivial solution of the system (2.4) is indicated, namely, the transformation of (2.4) into a Hill equation with a nonnegative periodic coefficient. Thus the method of estimating the characteristic constant is made applicable in the form proposed by Lyapunov (see [31]).

Article [8] is devoted to extending the above-mentioned Lyapunov method to systems of linear differential equations with periodic coefficients of an arbitrary order. A series of stability and instability criteria are presented there. In particular, the following theorem holds.

If in the vector equation

$$\ddot{y} + P(t)y = 0 \quad (P(t+T) \equiv P(t)) \tag{2.6}$$

all elements of the matrix $P(t)$ are nonpositive for $0 \leq t \leq T$, then the trivial solution of the system (2.6) is unstable.

In the scalar case we obtain the well-known theorem of Lyapunov (see [24']).

Another extension of the Lyapunov method of estimating the characteristic constant to systems of an arbitrary order is proposed by Yakubovich [46']. Among contemporary investigations in the theory of stability of motion the monographs by Chetaev [9'], Harris and Milles [14'], Kamenkov [17'], Krasovskii [20'], Lakshmikantham, *et al.* [23'], Malkin [26'], Martynyuk [29', 30'], Martynyuk and Gutowski [31'], Mel'nikov [32'], Rumyantsev [41'], and Zubov [48'] should be mentioned.

3 The Mathematical Theory of Parametric Resonance and Its Applications

The fundamentals of the mathematical theory of parametric resonance are presented in the basic investigations by Krein [21'], Malkin [27'], and Yakubovich (see [47'] and [C], Chapter V, and [A]). These principles are based on the development of methods of investigation of systems of linear Hamiltonian equations with periodic coefficients (see the survey by Yakubovich [46'] and [C], Chapter III, [20], and [22', 36']). For a parametric resonance in systems close to the canonical ones Starzhinskii [20] has established specific regions of the principal resonance which precede the regions of the fundamental and combined resonances, as well as specific regions of combined-difference resonances. The construction of these regions is based on the expressions of Yakubovich (see [8] and [C], Chapter IV) for the calculation of the characteristic indices.

We consider the vector equation

$$M\ddot{y} + \varepsilon Q(\vartheta t, \varepsilon)\dot{y} + [P_0 + \varepsilon P(\vartheta t, \varepsilon)]y = 0.$$

Here y is a k-dimensional vector; P and Q are $k \times k$ matrix functions which are analytic with respect to ε and 2π-periodic with respect to ϑt; M and P_0 are constant positive (in the sense of the quadratic form) symmetric matrices and $0 < \omega_1^2 \leq \cdots \leq \omega_k^2$ are the eigenvalues of the matrix $M^{-1}P_0$. We number its eigenvectors $a_1, \ldots, a_k$ as follows:

$$\frac{2}{\vartheta_0}\omega_\varkappa(Ma_\varkappa, a_\lambda) = \delta_{\varkappa\lambda} \quad (\varkappa, \lambda = 1, \ldots, k).$$

We determine the integers m_j and the numbers γ_{jh} and σ_{jh} according to the expressions

$$\omega_j = \omega_0 + m_j\vartheta_0; \quad \omega_{-j} = -\omega_j; \quad \gamma_{jh} = \delta_{jh}\,\text{sign}\,j;$$

$$\sigma_{jh} = -\frac{1}{\vartheta_0^2}\left(P_1^{(m_h - m_j)}a_{|j|}, a_{|h|}\right) + \frac{\mu\omega_j}{\vartheta_0^2}\gamma_{jh} - \frac{i\omega_j}{\vartheta_0^2}\left(Q_1^{(m_h - m_j)}a_{|j|}, a_{|h|}\right)$$

$$(j, h = j_1, \ldots, j_r),$$

where $j_1, \ldots, j_r$ are some of the numbers $\pm 1, \ldots, \pm k$; $P_1^{(m)}, Q_1^{(m)}$ are the Fourier matrix coefficients for the matrix function

$$P(\vartheta t, 0) \sim \Sigma e^{im\vartheta t}P_1^{(m)}; \quad Q(\vartheta t, 0) \sim \Sigma e^{im\vartheta t}Q_1^{(m)}.$$

We set up the equation

$$\det \|\sigma_{jh} + \nu\gamma_{jh}\|_{j_1, \ldots, j_r} = 0. \tag{3.1}$$

Principal Resonance. The class $\{\omega_{j_1}, \ldots, \omega_{j_r}\}$ consists of the single number ω_j; i.e. ω_j is incommensurable with the remaining $\omega_{-k}, \ldots, \omega_{-1}, \omega_1, \ldots, \omega_k$ to modulo ϑ_0. The region of instability is determined by the inequality

$$\text{Im}\,(\sigma_{jj}\,\text{sign}\,j) > 0.$$

The Fundamental and Combined Resonances. The class $\{\omega_{j_1}, \ldots, \omega_{j_r}\}$ consists of two numbers: ω_{j_1} and ω_{j_2}. Then

$$\vartheta_0 = \frac{1}{m}(\omega_{j_2} - \omega_{j_1}) \quad (m = m_{j_2}m_{j_1} > 0).$$

Equation (3.1) becomes

$$\vartheta_0^4 \nu^2 - 2(\alpha + i\beta)\vartheta_0^2 \nu - (\gamma + i\delta) = 0,$$

where

$$\alpha + i\beta = -\frac{1}{2}(\sigma_{j_1 j_1} \operatorname{sign} j_1 + \sigma_{j_2 j_2} \operatorname{sign} j_2)\vartheta_0^2;$$

$$\gamma + i\delta = (\sigma_{j_1 j_2}\sigma_{j_2 j_1} - \sigma_{j_1 j_1}\sigma_{j_2 j_2}\vartheta_0^4 \operatorname{sign}(j_1 j_2).$$

The regions of instability are determined by one of the inequalities

$$\beta < 0; \quad \beta = 0; \delta \neq 0; \quad \beta = \delta = 0; \quad \gamma < -\alpha^2,$$

or by the inequalities

$$\beta > 0; \quad \delta^2 + 4\alpha\beta\delta - 4\beta^2\gamma > 0.$$

The applications refer to stability of bending vibrations of coaxial shafts (see [20]).

In [20, 22] and [C] (Chapter VI) a series of problems concerning parametric resonance in mechanical systems is considered. Now we turn to the papers by Koroze and Starzhinskii [18′, 26] (see also [C], Section VI.5 and VI.6) dwelling on applications in physics. The boundary-value problem for the Helmholtz equation has the form

$$\Delta u + k^2 u = 0; \quad \left[\frac{\partial u}{\partial n} + \alpha u\right]_{\Pi} = 0 \quad (\alpha = \text{const}), \qquad (3.2)$$

where $u(x, y, z)$ is the velocity potential; k is the wavenumber; n is the outer normal to the surface of the waveguide Π. It is reduced by the Ritz–Kantorovich method [18′] to the canonical vector equation

$$J\frac{dh}{dz} = H(\vartheta z)h; \quad J = \begin{pmatrix} 0 & -I \\ I & 0 \end{pmatrix}.$$

We assume that the surface Π differs little from the surface of a right circular cylinder, i.e. it is given by the equation

$$r = a[1 + \varepsilon g(\vartheta z)] \quad (g(\vartheta z + 2\pi) \equiv g(\vartheta z), \ g_{av} = 0).$$

Then

$$H(\vartheta z) = H_0 + \varepsilon H_1(\vartheta z) + \ldots.$$

The blocking bands of the waveguide in the plane $\varepsilon\vartheta$ are determined according to the theorem of Yakubovich (see [47] and [C], Chapter V). These bands are adjacent to the axis points (critical frequencies)

$$\vartheta^0_{jhl} = \frac{\omega_j + \omega_h}{l} \quad (j, h = 0, 1, \ldots, s; \ l = 1, 2, \ldots)$$

and are given in the first approximation with respect to ε by the inequalities

$$\vartheta^0_{jhl} - \vartheta^1_{jhl}\varepsilon + \cdots < \vartheta < \vartheta^0_{jhl} + \vartheta^1_{jhl} + \cdots.$$

Here

$$\vartheta^1_{jhl} = \frac{2}{l}\left|(H^{(l)}_1 c_{-j}, c_h)\right|;$$

$H^{(l)}_1$ is the matrix coefficient of the Fourier series

$$H_1(\vartheta z) \sim \Sigma e^{i\lambda\vartheta z} H^{(\lambda)}_1;$$

$i\omega_p$ and $i\omega_{-p} = -i\omega_p$ are the eigenvalues of the matrix; $J^{-1}H_0$, while c_p and c_{-p} are the corresponding eigenvectors, normalized by the conditions

$$i(Jc_p, c_q) = \delta_{pq} \operatorname{sign} p \quad (p, q = \pm 0, \pm 1, \ldots, \pm s).$$

The blocking bands for acoustic waveguides with a periodic filling are determined in the same way in [C] (Section VI.5) and expressions of the second approximation are presented as well.

In [C] (Section VI.6) a problem on wave propagation in periodic electromagnetic waveguides is considered. Maxwell equations with the boundary conditions of M.A.Leontovich are the initial equations instead of (3.2). The resonances arising in this case are much more numerous than in the equivalent scalar case. The unsolved problems in the application of methods of the vibration theory to problems of wave propagation in guide structures are as follows.

Open problem 1. In the formulation of the problem (see [18']) the solution corresponds to the physical level of strictness; the problem of limit transition in the application of the Ritz–Kantorovich method is not discussed. It is of interest to apply to these problems the methods worked out by Yakubovich (see [C], Section VI.7), Fomin [11'], and Neimark [38'] for systems with an infinite number of degrees of freedom.

Open problem 2. For the problems thus formulated it is expedient to consider the case of almost-periodic coefficients and coefficients which are close to periodic [5', 10', 19', 45', 48'].

Open problem 3. Certain problems of wave propagation in guide structures involve nonlinear and, in particular, quasilinear systems. The application of asymptotic methods [2′, 3′, 33′ – 35′], and the method of parametric resonance in nonlinear systems (see [43′]) proves to be useful.

4. Oscillations in Substantially Nonlinear Systems

Substantially nonlinear systems are of considerable interest, as systems in which a pre-assigned small parameter is absent. An electric servodrive, subjected to the effect of a clearance and Coulomb friction, is an example of nonanalytic systems of such type. In the book by Starzhinskii [1] the method of integration by intervals of such systems was applied; the conditions for the emergence of self-excited vibrations were determined and their stability was investigated by the method of point transformations (see [1′]). Also forced vibrations of a follow-up electric drive were considered. The method of point transformations, developed by Neimark [8′, 38′], is effective for substantially nonlinear systems subjected to the forces with characteristics of Coulomb friction, clearance, and hysteresis type.

We proceed to analytical autonomous substantially nonlinear systems. What analytical methods can be proposed here in addition to the method of successive approximations?

In the first place we point out the method of Lyapunov (see [24′], Chapter II, and [25]) for the determination of periodic solutions, which was developed by Lyapunov when investigating the critical case of a pair of purely imaginary roots. The Lyapunov method for vibrations of systems close to Lyapunov systems is developed by Malkin [26′, 27′]. Kamenkov (see [17′], Vol. II) has presented a method for the construction of periodic solutions by means of the Lyapunov–Chetaev functions [24′, 9′], but for quasilinear systems.

Periodic solutions provided by the Lyapunov method can depend only on two arbitrary constants. However some classes are known (for example, the problem of energy jump) when the Lyapunov method does not work.

Starzhinskii (see [24, 25], and [B]) proposed a method of investigation of Lyapunov systems, which, however, reverts back to Lyapunov. Namely, a transformation of the original system of the $(2k+2)$-th order to a nonautonomous quasilinear system of the $2k$-th order is indicated, and linear and quadratic terms are written out with respect to the powers of a small parameter being the square root of the constant energy integral. The Poincaré method of determining periodic solutions, the method of averaging as well as asymptotic methods can be applied to the transformed system.

The proposed method turned out to be effective for the problem of energy jump. The general formulation of such problems proceeds from the determination of the vibrating chains (see [24], Part I and [B], Chapter II). Its first stage is to establish the initial periodic regime and to determine its regions of instability in the space of the system parameters on the basis of the mathematical theory of parametric resonance. The second stage consists of finding periodic regimes which arise in the case of critical values of the parameter and differ of course, from the initial regime. This stage is in fact based on the proposed method of determining periodic solutions. The third stage is the investigation of the process of transition from the initial periodic regime to that found in the second stage.

Among the unsolved problems on energy jump the stability of the periodic motion other then the initial motion has not been investigated. Apparently, in all known problems we deal with the critical case of stability of high order. If stability prevails, then the jump process is in essence a capture process.

In addition, if the instability of the periodic motion differs from the initial motion, then what is the probability of capture or return to the initial periodic regime, or motion for which neither will occur? How are the regions of conditional stability determined then? It may be advisable to investigate stability with respect to a part of variables.

We now go over to the application of the theory of perturbations for substantially nonlinear systems. We assume that the unperturbed nonlinear autonomous Lyapunov type system of the $(2k+2)$-th order is perturbed by analytical damping which is sufficiently small with respect to the norm (see [24], Part II and [B], Chapter IV). The perturbed system is transformed so that the unperturbed system can be transformed into a quasilinear autonomous system of the $2k$-th order. The solution of the latter is assumed as known for sufficiently small values of the system. For the first and the subsequent corrections of the corresponding (i.e. with the same initial conditions) solution of the perturbed system we set up the complete system of equations in variations with respect to the parameter, namely, a sequence of nonhomogeneous systems of linear differential equations of the $(2k+1)$-th order with variable coefficients. The complete system is written in operator form for the general finite-dimensional case of the analytical perturbation theory. If the general solution of the unperturbed system is known, then according to Poincaré the integration of the complete system is reduced to quadratures.

In conclusion, we consider the application of the theory of normal modes of analytical autonomous systems of ordinary differential equations to substantially nonlinear systems. We shall assume that this system under the assumption on the existence of simple elementary divisors of the matrix of its linear part (for the general case see [B], Chapter V.3.7), is brought to a diagonal form with respect to the linear part and is written in terms of symmetrical complex coefficients

$$\frac{dx_\nu}{dt} = \lambda_\nu x_\nu + \sum a_{jh}^\nu x_j x_h + \sum b_{jhk}^\nu x_j x_h x_k + \ldots \quad (\nu = 1, \ldots, n). \qquad (4.1)$$

Here and below the summation is carried out twice with respect to the input indices

taking the values $1,\ldots,n$, while the coefficients are symmetrical, i.e.
$$a_{hj}^\nu = a_{jh}^\nu; \quad b_{\{jhk\}}^\nu = \text{i.dem} \quad (\nu,\, j,\, h,\, k = 1,\ldots,n).$$
Besides $\{\alpha\beta\gamma\}$ is everywhere any permutation of the indices α, β and γ.

According to the fundamental theorem of Bryuno (see [6'], Part I), there exists the inverse (but generally non-single-valued and in certain cases diverging) normalizing transformation with complex coefficients
$$x_j = y_j + \sum \alpha_{lm}^j y_l y_m + \sum \beta_{lmp}^j y_l y_m y_p + \cdots \quad (j = 1,\ldots,n) \qquad (4.2)$$
$$(\alpha_{ml}^j = \alpha_{lm}^j; \quad \beta_{\{lmp\}}^j = \text{i.dem}; \quad j,\, l,\, m,\, p = 1,\ldots,n),$$
bringing the system (4.1) into the normal form
$$\frac{dy_\nu}{dt} = \lambda_\nu y_\nu + y_\nu \sum_{(\Lambda,Q)=0} g_{\nu Q}\, y^{q_1}\cdots y^{q_n} \quad (\nu = 1,\ldots,n). \qquad (4.3)$$
Here Λ and Q are vectors with the components $\lambda_1,\ldots,\lambda_n$ and $q_1,\ldots,q_n$, respectively. Moreover, the latter are the integers
$$q_\nu \geq -1; \quad q_j \geq 0\ (j \neq \nu); \quad q_1 + \cdots + q_n \geq 1.$$
In (4.3) the summation is only over the resonance terms which satisfy the resonance equation
$$(\Lambda, Q) \equiv \lambda_1 q_1 + \cdots + \lambda_n q_n = 0.$$
We symmetrize the coefficients of the normal form (4.3) and write it in the form
$$\frac{dy_\nu}{dt} = \lambda_\nu y_\nu + \sum \varphi_{lm}^\nu y_l y_m + \sum \chi_{lmp}^\nu y_l y_m y_p + \cdots \quad (\nu = 1,\ldots,n). \qquad (4.4)$$
In the representation (4.4) the nonzero terms φ_{lm}^ν, χ_{lmp}^ν are given by the form (3.3).

We introduce the symbols
$$\Delta_{lm}^\nu = \begin{cases} 1 & (\lambda_\nu = \lambda_l + \lambda_m); \\ 0 & (\lambda_\nu \neq \lambda_l + \lambda_m); \end{cases}$$
$$\Delta_{lmp}^\nu = \begin{cases} 1 & (\lambda_\nu = \lambda_l + \lambda_m + \lambda_p); \\ 0 & (\lambda_\nu \neq \lambda_l + \lambda_m + \lambda_p). \end{cases}$$
Then for the coefficients of the normal form (4.4) and the normalizing transformation (4.2) we obtain the expression (see [B], Section V.3)
$$\varphi_{lm}^\nu = \Delta_{lm}^\nu a_{lm}^\nu; \quad \chi_{lmp}^\nu = \Delta_{lmp}^\nu B_{lmp}^\nu;$$
$$\alpha_{lm}^\nu = \frac{1 - \Delta_{lm}^\nu}{\lambda_l + \lambda_m - \lambda_\nu} a_{lm}^\nu; \quad \beta_{lmp}^\nu = \frac{1 - \Delta_{lmp}^\nu}{\lambda_l + \lambda_m + \lambda_p - \lambda_\nu} B_{lmp}^\nu$$
$$(\nu,\, l,\, m,\, p = 1,\ldots,n),$$

where

$$B^\nu_{lmp} = b^\nu_{lmp} + \frac{2}{3}\sum_{j=1}^{n}[a^\nu_{jl}\alpha^j_{mp} + a^\nu_{jm}\alpha^j_{pl} + a^\nu_{jp}\alpha^j_{lm} - (\alpha^\nu_{jl}\varphi^j_{mp} - \alpha^\nu_{jm}\varphi^j_{pl} - \alpha^\nu_{jp}\varphi^j_{lm})],$$

and the symbols Δ are playing the part of a guard.

Indeed, for $\lambda_\nu \neq \lambda_l + \lambda_m$ or $\lambda_\nu \neq \lambda_l + \lambda_m + \lambda_p$ (nonresonance terms) we have $\varphi^\nu_{lm} = 0$ or $\chi^\nu_{lmp} = 0$. This shows that the expansions in the normal form contain only resonance terms. On the other hand, for $\lambda_\nu = \lambda_l + \lambda_m$ or $\lambda_\nu = \lambda_l + \lambda_m + \lambda_p$ (resonance terms) the expressions for α^ν_{lm} or β^ν_{lmp} yield the indeterminacy of $\frac{0}{0}$ type. This means that in a resonance case the coefficients of the normalizing transformation can be chosen arbitrarily (hence non-single-valuedness of the normalizing transformation). In vibration problems such coefficients are chosen either with respect to the continuity of the really entering parameters or are assumed to be zero.

In [46'], Section V.3 expressions are presented for the calculation of the coefficients of an arbitrary order. After the coefficients of the normal form and the normalizing transformation are calculated completely, there arises a problem of effective application of the normal forms to the problems of nonlinear vibrations.

First of all we singled out the class of problems in which the normal form contains only linear terms (the Poincaré theorem) and the representation of the Cauchy problem in the general form is determined by an effective inversion of the normalizing transformation. With respect to damped vibrating systems with analytical nonlinearities of the general form (see [24], Part II and [B], Chapter VI). Then investigated were the third order systems with two purely imaginary eigenvalues of the linear part and the third negative or zero one (see [24], Part III and [B], Chapter VII). Vibrations in electromechanical systems "with one and half degrees of freedom" are referred to these problems (cf. [44'] and [32']). Finally, normal modes and resonances were investigated in analytical autonomous systems of the fourth and sixth orders (see [24], Part II and [B], Chapter VIII) having respectively two and three pairs of different purely imaginary eigenvalues of the matrix of the linear part. Vibrations in gyroscopic systems considered by Ishlinskii (see [15'] and [24], Part II, and [B], Section VIII.2) and vibrations of a heavy body with a fixed point close to the lower equilibrium state (see [B], Chapter IX, and [49']) relate in these problems.

We note certain unsolved problems which are of considerable interest.

Open problem 4. The sufficient conditions of convergence and divergence of normalizing transformation (see Bryuno [6'], Part I) fail in the majority of problems of the theory of vibrations. Consequently, the construction of effective existence conditions for finite smooth transformations has assumed greater importance.

Open problem 5. Estimation of the accuracy of approximate integration in applications of the theory of normal modes to nonlinear vibration problems.

Open problem 6. Normal modes, of course, do not exhaust the potentialities of the local method. The subsequent development of the local method (seminormal modes and related integral manifolds (see [4']) as well as its interpretation in nonlinear vibration problems are essential.

Books

[A] Yakubovich, V.A. (1987). *Parametric Resonance in Linear Systems.* Nauka, Moscow (Russian).
[B] (1977). *Applied Methods of Nonlinear Oscillations.* Nauka, Moscow (Russian).
[C] Yakubovich, V.A. (1972). *Linear Differential Equations with Periodic Coefficients and Their Applications.* Nauka, Moscow (Russian).

Articles

[1] (1948). The effect of clearance and friction on the motion of follow-up electric drive. ONTI NII PSSM.
[2] (1952). Sufficient stability conditions for a mechanical system with one degree of freedom. *Prikl. Mat. Mekh.*, **16**(3), 369–374 (Russian).
[3] (1953). On stability of a mechanical system with one degree of freedom. *Prikl. Mat. Mekh.*, **17**(1), 117–122 (Russian).
[4] (1954). A review of work on the conditions of stability of the trivial solution of a system of linear differential equations with periodic coefficients. *Prikl. Mat. Mekh.*, **18**(4), 469–510.
[5] (1954). On stability of trivial solution of second order differential equation with periodic coefficients. *Inzhen. Sbornik*, **18**, 119–138 (Russian).
[6] (1955). A remark to the stability investigation of periodic motions. *Prikl. Mat. Mekh.*, **19**(1), 119–120 (Russian).
[7] (1957). On stability of periodic motions in a special case. *Prikl. Mat. Mekh.*, **21**(5), 720–722 (Russian).
[8] (1958). On stability of trivial solution of linear systems with periodic coefficients. *Prikl. Mat. Mekh.*, **22**(5), 646–656 (Russian).
[9] (1958, 1959). On stability of periodic motions. *Bul. Inst. Politehn., Din. Iasi*, Serie noua 4–8, Part I, no. 3–4, 19–68; Part II, 5(9), no. 1–2, 51–100 (Russian).
[10] (1959). To the problem on boundedness of solution to system of linear differential equations with periodic coefficients. *Trudy 3 Vsesouzn. Mat. S'ezda*, **4**, 37–39 (Russian).
[11] (1959). Torsion oscillations of loom crankshafts. *Nauchn. Dokl. Vysshei shkoly mashyn. i priborostroen.*, **1**, 51–57 (Russian).
[12] (1959). On Liapunov's method of estimating characteristical constant. *Izd. Akad. Nauk USSR, OTN, Mekh. Mashinostroen.*, **4**, 46–55 (Russian).
[13] (1960). On stability of trivial solution of system of two linear differential equations with periodic coefficients. *Prikl. Mat. Mekh.*, **24**(3), 578–581 (Russian).
[14] (1960). The stability of periodic motions in a special case. *Royal Aircraft Establishment*, no. 883, 2–4.

[15] (1961). Free not entirely elastic oscillating chains. *Izd. Akad. Nauk USSR, OTN, Mekh. Mashinostroen.*, **6**, 68–73 (Russian).
[16] (1962). Free entirely elastic oscillating chains. *Prikl. Mat. Mekh.*, **26**(1), 172–181 (Russian).
[17] (1963). Oscillating chains. *Proceedings of International Symposium on Nonlinear Vibrations*, Vol. 1., Naukova Dumka, Kiev, 446–455 (Russian).
[18] (1963). On the stability of periodic motion. *Amer. Math. Soc. Trans.*, Ser.2, Part I, **33**, 59–121; Part II, i.d. 123–187.
[19] (1964). On the stability of periodic regimes. *Nonlin. Vibrations Problems*, PWN, Warszawa, **5**, 360–369.
[20] (1967). Parametric resonance in systems close to the canonical ones. *Inzhen. Zh. Mekh. Tver. Tela*, **3**(3), 174–180 (Russian).
[21] (1968). On one version of the method of determining periodic solutions. *Inzhen. Zh. Mekh. Tver. Tela.*, **4**(6), 67–71 (Russian).
[22] (1968). To the theory of parametric resonance. *Proc. of the Fourth Conf. of Nonlin. Oscil.*, Prague, 475–480.
[23] (1968). Dynamic stability of thin-walled rods loaded with longitudinal periodic forces. *Proc. of the Fourth Conf. of Nonlin. Oscil.*, Prague, 467–474.
[24] (1970, 1971, 1972). *On the Theory of Non-Linear Vibration.* Part 1, Part 2, Part 3, Izd. Moscow University (Russian).
[25] (1973). Certain problems of the theory of nonlinear vibrations. *Izv. Yassk. Politekh. Inst.*, Part 1, **19**(**23**), Nos. 1–2, 113–120; Part 2, **19**(**23**), Nos. 3–4, 127–134.
[26] Koroza, V.I. (1969, 1970, 1971). Theory of periodic waveguides. *Izv. Yassk. Politekh. Inst.*, Part 1, **15**(**19**), Nos. 3–4, 7–16; Part 2, **16**(**20**), Nos. 3–4, 21–30; Part 3, **17**(**21**), Nos. 3–4, 31–37.
[27] (1971). Anwendung der theorie linearer differentialgleichungen mit periodischen koeffizienten in der mechanik. *Mitteilungen der Math.*, Gesselschaft DDR, **1**, 53–65.
[28] (1971). Interruption of spring oscillations of the mathematical pendulum. *Inzhen. Zh. Mekh. Tver. Tela*, **7**(2), 154–156 (Russian).
[29] Bairoiter, I. (1966). Parameterresonanz in fastkanonishen Systemen. *ZAMM*, **46**(7), 459–464.
[30] (1976). Orbital stability in a partial case of the problem of n bodies. *Izbr. Voprosy Dinam.*, Nauka, Moscow, 7–11 (Russian).
[31] (1971, 1973) Einige probleme nichtlinearer Schwingungen. *ZAMM*, Part 1, **51**(6), 455–469; Part 2, **53**(8), 453–462.

Supplementary References

[1'] Andronov, A.A., Vitt, A.A. and Khaikin, S.E. (1959). *Theory of Vibrations.* Fizmatgiz, Moscow (Russian).
[2'] Bogolyubov, N.N. (1969). *Selected Works*, **I**. Naukova Dumka, Kiev (Russian).
[3'] Bogolyubov, N.N. and Mitropol'skii, Yu.A. (1974). *Asymptotic Methods in the Theory of Nonlinear Vibrations.* Nauka, Moscow (Russian).
[4'] Bogolyubov, N.N. and Mitropol'skii, Yu.A. (1963). The method of integral manifolds in nonlinear mechanics. *Proc. of Int. Symp. on Nonlinear Vibrations*, **3**, Naukova Dumka, Kiev, 93–154 (Russian).

[5'] Bogolyubov, N.N., Mitropol'skii, Yu.A. and Samoilenko, A.M. (1969). *The Method of Accelerated Convergence in Nonlinear Mechanics*. Naukova Dumka, Kiev (Russian).
[6'] Bryuno, A.D. (1971, 1972). The analytical form of differential equations. *Trudy Mosk. Mat. Ob-va*, Part I, **25**, 119–262; Part II, **26**, 199–239 (Russian).
[7'] Bulgakov, B.V. (1954). *Vibrations*. Gostechizdat, Moscow (Russian).
[8'] Butenin, N.V., Neimark, Yu.I. and Fufaev, N.A. (1976). *An Introduction to the Theory of Nonlinear Vibrations*. Nauka, Moscow (Russian).
[9'] Chetaev, N.G. (1962). *Stability of Motion. Papers on Analytical Mechanics*. Akad. Nauk USSR, Moscow (Russian).
[10'] Fink, A.M. (1974). *Almost Periodic Differential Equations*, Springer-Verlag, Berlin, etc.
[11'] Fomin, V.N. (1972). *The Mathematical Theory of Parametric Resonance in Linear Distributed Systems*. Leningrad. University, Leningrad (Russian).
[12'] Grebenikov, E.A. and Ryabov, Yu.A. (1971). *New Qualitative Methods of Celestial Mechanics*. Nauka, Moscow (Russian).
[13'] Grebenikov, E.A. and Ryabov, Yu. A. (1982). *Metoda Usrednienia w Mechanice Nieliniowej*. PWN, Warszawa (Polska).
[14'] Harris, C.J. and Milles, J.F. (1980). *Stability of Linear Systems: Some Aspects of Kinematic Similarity*. Academic Press, London, etc.
[15'] Ishlinskii, A.Yu. (1963). *Mechanics of Gyroscopic Systems*. AN SSSR, Moscow (Russian).
[16'] Kalinin, S.V. (1972). *Stability of Periodic Motions in Critical Cases*. Mosc. University, Moscow (Russian).
[17'] Kamenkov, G.V. (1971). *Selected Works*, **I, II**. Nauka, Moscow (Russian).
[18'] Koroza, V.I. (1967). The Ritz-Kantorovich method in the problem of wave propagation in electromagnetic waveguides. *Uskoriteli*, **11**, 88–92 (Russian).
[19'] Krasnosel'skii, M.A., Burd, V.Sh. and Kolesov, Yu.S. (1970). *Nonlinear Almost-Periodic Vibrations*. Nauka, Moscow (Russian).
[20'] Krasovskii, N.N. (1959). *Certain Problems of Stability Theory of Motion*. Fizmatgiz, Moscow (Russian).
[21'] Krein, M.G. (1955). The basis premises of the theory of λ zones of stability of a canonical system of linear differential equations with periodic coefficients. *Collection in Remembrance of A.A. Andronov*. Fizmatgiz, Moscow (Russian).
[22'] Krein, M.G. and Yakubovich, V.A. (1963). Hamiltonian Systems of linear differential equations with periodic coefficients. *Proc. of Int. Symp. on Nonlinear Vibrations*, **I**. Naukova Dumka, Kiev, 277–305 (Russian).
[23'] Lakshmikantham, V., Leela, S. and Martynyuk, A.A. (1989). *Stability Analysis of Nonlinear Systems*. Marcel Dekker, Inc., New York.
[24'] Lyapunov, A.M. (1956) The general problem of stability of motion. *Collected Works*, **2**. Akad. Nauk USSR, 5–263 (Russian).
[25'] Lyapunov, A.M. (1902). Sur une serie dans la theorie des equations differentielles lineaires du second ordre a coefficients periodiques. *Zap. Akad. Nauk Fiz.- Mat. Otd.*, Ser. 8, **13**(2), 1–70 (Russian).
[26'] Malkin, I.G. (1966). *The Theory of Stability of Motion*. Nauka, Moscow (Russian).
[27'] Malkin, I.G. (1949). *Methods of Lyapunov and Poincaré in the Theory of Nonlinear Vibrations*. Gostechizdat, Moscow (Russian).
[28'] Marsden, J.E. and McCracken, M.F.(1976). *The Hopf bifurcation and its applications*. Appl. Math. Sciences, **19**, Springer-Verlag, New York.

[29'] Martynyuk, A.A. (1995). *Stability Analysis: Nonlinear Mechanics Equations*. Gordon and Breach Publisher, New York.
[30'] Martynyuk, A.A. (1998). *Stability by Liapunov's Matrix Function Method with Applications*. Marcel Dekker, Inc., New York.
[31'] Martynyuk, A.A. and Gutowski, R. (1979). *Integral Inequalities and Stability of Motion*. Naukova Dumka, Kiev (Russian).
[32'] Mel'nikov, G.I. (1975). *Dynamics of Nonlinear Mechanical and Electromechanical Systems*. Mashinostroenie, Leningrad (Russian).
[33'] Mitropol'skii, Yu.A. (1964). *Problems of Asymptotic Theory of Nonstationary Vibrations*. Nauka, Moscow (Russian).
[34'] Mitropol'skii, Yu.A. (1995). *Nonlinear Mechanics: Asymptotic Methods*. Inst. of Math., Kiev (Russian).
[35'] Moiseev, N.N. (1972). *Asymptotic Methods in the Theory of Nonlinear Vibrations*. Nauka, Moscow (Russian).
[36'] Moser, J. (1981). *Integrable Hamiltonian Systems and Spectral Theory*. Accademia Nazionale dei Lincei, Pisa.
[37'] Nayfeh, A.H. and Mook, D.T. (1979). *Nonlinear Oscillations*. Wiley-Interscience, New York.
[38'] Neimark, Yu.I. (1972). *The Point-Mapping Method in the Theory of Nonlinear Vibrations*. Nauka, Moscow (Russian).
[39'] Proskuryakov, A.P. (1977). *The Poincaré Method in the Theory of Nonlinear Vibrations*. Nauka, Moscow (Russian).
[40'] Poincaré, A. (1971, 1972). *Selected Works*, **I, II**. Nauka, Moscow (Russian).
[41'] Rumyantsev, V.V. (1967). Stability of stationary motions of satellites. *Mat. Metody Dinam. Kosm. Appar.*, **4**, 3–141 (Russian).
[42'] Sanders, J.A. and Verhulst, F. (1985). *Averaging Methods in Nonlinear Dynamical Systems*. Springer-Verlag, New York.
[43'] Schmidt, G. (1978). *Parametric Vibrations*. Inostr. Literatura, Moscow (Russian).
[44'] Stoker, J.J. (1950). *Nonlinear Vibrations in Mechanical and Electrical Systems*. Inostr. Literatura, Moscow (Russian).
[45'] Verhulst, F. (1990). *Nonlinear Differential Equations and Dynamical Systems*. Springer-Verlag, Berlin.
[46'] Yakubovich, V.A. (1957). Extension of certain results of A.M. Lyapunov to linear canonical systems with periodic coefficients. *Prikl. Mat. Mech.*, **21**, 707–713 (Russian).
[47'] Yakubovich, V.A. (1966). Regions of dynamic instability of Hamiltonian systems. *Metody Vychisl.*, **3**, 51–59 (Russian).
[48'] Zubov, V.I. (1979). *The Theory of Vibrations*. Vysshaja Shkola, Moscow (Russian).
[49'] Zubov, V.I. (1970). *Analytical Dynamics of Gyroscopic Systems*. Sudostroenie, Leningrad (Russian).

3.2 IMPLICATIONS OF THE STABILITY OF AN ORBIT FOR ITS OMEGA LIMIT SET*

J.S. MULDOWNEY

Department of Mathematical Sciences, University of Alberta, Edmonton, Canada

1 Introduction

In dynamics, it is natural to enquire how much of the nontransient behaviour can be detected from an analysis of an individual orbit and its relationship with its neighbours. Can the dynamical properties of some attractors be deduced from a single trajectory? In differential equations, even when the algebraic equations yielding the equilibria cannot be solved explicitly, is it possible to detect a stable equilibrium in this way? Similar questions regarding periodic orbits and more complicated structures are of interest.

For autonomous 2-dimensional differential equations, the Poincaré-Bendixson theory [1, 2] shows that a bounded orbit which does not get close to any equilibrium has a periodic orbit as its omega limit set. Similarly, Massera's Theorem [3] infers the existence of a periodic solution to a nonautonomous time-periodic scalar differential equation from the existence of a bounded solution. In this spirit, Sell [4] shows for a general semiflow on a metric space that a Lagrange stable orbit has as its omega limit set a phase asymptotically stable periodic orbit if it is itself phase asymptotically stable. Good expositions of Sell's results may be found in Saperstone [5] Chapter III and Cronin [6] Chapter 6. Yoshizawa [7] also discusses these results and extends the applications to functional differential equations. Pliss, in [8] Theorem 1.6, establishes a closely related result for autonomous differential equations in $\mathbb{R}^n$ where the stability requirements are somewhat different from those of Sell: Lyapunov stability is not required but a certain uniformity is imposed on the manner in which the orbit attracts its neighbours. More recently these results

---

Advances in Stability Theory (Ed.: A.A. Martynyuk). Stability and Control: Theory, Methods and Applications, Taylor & Francis, London, **13** (2003) 217-229.

have been extended by Li and Muldowney [9] and in Muldowney [10] with greatly simplified proofs.

There are many papers which deduce the existence of a periodic orbit solely from the existence of a bounded orbit. These do so without any a priori restriction on the attraction of this orbit for its neighbours; there is however always some form of attraction inherent in the dynamical requirements as in the Poincaré-Bendixson theory when the orbit and its limit cycle are both orbitally stable at least from one side. For example, Hirsch [11] and H.L. Smith [12] show that 3-dimensional order-preserving flows have this classical Poincaré-Bendixson property. Mallet-Paret and H.L. Smith develop the theory for monotone cyclic feedback systems in [13]. R.A. Smith [14–16] is the author of a higher dimensional theory based on guiding functions to show that systems which behave asymptotically in a sufficiently 2-dimensional fashion also have the Poincaré-Bendixson property. This paper gives an exposition and extension of those results that guarantee the existence of stable equilibria or periodic orbits in the spirit of [4, 8–10]. The general approach is to express the orbital stability for an equilibrium or periodic orbit in such a way that, when this definition of stability is applied instead to a bounded orbit, its omega limit set is a similarly stable equilibrium or periodic orbit. In Section 2, such results are explored for discrete and continuous semiflows on a metric space. Sections 3, 4 and 5 deal with hyperbolic stable equilibria for smooth systems in $\mathbb{R}^n$.

2 Discrete and Continuous Semiflows

Let $\{X, d\}$ be a metric space and let $\mathbb{T}_+ = \mathbb{Z}_+$ or $\mathbb{R}_+$, the nonnegative integers or real numbers respectively. A map ϕ with domain $\mathbb{T}_+ \times X$ and range in X is a *semiflow* on $\{X, d\}$ if, for each $x \in X$ and $t, s \in \mathbb{T}_+$,

(i) $\phi(0, x) = x$
(ii) $\phi(t + s, x) = \phi(t, \phi(s, x))$
(iii) $(t, x) \mapsto \phi(t, x)$ is continuous.

The *semiflow* is *discrete* if $\mathbb{T}_+ = \mathbb{Z}_+$ and *continuous* if $\mathbb{T}_+ = \mathbb{R}_+$. In the case of a discrete semiflow, $\phi(t, x) = \phi^t(x)$, where $\phi(x) = \phi(1, x)$, $\phi^0(x) = x$ and $\phi^{t+1}(x) = \phi(\phi^t(x))$, $t = 0, 1, 2, \ldots$.

Definition 2.1

(a) For any $x \in X$, the *(positive) orbit* of x is $C_+(x) = \{\phi(t, x) \colon t \in \mathbb{T}_+\}$ and the *omega limit set* is $\Omega(x) = \bigcap_{t \in \mathbb{T}_+} c\ell\, C_+(\phi(t, x))$, where $c\ell$ denotes the topological closure.
(b) $C_+(x)$ is a *periodic orbit* of period ω and x is a ω-*periodic point* if $\phi(\omega, x) = x$ for some $0 < \omega \in \mathbb{T}_+$. It is an *equilibrium* if $\phi(\omega, x) = x$ for all $\omega \in \mathbb{T}_+$.
(c) The *semiflow* is *Lagrange stable* at x if $c\ell\, C_+(x)$ is compact.

(d) The *semiflow* is *Lyapunov stable* at $S \subset X$ if, for each $\varepsilon > 0$, there exists $\delta > 0$ such that $x_0 \in S$ and $d[x_0, x] < \delta$ implies $d[\phi(t, x_0), \phi(t, x)] < \varepsilon$ for all $t \in \mathbb{T}_+$. When $S = C_+$, an orbit, this is the usual concept of uniform Lyapunov stability of C_+.

(e) The *semiflow* is *asymptotic* at $S \subset X$ if there exists $\rho > 0$ such that $x_0 \in S$ and $d[x_0, x] < \rho$ implies $\lim_{t \to \infty} d[\phi(t, x_0), \phi(t, x)] = 0$.

(f) The *semiflow* is *phase asymptotic* at $S \subset X$ if there exist $\rho, \eta > 0$ such that $x_0 \in S$ and $d[x_0, x] < \rho$ implies there is a real-valued function $(x_0, x) \mapsto h(x_0, x)$, with $\eta > |h(x_0, x)| \in \mathbb{T}_+$, such that $\lim_{t \to \infty} d[\phi(t, x_0), \phi(t + h, x)] = 0$. While the *phase* h depends on (x_0, x) in general, the dependence will frequently be suppressed in the notation.

Lemma 2.1 *Suppose the semiflow ϕ is Lagrange stable at x_*. Then ϕ is Lyapunov stable, asymptotic or phase asymptotic at $C_+(x_*)$ if and only if it has the same property at $\Omega(x_*)$.*

Proof We prove the statement with respect to the phase asymptotic property. Proofs for the other two properties are similar. Suppose ϕ is phase asymptotic at $C_+(x_*)$ and $x_0 \in \Omega(x_*)$. There exists $x_1 \in C_+(x_*)$ such that $d[x_1, x_0] < \rho/2$. If $d[x_0, x] < \rho/2$, then $d[x_1, x] < \rho$ and $\lim_{t \to \infty} d\big[\phi(t, x_1), \phi(t + h(x_1, x), x)\big] = 0$ so that

$$\lim_{t \to \infty} d\big[\phi(t, x_0), \phi\big(t + h(x_1, x) - h(x_1, x_0), x\big)\big]$$
$$= \lim_{t \to \infty} d\big[\phi\big(t + h(x_1, x_0), x_0\big), \phi\big(t + h(x_1, x), x\big)\big] = 0,$$

and

$$\lim_{t \to \infty} d[\phi(t, x_0), \phi(t + h, x)] = 0,$$

where $h = h(x_1, x) - h(x_1, x_0)$, $|h| \leq 2\eta$. Thus ϕ is phase asymptotic at $\Omega(x_*)$ with ρ, η replaced by $\rho/2, 2\eta$ respectively. A similar argument shows conversely that ϕ is phase asymptotic at $C_+(x_*)$ if it is phase asymptotic at $\Omega(x_*)$.

In the following theorems the phrase in square brackets may be included or omitted throughout.

Theorem 2.1 *Suppose that $(t, x) \mapsto \phi(t, x)$ is a semiflow which is Lagrange stable at x_*. Then (a) and (b) are equivalent.*

(a) *ϕ is asymptotic [and Lyapunov stable] at $C_+(x_*)$.*

(b) *$\Omega(x_*)$ is a periodic orbit at which ϕ is asymptotic [and Lyapunov stable]. The periodic orbit is an equilibrium if the semiflow is continuous.*

Proof Suppose that (a) is satisfied. We may choose $x_1, x_2 \in C_+(x_*)$ so that $d[x_1, x_2] < \rho$ and $x_2 = \phi(\omega, x_1)$, $\omega > 0$. Then $\lim_{t \to \infty} d[\phi(t, x_1), \phi(\omega, \phi(t, x_1))] = \lim_{t \to \infty} d[\phi(t, x_1), \phi(t, x_2)] = 0$, since ϕ is asymptotic at $C_+(x_*)$. If $x_0 \in \Omega(x_*)$, choose $t = t_n \to \infty$, so that $\phi(t_n, x_1) \to x_0$, $n \to \infty$. We find $d[x_0, \phi(\omega, x_0)] = 0$ so that $x_0 = \phi(\omega, x_0)$ and x_0 is ω-periodic. The lemma implies that the semiflow is asymptotic at $C_+(x_0) \subset \Omega(x_*)$. It therefore attracts all nearby orbits including $C_+(x_*)$ so that $C_+(x_0) = \Omega(x_*)$. Then Lemma 2.1 also shows that (b) implies (a) and that the statement on Lyapunov stability may be included. Finally, if the semiflow is continuous, we may choose any ω in $[0, \varepsilon]$ in the preceding argument and $x_0 = \phi(\omega, x_0)$, $0 \leq \omega \leq \varepsilon$, implies $x_0 = \phi(t, x_0)$ for all $t \geq 0$.

Theorem 2.2 *Suppose ϕ is a semiflow which is Lagrange stable at x_*. Then (a) and (b) are equivalent*

(a) *ϕ is phase asymptotic [and Lyapunov stable] at $C_+(x_*)$*

(b) *$\Omega(x_*)$ is a periodic orbit at which ϕ is phase asymptotic [and Lyapunov stable].*

Proof If (a) is satisfied, then we may choose $x_1, x_2 \in C_+(x_*)$ such that $d[x_1, x_2] < \rho$ and $x_2 = \phi(t_1, x_1)$, where t_1 may be arbitrarily large, in particular $t_1 > \eta$. Then, with $\omega = t_1 + h > 0$ since $|h| < \eta$, $\lim_{t \to \infty} [\phi(t, x_1), \phi(\omega, \phi(t, x_1))] = \lim_{t \to \infty} d[\phi(t, x_1), \phi(t + h, x_2)] = 0$. If $x_0 \in \Omega(x_*)$, choose $t = t_n \to \infty$ so that $\phi(t_n, x_1) \to x_0$, $n \to \infty$. It follows that $x_0 = \phi(\omega, x_0)$ and x_0 is ω-periodic. Since the semiflow is phase asymptotic at $C_+(x_0) \subset \Omega(x_*)$ from the lemma, it attracts all orbits nearby including $C_+(x_*)$. Therefore $C_+(x_0) = \Omega(x_*)$ and (b) is established. Again Lemma 2.1 allows us to deduce that (b) implies (a) and to include the statement on Lyapunov stability.

Remark 2.1

(i) Equally brief proofs of Theorems 2.1, 2.2 may be based on the observation that a nonempty omega limit set at which the semiflow is phase asymptotic consists of exactly one orbit.

(ii) Theorem 2.2 is due to Li and Muldowney [9]. It generalizes Pliss [8] Theorem 1.6 and Sell [3] Theorem 1. Pliss' result is proved for flows in $\mathbb{R}^n$ and the condition imposed is equivalent to the phase asymptotic concept used here with an additional requirement of uniformity on the rate at which an orbit attracts its neighbours. Sell's theorem has Lyapunov stability as a requirement rather than an option.

(iii) A theorem of Deysach and Sell shows that if, in the terminology of this paper, the semiflow is Lagrange stable at x_* and Lyapunov stable at $C_+(x_*)$, then $\Omega(x_*)$ is a minimal set of almost periodic motions. It is interesting to speculate on conditions which ensure these motions are quasi-periodic.

3 Discrete Semiflows in $\mathbb{R}^n$

We are now in a position to characterize in terms of their stability properties those Lagrange stable motions in a discrete C^1 semiflow in $\mathbb{R}^n$ which limit to a stable hyperbolic periodic orbit. Consider the semiflow

$$(t,x) \mapsto \phi(t,x) = \phi^t(x), \quad t \in \mathbb{Z}_+, \quad x \in \mathbb{R}^n, \tag{3.1}$$

where $\phi(\cdot)$ is a C^1 function and $\phi^0(x) = x$, $\phi^{t+1}(x) = \phi(\phi^t(x))$. The linearization of the flow with respect to the motion at x_0 is the nonautonomous recursion

$$y_{t+1} = \frac{\partial \phi}{\partial x}(x_t) y_t, \quad x_t = \phi^t(x_0), \quad t \in \mathbb{Z}_+. \tag{3.2}$$

A solution y_t of (3.2) is uniquely determined by y_s and satisfies

$$y_t = \frac{\partial \phi}{\partial x}(x_{t-1}) \frac{\partial \phi}{\partial x}(t_{t-2}) \cdots \frac{\partial \phi}{\partial x}(x_s) y_x = \frac{\partial}{\partial x} \phi^{t-s}(x_s) y_s, \quad t \geq s. \tag{3.3}$$

However y_s is not uniquely determined by y_t if $\frac{\partial \phi}{\partial x}(x_j)$ is singular for some j, $s \leq j < t$. The recursion (3.2) is uniformly asymptotically stable if there exist constants $1 \leq K$, $0 < \alpha < 1$ such that solutions to (3.2) satisfy $|y_t| \leq |y_s| K \alpha^{t-s}$, $s \leq t$, which from (3.3) is equivalent to $|\frac{\partial}{\partial x} \phi^{t-s}(x_s)| \leq K \alpha^{t-s}$, $t \geq s$, or

$$\left| \frac{\partial}{\partial x} \phi^t(x) \right| \leq K \alpha^t, \tag{3.4}$$

if $t \geq 0$, $x \in C_+(x_0)$.

The recursion (3.2) is autonomous if $x_0 = \phi(x_0)$, an equilibrium of (3.1). The equilibrium is said to be *stable hyperbolic* if every eigenvalue λ of $\frac{\partial \phi}{\partial x}(x_0)$ satisfies $|\lambda| < 1$. This is equivalent to the (uniform) asymptotic stability of (3.2) and implies that ϕ is Lyapunov stable and asymptotic at x_0.

Similarly, a ω-periodic point $x_0 = \phi^\omega(x_0)$ is an equilibrium of $\phi^\omega(\cdot)$ and is said to be stable hyperbolic if every eigenvalue λ of $\frac{\partial}{\partial x} \phi^\omega(x_0)$ satisfies $|\lambda| < 1$ or equivalently $|\frac{\partial}{\partial x} \phi^{\omega s}(x_0)| \leq M \beta^s$, $s \in \mathbb{Z}_+$, for constants $0 < M$, $0 < \beta < 1$; see Szlenk [17, p.39]. It is an exercise to show that, since $C_+(x_0)$ is a finite set, this is equivalent to (3.4) so that the periodic orbit $C_+(x_0)$ is stable hyperbolic if and only if (3.2) is uniformly asymptotically stable.

Lemma 3.1 *Suppose the discrete semiflow is Lagrange stable at x_* and (3.4) is satisfied if $t \geq 0$, $x \in C_+(x_*)$. If $L > K$ and $\alpha < \gamma < 1$, there exists a neighbourhood U of $\operatorname{cl} C_+(x_*)$ and $\delta > 0$ such that $y \in U$, $|y - z| < \delta$ implies $|\phi^t(y) - \phi^t(z)| \leq L \gamma^t |y - z|$ for all $t \in \mathbb{Z}_+$. In particular, the semiflow is asymptotic and Lyapunov stable at U.*

Proof Since ϕ is C^1, (3.4) is satisfied if $t \geq 0$, $x \in \operatorname{cl} C_+(x_*)$. Choose β, $\alpha < \beta < \gamma$, and a positive integer s such that $L(\beta/\gamma)^s < 1$. There exists $\delta > 0$

such that, if $x_0 \in c\ell\, C_+(x_*)$ and $|x - x_0| < 2\delta$, then $|\frac{\partial}{\partial x} \phi^t(x)| \leq L\beta^t$, $0 \leq t \leq s$. It follows that, if $|y - x_0| < \delta$, $|y - z| < \delta$ and $x(\lambda) = (1 - \lambda)z + \lambda y$, $0 \leq \lambda \leq 1$ then

$$|\phi^t(y) - \phi^t(z)| = \left| \int_0^1 \frac{d}{d\lambda} \phi^t(x(\lambda))\, d\lambda \right|$$

$$= \left| \int_0^1 \frac{\partial}{\partial x} \phi^t(x(\lambda))(y - z)\, d\lambda \right| \leq L\beta^t |y - z|,$$

$$0 \leq t \leq s.$$

In particular $|\phi^s(y) - \phi^s(x_0)| < \delta$, $|\phi^s(z) - \phi^s(x_0)| < \delta$ and by induction $|\phi^{ks}(y) - \phi^{ks}(x_0)| < \delta$, $|\phi^{ks}(z) - \phi^{ks}(x_0)| < \delta$. Therefore $|\phi^{t+ks}(y) - \phi^{t+ks}(z)| \leq L\beta^t |\phi^{ks}(y) - \phi^{ks}(z)|$, $0 \leq t \leq s$. Thus, if $ks \leq t \leq (k+1)s$,

$$|\phi^t(y) - \phi^t(z)| \leq L\beta^{t-ks} L^k \beta^{ks} \leq L\gamma^t |y - z|$$

since $(L\beta^s)^k \leq \gamma^{ks}$ and $\beta^{t-ks} \leq \gamma^{t-ks}$.

Theorem 3.1 *Let* $(t, x) \mapsto \phi^t(x)$ *be a discrete* C^1 *semiflow on* $\mathbb{R}^n$ *which is Lagrange stable at* x_*. *Then (a) and (b) are equivalent.*

(a) *The linearization (3.2) of the semiflow with respect to the motion at x_* is uniformly asymptotically stable.*
(b) $\Omega(x_*)$ *is a stable hyperbolic periodic orbit.*

Proof Suppose that (a) is satisfied. Lemma 3.1 shows that the semiflow is asymptotic and Lyapunov stable at $c\ell\, C_+(x_*)$ from which we infer that $\Omega(x_*)$ is a periodic orbit at which the semiflow is asymptotic and Lyapunov stable, from Theorem 2.1. Indeed Lemma 3.1 implies $|\frac{\partial}{\partial x} \phi^t(x)| \leq L\gamma^t$ for all x in a neighbourhood U of $c\ell\, C_+(x_*) \supset \Omega(x_*)$. Thus $\Omega(x_*)$ is a stable hyperbolic periodic orbit as asserted in (b). Conversely if (b) is satisfied, the lemma again implies an inequality of the form (3.4) holds for x in a neighbourhood U of $\Omega(x_*)$ and $t \geq 0$. We may assume $C_+(x_*) \subset U$ so that the assertion (a) holds.

Remark 3.1

(i) Less general forms of Theorem 3.1 are proved in Li and Muldowney [9] Theorem 5.1 and Muldowney [10] Theorem 2.1, where it is assumed that $\frac{\partial \phi}{\partial x}$ is invertible. Those proofs use the variation of parameters formula for difference equations and a discrete version of Gronwall's inequality.
(ii) Theorem 3.1 may be stated equivalently as follows: In a smooth discrete dynamical system on $\mathbb{R}^n$, the set of orbits that limit to a periodic orbit which attracts its neighbours exponentially consists precisely of those orbits that are bounded and attract their neighbours exponentially.

4 Periodic Differential Equations in $\mathbb{R}^n$

Let $(t,x) \mapsto f(t,x)$ be a continuous function from $\mathbb{R}^{n+1}$ to $\mathbb{R}^n$ such that $f(t+\omega,x) = f(t,x)$ for each $t \in \mathbb{R}$, $x \in \mathbb{R}^n$, where $\omega > 0$. Suppose that a solution $x(t)$ of

$$\dot{x} = f(t,x) \tag{4.1}$$

is uniquely determined for $t \geq s$ by $x(s)$ if $s \geq 0$. Massera [3] Theorem 1 shows that the existence of a bounded solution implies that of a periodic solution when $n = 1$. When $n > 1$, this result is no longer valid. When $n = 2$ however, it is shown in [3] Theorem 2 that the conclusions still holds if the additional assumption is satisfied that all solutions exist on rays of the form $[s,\infty)$. But this assumption is no longer sufficient for $n > 2$ except for special equations. For example, Massera shows in [3] Theorem 4 that when $f(t,x) = A(t)x + b(t)$, where $A(\cdot)$, $b(\cdot)$ are $n \times n$, $n \times 1$ ω-periodic matrix-valued functions respectively, the existence of a bounded solution implies the existence of a ω-periodic solution. Yoshizawa gives a comprehensive discussion of Massera's work in [18] Chapter VII. Yoshizawa also surveys other work in this area in [7] and gives useful brief descriptions of the results; in particular there is a discussion of the work of Sell for (4.1) and extension of those results to functional equations. Saperstone [5] Chapter III also has a good exposition on Sell's results as does Cronin [6] Chapter 6.

Let $\phi(t;s,x_0)$ denote the solution $x(t)$ of (4.1) such that $x(s) = x_0$.

Definition 4.1

(a) A *solution $x(t)$* of (4.1) is *uniformly Lyapunov stable* if, for each $\varepsilon > 0$, there exists $\delta > 0$ such that $s \geq 0$, $|x_0 - x(s)| < \delta$ implies $|\phi(t;s,x_0) - x(t)| < \varepsilon$ for all $t \geq s$.

(b) A *solution $x(t)$* is *asymptotic* if there exists $\delta_0 > 0$ such that $s \geq 0$, $|x_0 - x(s)| < \delta_0$ implies $|\phi(t;s,x_0) - x(t)| \to 0$ as $t \to \infty$.

(c) A *solution $x(t)$* is *uniformly asymptotically stable* if it is uniformly Lyapunov stable and asymptotic.

(d) A *solution $x(t)$* is *harmonic* if it is periodic of period $m\omega$, where m is a positive integer.

Sell [4] Theorem 4 shows that, if there exists a solution $x(t)$ of (4.1) which is bounded and uniformly asymptotically stable, then there is a harmonic solution which is also uniformly asymptotically stable.

By considering the Poincaré map $x \mapsto \phi(x) = \phi(\omega;0,x)$, we see that $(t,x) \mapsto \phi^t(x)$, $t \in \mathbb{Z}_+$ defines a discrete semiflow on $X \subset \mathbb{R}^n$, where X is the set of x such that $\phi(t;0,x)$ exists for all $t \geq 0$. Evidently a solution $x(t)$ is bounded if and only if the discrete semiflow is Lagrange stable at $x(0)$. Moreover $x(t)$ is uniformly Lyapunov stable or asymptotic if and only if $C_+(x(0))$ is Lyapunov stable or asymptotic respectively with respect to ϕ^t. Finally $x(t)$ is ω-periodic if and only if

$x(0)$ is an equilibrium of the discrete flow and it is harmonic if and only if $x(0)$ is periodic. Theorem 2.1 therefore shows that the uniform Lyapunov stability of $x(t)$ in Sell's theorem is not an essential requirement for the existence of a harmonic solution. It is the asymptotic property which establishes the existence of a harmonic solution and the Lyapunov stability may be either included or excluded in both the hypothesis and conclusion.

Theorem 4.1 *Suppose that $x(t)$ is a bounded solution of (4.1). Then $x(t)$ is asymptotic [and uniformly Lyapunov stable] if and only if there exists a harmonic solution $x_0(t)$ which is asymptotic [and uniformly Lyapunov stable] and such that $\lim_{t \to \infty} |x(t) - x_0(t)| = 0$.*

When $f(t, \cdot)$ is C^1 the Poincaré map ϕ is also smooth. The matrix $\frac{\partial}{\partial x} \phi(t; s, x(s))$ is a fundamental matrix for the linearized system

$$\dot{y} = \frac{\partial f}{\partial x}(t, x(t))y. \tag{4.2}$$

This equation is said to be *uniformly asymptotically stable* if there exist constants $K, \beta > 0$ such that $|\frac{\partial}{\partial x} \phi(t; s, x(s))| \le K e^{-\beta(t-s)}$, $0 \le s \le t$; see Coppel [19, p.54]. This condition is equivalent to

$$\left|\frac{\partial}{\partial x} \phi(t; 0, x)\right| \le K e^{-\beta t}, \quad 0 \le t \tag{4.3}$$

if $x \in \{x(s): 0 \le s < \infty\}$ which in turn implies $|\frac{\partial}{\partial x} \phi(k\omega; 0, x_j)| \le K\alpha^k$, $0 \le k, j$, where $x_j = x(j\omega)$. This is the condition (3.4) for the Poincaré map and, from Theorem 3.1, if $x(t)$ is bounded, $\Omega(x_0)$ is a stable hyperbolic m-periodic orbit for this map and some $m > 0$. It follows that, if $x \in \Omega(x_0)$, $x(t) = \phi(t; 0, x)$ is a harmonic solution of (4.1). Since $\varphi(t; s, \cdot)$ is C^1, (4.3) is also satisfied if $x(t)$ is the harmonic solution and so (4.2) is uniformly asymptotically stable in this case also. Finally, it may be argued as in the proof of Lemma 3.1 that when $x(t)$ is a bounded solution of (4.1) such that (4.2) is uniformly asymptotically stable, there is a neighbourhood U of $\{x(s): 0 \le s < \infty\}$ and $K, \beta, \delta > 0$ such that $x_0 \in U$, $|x - x_0| < \delta$ implies $|\phi(t; s, x) - \phi(t; s, x_0)| \le K e^{-\beta(t-s)} |x - x_0|$. We can now synthesize these observations in the following theorem.

Theorem 4.2 *Suppose that $x(t)$ is a bounded solution of the periodic system (4.1), where $f(t, \cdot)$ is C^1. Then there exists a harmonic solution $x_0(t)$ such that $\lim_{t \to \infty} |x(t) - x_0(t)| = 0$ and constants $K, \beta, \delta > 0$ for which*

$$|\phi(t; s, x) - x_0(t)| \le K e^{-\beta(t-s)} |x - x_0(s)|, \quad 0 \le s \le t,$$

when $|x - x_0(s)| < \delta$, if and only if the linearized equation (4.2) is uniformly asymptotically stable.

Earlier forms of both results in this section appear in [9].

5 Autonomous Differential Equations

We now consider an autonomous differential equation

$$\dot{x} = f(x), \tag{5.1}$$

where $x \mapsto f(x)$ is a C^1 function from $\mathbb{R}^n$ to $\mathbb{R}^n$. This defines a semiflow $(t, x) \mapsto \phi(t, x)$ on the set X of points x in $\mathbb{R}^n$ such that the solution $x(t) = \phi(t, x)$ satisfying $x(0) = x$ exists for all $t \geq 0$. Necessary and sufficient conditions are given for an omega limit set to be a stable hyperbolic equilibrium or a phase stable hyperbolic periodic orbit.

The linearization of (5.1) with respect to a solution $x(t)$ is

$$\dot{y} = \frac{\partial f}{\partial x}(x(t))y. \tag{5.2}$$

This equation is uniformly asymptotically stable if and only if there exist constants $K, \alpha > 0$ such that

$$\left|\frac{\partial}{\partial x}\phi(t,x)\right| \leq Ke^{-\alpha t}, \quad t \geq 0, \tag{5.3}$$

if $x \in C_+(x(0))$.

Definition 5.1 An equilibrium x_0, $f(x_0) = 0$, is *stable hyperbolic* if Re $\lambda < 0$ for every eigenvalue λ of $\frac{\partial f}{\partial x}(x_0)$. This is equivalent to the uniform asymptotic stability of (5.2) with $x(t) = x_0$.

Theorem 5.1 *Suppose $x(t)$ is a bounded solution of (5.1) then $\lim_{t\to\infty} x(t) = x_0$, where x_0 is a hyperbolic stable equilibrium, if and only if the linearization (5.2) of (5.1) with respect to $x(t)$ is uniformly asymptotically stable.*

Proof This may be established using Theorem 2.1 but a more direct proof is possible. First observe that $y(t) = \dot{x}(t) = f(x(t))$ is a solution of (5.2). Suppose now that (5.2) is uniformly asymptotically stable. Then $\lim_{t\to\infty} f(x(t)) = 0$ and x_0 is an equilibrium if $x_0 \in \Omega(x(0))$. Thus an inequality of the form (5.3) is satisfied with $x = x_0$; this equilibrium is stable hyperbolic and therefore isolated. In particular $\lim_{t\to\infty} x(t) = x_0$. Conversely, if x_0 is a stable hyperbolic equilibrium, then (5.4) is satisfied with $x = x_0$. It may be argued as in the proof of Lemma 3.1 that this implies the existence of a neighbourhood U of x_0 and constants $L > K$, $0 < \gamma < \alpha$ such that $\left|\frac{\partial}{\partial x}\phi(t,x)\right| \leq Le^{-\gamma t}$, if $t \geq 0$, $x \in U$. Thus (5.2) is uniformly asymptotically stable if $\lim_{t\to\infty} x(t) = x_0$.

We conclude this section with a discussion of a result of Li and Muldowney [9], extended in [10] on the existence of a stable hyperbolic limit cycle for (5.1). When $x(t)$ is a ω-periodic solution, the linear equation (5.2) has ω-periodic coefficient matrix. A nontrivial periodic solution of (5.2) is $y_1(t) = \dot{x}(t)$ so the system has a Floquet multiplier $\lambda_1 = 1$; see Guckenheimer and Holmes [20, p.25]. The remaining Floquet multipliers $\lambda_2, \ldots, \lambda_n$ are the eigenvalues of the linearized Poincaré map $x \mapsto \phi(\omega, x)$ at a point x in the periodic orbit $C_+(x(0))$. When $|\lambda_i| < 1$, $i = 2, \ldots, n$, the periodic orbit is said to be stable hyperbolic. In that case, there exist constants $K, \alpha, \delta > 0$ such that $x_0 \in C_+(x(0))$, $|x - x_0| < \delta$ implies that for some h

$$|\phi(t+h, x) - \phi(t, x_0)| \leq K e^{-\alpha t} |x - x_0|. \tag{5.4}$$

In particular, the semiflow is Lyapunov stable and phase asymptotic at $C_+(x(0))$. Details may be found in Coppel [19, p.82] Theorem 14.

We now reformulate the condition of hyperbolic stability of a periodic orbit so that it can be applied to a general bounded orbit and, in the spirit of the present discussion, provide a criterion for the omega limit set to be a stable hyperbolic periodic orbit. The second compound equation of (5.2) is

$$\dot{z} = \frac{\partial f^{[2]}}{\partial x}(x(t))z, \tag{5.5}$$

the equation whose solution set is the space spanned by the exterior products $z(t) = y_1(t) \wedge y_2(t)$, where $y_1(t)$, $y_2(t)$ are solutions of (5.2). The matrix $\frac{\partial f^{[2]}}{\partial x}$ is $N \times N$, $n = \binom{n}{2}$, the second additive compound matrix of $\frac{\partial f}{\partial x}$. A discussion of compound matrices may be found in Fiedler [21], Marshall and Olkin [22]; see also the Appendix. The relevance of (5.5) to (5.1) is discussed in [23]. When $x(t)$ is ω-periodic, (5.5) is also a ω-periodic system and its Floquet multipliers are the products $\lambda_i \lambda_j$, $i \neq j$, where $\lambda_1, \ldots, \lambda_n$ are the Floquet multipliers of (5.2). Since $\lambda_1 = 1$ is a multiplier, it follows that $\lambda_j = \lambda_1 \lambda_j$, $j = 2, \ldots, n$ are also Floquet multipliers of (5.5) and $|\lambda_j| < 1$, $j = 2, \ldots, n$ is equivalent to the uniform asymptotic stability of (5.5). We conclude that the periodic solution $x(t)$ of (5.1) is stable hyperbolic if and only if (5.5) is uniformly asymptotically stable. This generalizes the Poincaré condition for stability when $n = 2$: $\int_0^\omega \operatorname{div} f(x(t)) dt < 0$. The Poincaré condition is equivalent to the uniform asymptotic stability of the Liouville equation $\dot{z} = \operatorname{div} f(x(t)) z$, which is (5.5) when $n = 2$.

The following theorem follows from Li and Muldowney [9] Theorem 4.1 and Muldowney [10] Theorem 2.2.

Theorem 5.2 *Suppose that $x(t)$ is a bounded solution of (5.1) whose omega limit set Ω contains no equilibrium. Then Ω is a stable hyperbolic periodic orbit if and only if the second compound equation (5.5) is uniformly asymptotically stable.*

The sufficiency of the stability of (5.5) is proved in [9] and the necessity in [10].

6 Discussion

It is interesting to speculate on the type of stability conditions which would imply the existence of more complex dynamical objects than equilibria and periodic orbits. For example, the k-th compound equation of (5.2), $\dot{z} = \frac{\partial f}{\partial x}^{[k]}(x(t))z$, is the equation whose solution set is the span of all exterior products $z(t) = y+1(t) \wedge \cdots \wedge y_k(t)$, where $y_1(t), \ldots, y_k(t)$ are solutions of (5.1). Li and Muldowney [24] show that, if $x(t)$ is a quasi-periodic solution of (5.1) with m basic frequencies, then it is orbitally asymptotically stable with asymptotic phase provided the $(m+1)$-th compound equation of (5.2) is uniformly asymptotically stable. Comparing this with Theorem 5.2 a natural question arises: when $x(t)$ is simply a bounded solution of (5.1), does stability of this compound equation imply that the omega limit set is the orbit closure of a stable quasi-periodic solution? This appears to be not necessarily the case; some additional conditions are required. However it appears to be a promising direction for investigation.

Appendix

If $A = [a_{ij}]$ is an $n \times n$ matrix, its second additive compound $A^{[2]}$ is the $\binom{n}{2} \times \binom{n}{2}$ matrix defined as follows. For any integer $i = 1, \ldots, \binom{n}{2}$, let $(i) = (i_1, i_2)$ be the i-th member in the lexicographic ordering of the integer pairs (i_1, i_2) such that $1 \leq i_1 < i_2 \leq n$. Then the element in the i-row and j-column of $A^{[2]}$ is

$$\begin{cases} a_{i_1 i_1} + a_{i_2 i_2}, & \text{if } (j) = (i) \\ (-1)^{r+s} a_{i_r j_s}, & \text{if exactly one entry } i_r \text{ or } (i) \text{ does not occur in } (j) \\ & \text{and } j_s \text{ does not occur in } (i) \\ 0, & \text{if neither entry from } (i) \text{ occurs in } (j). \end{cases}$$

Table 1 gives $A^{[2]}$ in the cases $n = 2, 3, 4, 5$.

$n = 2:$ $(1) = (1, 2)$

$$A^{[2]} = [a_{11} + a_{22}] = \operatorname{tr} A$$

$n = 3:$ $\begin{array}{l}(1) = (1,2) \\ (2) = (1,3) \\ (3) = (2,3)\end{array}$

$$A^{[3]} = \begin{bmatrix} a_{11}+a_{22} & a_{23} & -a_{13} \\ a_{32} & a_{11}+a_{33} & a_{12} \\ -a_{31} & a_{21} & a_{22}+a_{33} \end{bmatrix}$$

$n = 4:$ $\begin{array}{l}(1) = (1,2) \\ (2) = (1,3) \\ (3) = (1,4) \\ (4) = (2,3) \\ (5) = (2,4) \\ (6) = (3,4)\end{array}$

$$A^{[4]} = \begin{bmatrix} a_{11}+a_{22} & a_{23} & a_{24} & -a_{13} & -a_{14} & 0 \\ a_{32} & a_{11}+a_{33} & a_{34} & a_{12} & 0 & -a_{14} \\ a_{42} & a_{43} & a_{11}+a_{44} & 0 & a_{12} & a_{13} \\ -a_{31} & a_{21} & 0 & a_{22}+a_{33} & a_{34} & -a_{24} \\ -a_{41} & 0 & a_{21} & a_{43} & a_{22}+a_{44} & a_{23} \\ 0 & -a_{41} & a_{31} & -a_{42} & a_{32} & a_{33}+a_{44} \end{bmatrix}$$

$n = 5:$ $\begin{array}{l}(1) = (1,2) \\ (2) = (1,3) \\ (3) = (1,4) \\ (4) = (1,5) \\ (5) = (2,3) \\ (6) = (2,4) \\ (7) = (2,5) \\ (8) = (3,4) \\ (9) = (3,5) \\ (10) = (4,5)\end{array}$

$$A^{[5]} = \begin{bmatrix} a_{11}+a_{22} & a_{23} & a_{24} & a_{25} & -a_{13} & -a_{14} & -a_{15} & 0 & 0 & 0 \\ a_{32} & a_{11}+a_{33} & a_{34} & a_{35} & a_{12} & 0 & 0 & -a_{14} & -a_{15} & 0 \\ a_{42} & a_{43} & a_{11}+a_{44} & a_{45} & 0 & a_{12} & 0 & a_{13} & 0 & -a_{15} \\ a_{52} & a_{53} & a_{54} & a_{11}+a_{55} & 0 & 0 & a_{12} & 0 & a_{13} & a_{14} \\ -a_{13} & a_{21} & 0 & 0 & a_{22}+a_{33} & a_{34} & a_{35} & -a_{24} & -a_{25} & 0 \\ -a_{41} & 0 & a_{21} & 0 & a_{43} & a_{22}+a_{44} & a_{45} & a_{23} & 0 & -a_{25} \\ -a_{51} & 0 & 0 & a_{21} & a_{53} & a_{54} & a_{22}+a_{55} & 0 & a_{23} & a_{24} \\ 0 & -a_{41} & a_{31} & 0 & -a_{42} & a_{32} & 0 & a_{33}+a_{44} & a_{45} & -a_{35} \\ 0 & -a_{51} & 0 & a_{31} & -a_{52} & 0 & a_{32} & a_{54} & a_{33}+a_{55} & a_{34} \\ 0 & 0 & -a_{51} & a_{41} & 0 & -a_{52} & a_{42} & -a_{53} & a_{43} & a_{44}+a_{55} \end{bmatrix}$$

Table 1 The matrix $A^{[n]}$, $n = 2, 3, 4, 5$

References

[1] Bendixson, I. (1901). Sur les curbes définiés des équations différentielle. *Acta Math.*, **24**, 1–88.

[2] Hale, J.K. (1969). *Ordinary Differential Equations*. Wiley-Interscience, New York.

[3] Massera, J.L. (1950). The existence of periodic solutions of systems of differential equations. *Duke Math. J.*, **17**, 457–475.

[4] Sell, G.R. (1966). Periodic solutions and asymptotic stability. *J. Diff. Eqns*, **2**, 143–157.

[5] Saperstone, S.H. (1981). *Semidynamical Systems in Infinite Dimensional Spaces*. Applied Mathematical Sciences, Vol. 37, Springer-Verlag, New York.

[6] Cronin, J. (1994). *Differential Equations, Introduction and Qualitative Theory*. Second Ed., Marcel Dekker, New York.

[7] Yoshizawa, T. (1967). Stability and existence of periodic and almost periodic solutions. *Proceedings United States-Japan Seminar on Differential and Functional Equations* (Eds.: W.A. Harris and Y. Sibuya). Benjamin, New York, 411–427.

[8] Pliss, V.A. (1966). *Nonlocal Problems of the Theory of Oscillations*. Academic Press, New York.

[9] Li, M.Y. and Muldowney, J.S. (1996). Phase asymptotic semiflows, Poincaré's condition and the existence of stable limit cycles. *J. Diff. Eqns*, **124**, 425–448.

[10] Muldowney, J.S. (1996). Stable hyperbolic limit cycles. In: *Differential Equations and Applications to Biology and to Industry* (Eds.: M. Martelli and K. Cooke). World Scientific, Singapore.

[11] Hirsch, M.W. (1982). Systems of differential equations which are competitive or cooperative I: limit sets. *SIAM J. Math. Anal.*, **13**, 432–439.

[12] Smith, H.L. (1988). Systems of ordinary differential equations which generate an order preserving semiflow. *SIAM Review*, **30**, 87–113.

[13] Mallet-Paret, J. and Smith, H.L. (1990). The Poincaré-Bendixson Theorem for monotone cyclic feedback systems. *J. Dynamics and Diff. Eqns*, **2**, 367–421.

[14] Smith, R.A. (1979). The Poincaré-Bendixson Theorem for certain differential equations of higher order. *Proc. Roy. Soc. Edinburgh Sect. A*, **83**, 63–75.

[15] Smith, R.A. (1980). Existence of periodic orbits of autonomous ordinary differential equations. *Proc. Roy. Soc. Edinburgh Sect. A*, **85**, 153–172.

[16] Smith, R.A. (1986). Massera's convergence theorem for periodic nonlinear differential equations. *J. Math. Anal. Applic.* **120**, 679–708.

[17] Szlenk, W. (1984). *An Introduction to the Theory of Smooth Dynamical Systems*. Wiley, Chichester.

[18] Yoshizawa, T. (1966). *Stability Theory of Liapunov's Second Method*. Math. Soc. Japan, Tokyo.

[19] Coppel, W.A. (1965). *Stability and Asymptotic Behavior of Differential Equations*. Heath, Boston.

[20] Guckenheimer, J. and Holmes, P. (1983). *Nonlinear Oscillations, Dynamical Systems and Bifurcations of Vector Fields*. Springer, New York.

[21] Fiedler, M. (1974). Additive compound matrices and inequality for eigenvalues of stochastic matrices. *Czech. Math. J.*, **24** (99), 392–402.

[22] Marshall, A.W. and Olkin, I. (1979). *Inequalities: Theory of Majorization and its Applications*. Academic Press, New York.

[23] Muldowney, J.S. (1990). Compound matrices and ordinary differential equations. *Rocky Mountain J. Math.*, **20**, 857–871.

[24] Li, M.Y. and Muldowney, J.S. (1998). Poincaré's stability condition for quasi-periodic orbits. *Canadian Applied Mathematics Quarterly*, **6**, 367–381.

3.3 SOME CONCEPTS OF PERIODIC MOTIONS AND STABILITY ORIGINATED BY ANALYSIS OF HOMOGENEOUS SYSTEMS*

V.N. PILIPCHUK

Department of Mechanical Engineering, Wayne State University, Detroit, USA

1 Introduction

Theory of periodic motions of nonlinear mechanical systems is a quite general field in pure and applied Mathematics and Mechanics. The related treatments and references can be found in survey [1]. As follows from the Floquet's theory [2] the periodic solutions have a special meaning for linear differential equations with periodic coefficients. As a rule, the periodic solutions separate regions of stability and instability in a space of parameters, and hence give us a complete enough information about the space structure. The present paper is focused on some ideas concerning periodic solutions initiated by the investigation of homogeneous oscillators. Probably A.M.Lyapunov was the first who has paid his attention for special meaning nonlinear oscillator with homogeneous power form characteristic. He obtained the homogeneous oscillator when dealing with stability problems in general formulation and considering so-called degenerated case [3] (see also in [4]). Later on, within approximately last 35 years (the references are given below in the text), the homogeneous systems were under consideration from different points of view in applied mathematics, mechanics and theoretical physics. These studies generated an interesting circle of ideas, some of which came back into the stability theory.

---

*Advances in Stability Theory (Ed.: A.A. Martynyuk). Stability and Control: Theory, Methods and Applications, Taylor & Francis, London, **13** (2003) 231–242.

2 Homogeneous Oscillator

Let us consider a family of one degree of freedom oscillators described by the differential equation

$$\ddot{x} + x^{2n-1} = 0; \quad x \in R, \qquad (1)$$

where n is positive integer.

When $n = 1$ one has the simplest linear (harmonic) oscillator, however when $n > 1$ the system becomes essentially nonlinear and can not be linearized in the class of vibrating systems. From a point of view of mechanics this situation may correspond to a critical loading applied to an elastic system. For example, in appropriate variables, vibration of a one-mode model of linear beam on nonlinear elastic foundation loaded by axial force T is described by the differential equation of motion with respect to the modal coordinate $x(t)$ as

$$\ddot{x} + \left(1 - \frac{T}{T_*}\right)x + x^3 = 0,$$

where T^* is a certain critical (by Euler) value of the axial force.

At critical value $T = T^*$ the linear term in the last equation vanish, and one obtains oscillator (1) with cubic characteristic ($n = 2$). Oscillator (1) can be found in physical literature and also in different fields of applied mathematics and mechanics [5–13]. On the other hand, there is a purely mathematical reason to consider equation (1) as a family of oscillators with power form characteristic including both linear and strongly nonlinear cases.

For arbitrary positive integer n, general solution of equation (1) can be expressed in terms of special Lyapunov's function [4] such as $cn\theta$ (another version of special functions for equation (1) was considered in [8]). The pair of periodic functions, $sn\theta$ and $cn\theta$, are defined by expressions

$$\theta = \int_0^{sn\theta} \left(1 - nz^2\right)^{\frac{1-2n}{2n}} dz, \quad cn^{2n}\theta + n\,sn^2\theta = 1$$

and possesses the properties

$$cs0 = 1, \quad sn0 = 0, \quad \frac{dsn\theta}{d\theta} = cn^{2n-1}\theta, \quad \frac{dcn\theta}{d\theta} = -sn\theta.$$

Now, the general solution of equation (1) can be written as $x = Acn\left(A^{n-1}t + \alpha\right)$, where A and α are arbitrary constants. For $n = 1$ the functions $sn\theta$ and $cn\theta$ give the standard pair of trigonometric functions $\sin\theta$ and $\cos\theta$ respectively. Interestingly enough, the strongly nonlinear limit $n \to \infty$ gives a quite simple pair of periodic functions too. This case can be interpreted by means of the first integral of motion

$$\frac{\dot{x}^2}{2} + \frac{x^{2n}}{2n} = \frac{1}{2}, \qquad (2)$$

where the right-hand side (total energy) is given by the initial conditions $x(0) = 0$ and $\dot{x}(0) = 1$. Taking into account that the coordinate of the oscillator reaches its amplitude value at zero kinetic energy, one obtains $x(t) \in \left[-n^{1/(2n)}, n^{1/(2n)}\right]$ at any time t. Since $n^{1/(2n)} \to 1$, when $n \to \infty$, the oscillator's motion will be restricted by the interval $[-1, 1]$. Inside of this interval, the second term at the left-hand side of expression (2) vanishes and hence, $\dot{x} = \pm 1$ or $x = \pm t + \alpha_\pm$ where $\alpha_\pm$ are constants. Manipulating with the constants one can construct a periodic piece-wise linear sawtooth function as (sawtooth sine)

$$\tau(t) = \begin{cases} t, & -1 \leq t \leq 1, \\ -t+2, & 1 \leq t \leq 3, \end{cases} \qquad \tau(t) \stackrel{\forall t}{=} \tau(4+t). \qquad (3)$$

One should expect that $cnt \stackrel{\forall t}{\longrightarrow} \tau(t+1)$ when $n \to \infty$. However there is a question remains in terms of the differential equation, namely, will the sawtooth function, $\tau(t)$, satisfy the differential equation (1) when $n \to \infty$ and what sense one should provide the limit solution with? Indeed, one has $\tau(t) \in C(R)$, whereas a classic solution of the differential equation (1) has to be of class $C^2(R)$. The problem is that a definition of solution in terms of distributions (generalized functions) is well formulated in the linear case, but it requires a more special treatment for nonlinear differential equations [15]. This kind of problems will not be considered here. From the point of view of the below treatments, the most important remark is that the family of oscillators (1) includes the two simple cases associated with the boundaries of the interval $1 \leq n < \infty$. Respectively, one has the two pair of periodic functions

$$\{x, \dot{x}\} = \{\sin t, \cos t\}, \quad \text{if} \quad n = 1 \qquad (4)$$

and

$$\{x, \dot{x}\} \to \{\tau(t), \dot{\tau}(t)\}, \quad \text{if} \quad n \to \infty, \qquad (5)$$

where $\dot{\tau}(t)$ is a generalized derivative of the sawtooth sine and will be named as a rectangular cosine. It is an important to note that the piece-wise linear periodic functions (5) did not come as an abstract artificial construction. It is treated as a special representative of the family of periodic functions generated by the homogeneous oscillator (1). From that point of view functions (5) and (4) possess the "equal rights."

3 Sawtooth Time Variable

There are many mathematical techniques and methods exist based on suitable mathematical properties of trigonometric pair (4). It would be enough to mention the Fourier analysis. Now, in addition to (4), is it possible to employ the obvious simplicity of the sawtooth sine and its first generalized derivative (5) in order to develop an analytical technique for oscillating processes? The reason of construction of alternative techniques to the Fourier analysis is clear. Indeed, if a process to be investigated has a sawtooth time history, say $x = \tau(t)$, then one should keep a large number of terms in the related Fourier expansion when trying to describe the discontinuities of slope. This well known problem of the Fourier analysis initiated a series of investigations aimed on construction of expansions for a local analysis of functions. One should mention the Haar system [16] and the relatively new wavelets theory (an introductory articles can be found in [17]). An advantage of this kind of analysis is that any localized perturbation of the function affects the only few number of coefficients of the expansion. The local expansions appeared to be very suitable for different problems of signal processing while its applications to differential still remains quite limited. In fact, a broad applicability of the Fourier series to partial and ordinary differential equations is provided, first of all, by the remarkable property of the Fourier basic functions to be eigen-functions of the differentiating operator, $\left(e^{ik\omega t}\right)'_t = ik\omega e^{ik\omega t}$. Unfortunately, as a rule, it is not so in the above mentioned theory of local expansions. Differentiation of the local basic functions may even bring them into a worse class of functions. This is a fact of quite general nature, namely, since a function becomes localized its derivative is getting worse. Now, it will be shown that there is another way to adopt the non-smooth basic functions such as (5) for different problems of mechanics described by the differential equations. Our attempt is inspired by the fact that functions (5) itself generated by the special limiting case of the differential equation (1). All illustrations will be done on the periodic processes (some generalizations can be found in the references).

First, note that the period of the pair (5) is normalized to four in order to provide the following suitable relationship for the generalized derivative $\dot{\tau}(t) = e(t)$

$$[e(t)]^2 = 1 \quad \text{for almost all} \quad t. \tag{6}$$

Now, let us reproduce the following proposition [18]:

Proposition 3.1 *Any periodic function $x(t)$ whose period has been normalized to $T = 4$, can be represented as*

$$x(t) = X[\tau(t)] + Y[\tau(t)]e(t), \tag{7}$$

where $X(\tau) = [x(\tau) + x(2-\tau)]/2$ and $Y(\tau) = [x(\tau) - x(2-\tau)]/2$.

Example 3.1

$$C_1 \sin \frac{\pi t}{2} + C_2 \cos \frac{\pi t}{2} = C_1 \sin \left[\frac{\pi}{2}\tau(t)\right] + C_2 \cos \left[\frac{\pi}{2}\tau(t)\right] e(t).$$

Expression (7) can be understood as a "complex" value with real X and "imaginary" Y components, where basic "imaginary" element e creates a circling group: $e^2 = 1$, $e^3 = e$, $e^4 = 1$, ... This remark significantly simplifies all operations with the representation, when transforming a differential equation of motion. For example, one has relationship

$$\exp(X + Ye) = \exp(X)[\cosh Y + e \sinh Y] \tag{8}$$

which follows from the more general one

$$f(X + Ye) = \frac{1}{2}[f(X+Y) + f(X-Y)] + \frac{1}{2}[f(X+Y) - f(X-Y)]e.$$

An important feature is that the algebraic structure of the representation (7) is not changed after differentiation if necessary conditions of continuity for function under the differentiation is provided. For first two derivative the result of differentiation and the conditions related are

$$\dot{x} = Y' + X'e \quad \text{if} \quad Y|_{\tau=\pm 1} = 0, \tag{9}$$

$$\ddot{x} = X'' + Y''e \quad \text{if also} \quad X'|_{\tau=\pm 1} = 0, \tag{10}$$

where prime denotes differentiation with respect to τ.

The above listed relations enables one to introduce an oscillating time parameter $t \to \tau$ into differential equations on the manifold of periodic solutions. In this case, the inverted transformation of time over the period $-1 \leq t \leq 3$ is

$$t = 1 + (\tau - 1)e. \tag{11}$$

A role of new unknown functions is played by the pair $X(\tau)$ and $Y(\tau)$. In many cases one of the two components can be identically equal to zero. Manipulating with the transformed equations using different successive approximations techniques, one can construct periodic solutions of linear and nonlinear differential equations [19, 20]. A special case based on the one-component (X) representation can be found in [10], where an approximate general solution of the homogeneous differential equation (1) for arbitrary n was obtained in a power series form with respect to the oscillating time τ.

To this end let us reproduce two alternative expansions. The first one is simply the Fourier series for the sawtooth sine, i.e. it expresses the sawtooth functions in terms of the trigonometric ones, whereas the second expansion represents the trigonometric sine in terms of the sawtooth function:

$$\tau(t) = \frac{8}{\pi^2}\left(\sin\frac{\pi t}{2} - \frac{1}{3^2}\sin\frac{3\pi t}{2} + \frac{1}{5^2}\sin\frac{5\pi t}{2} - \cdots\right), \qquad (12)$$

$$\sin t = \sin\frac{\pi\tau}{2} = \frac{\pi\tau}{2} - \frac{1}{3!}\left(\frac{\pi\tau}{2}\right)^3 + \frac{1}{5!}\left(\frac{\pi\tau}{2}\right)^5 - \cdots, \quad \tau = \tau\left(\frac{2t}{\pi}\right). \qquad (13)$$

So one has the two different kind of expansions associated with the boundaries of interval $1 \leq n < \infty$, where n is the parameter of the homogeneous oscillator (1). Any truncated series of the expansions (12) is smooth function, whereas approximated function $\tau(t)$ is non-smooth. On other hand, any truncated series of the expansion (13) is non-smooth, whereas the approximated function $\sin t$ is smooth.

4 Differential Equations with Periodic Coefficients

Let us show what relation the above illustrated sawtooth transformation of time may have to the problem of stability.

Consider the following second order differential equation with periodic coefficient

$$\ddot{x} + q(t)x = 0; \quad x \in R, \quad q(t+T) \stackrel{\forall t}{=} q(t). \qquad (14)$$

Let us introduce the oscillating time parameter as $\tau = \tau(\omega t)$, were $\omega = 4/T$ (the numerical factor 4 appeared instead of 2π due to normalization for the period of the sawtooth sine). Being interested in periodic solutions of the period T and taking into account the remarks of Section 1, the periodic coefficient of the equation and the solution to be found are represented respectively as

$$q(t) = Q[\tau(\omega t)] + P[\tau(\omega t)]e(\omega t) \quad \text{and}$$
$$x(t) = X[\tau(\omega t)] + Y[\tau(\omega t)]e(\omega t), \qquad (15)$$

where the components Q and P are defined by (7), X and Y are the components of unknown function to be found.

Substituting (15) into differential equation (14), and taking into account the "multiplication table" $e^2 = 1$ and the differentiation rules (9)–(10), one obtains

$$\omega^2 X''(\tau) + Q(\tau)X(\tau) + P(\tau)Y(\tau) + [\omega^2 Y''(\tau) + P(\tau)X(\tau) + Q(\tau)Y(\tau)]e = 0.$$

Equating separately the "real" and "imaginary" parts of this expression to zero, one keeping in mind (9)–(10), one obtains the homogeneous boundary value problem

$$\omega^2 X'' + QX + PY = 0, \quad X'|_{\tau=\pm 1} = 0,$$
$$\omega^2 Y'' + PX + QY = 0, \quad Y|_{\tau=\pm 1} = 0. \qquad (16)$$

Now the new independent variable is bounded as $|\tau| \le 1$, and hence one can seek the general solutions in the power series form with respect to τ after the coefficient $Q(\tau)$ and $P(\tau)$ have been expressed by its Maclaurin series. Substituting the solutions into the boundary conditions in (16) one will get the relations for parameters at which the periodic solutions exist. Note that the system becomes decoupled with respect to X and Y if $P \equiv 0$. The above transformation will be still valid in the multidimensional case, when $x \in R^n$ and $q(t)$ is $n \times n$-matrix.

The most reasonable cases to apply the sawtooth transformation of time are those when the periodic coefficient $q(t)$ is expressed through the functions τ and e in a simple manner.

For example, consider the case of parametric step-wise periodic excitation of the one degree of freedom harmonic oscillator

$$\ddot{x} + \omega_0^2 [1 + \varepsilon e(\omega t)] x = 0, \qquad (17)$$

where ω_0, ω and ε are constants and the parameter ε is not necessarily small.

In this case $Q = \omega_0^2$ and $P = \varepsilon \omega_0^2$, and one obtains the boundary value problem for linear system with constant coefficients

$$\begin{pmatrix} X \\ Y \end{pmatrix}'' + r^2 \begin{pmatrix} 1 & \varepsilon \\ \varepsilon & 1 \end{pmatrix} \begin{pmatrix} X \\ Y \end{pmatrix} = 0, \qquad (18)$$

where $r = \omega_0/\omega$.

More details and examples handled by this kind of analysis can be found in previous work [21].

One should note that the above transformation can be implemented for a broad class of periodic systems and the transformed equations possess a relatively simple structure due to some special properties of the pair (5). An essential physical feature of the transformation is that the oscillating time, τ, is introduced in such a manner that it does not affect the metrical properties of the time, namely, the metric remains constant after the transformation: $[d\tau(t)]^2 = dt^2$ due to (6). This defines a special role of the saw-tooth transformation among other periodic transformations of time. Due to this property the transformed differential equations of motion remain in the framework of Newton's formulation. Another important feature is that one has an inverted transformation of time in a simple form (11) of the algebraic structure with the basic elements $\{1, e\}$. Finely, the transformation is suitable to apply to different non-autonomous cases of the non-smooth external excitation.

Now let us discuss a possibility of introduction of an oscillating time parameter, say z, in a more general manner as $z = z(t)$, where $z(t)$ is periodic function, not necessarily the saw-tooth sine. One could consider trigonometric sine, elliptic sine, and others specially selected functions dependently on a problem investigated. The transformation becomes physically reasonable if the variable coefficients appear to be functions of the new time parameter z over all period of the oscillation. For example, let us consider equation (14), when $q(t) = a + b\cos 2t$ (the Matheu's equation). In this case one can introduce the new independent variable $z = \sin t$. Indeed, $q(t) = a + b\cos 2t = a + b - 2bz^2$, and the equation can be written with respect to $x = X(z(t))$ as

$$(1 - z^2)\frac{d^2 X}{dz^2} - z\frac{dX}{dz} + (a + b - 2bz^2)X = 0, \quad |z| \leq 1. \tag{19}$$

This transformation is known as Ince algebraization [22]. It can be realized in those cases when the periodic coefficient is represented as a function of $\sin t$, i.e. $q(t) = Q(\sin t)$. Future investigation is based on the fact that the new independent variable is bounded [23].

5 Stability of Normal Modes in Two Degrees of Freedom Homogeneous System

The above sections show an interesting role of homogeneous oscillator (1) from the point of view theory of stability.

In this section, an attention will brought to the nonlinear normal modes (NNMs) theory from the point of view special meaning of the homogeneous systems for the theory. Extending the classic concept of linear normal modes on nonlinear situation, the NNMs can be understood as synchronous periodic particular solutions of the nonlinear equations of motion. The linear superposition principle becomes inapplicable, however the NNMs should be understood as a natural motions of a nonlinear system and that is why this kind of particular solutions remains still important in nonlinear theory. A quasi-linear formulation of the NNMs theory started from the classic Lyapunov's works [24, 25]. However, is there any possible way exist to make use the idea of normal modes when rigidities of a many degrees of freedom system do not include linear components at all? H. Kauderer [26] showed that a common language for both linear and nonlinear systems is given by geometry of the configurational space. Namely, one can focus an analysis on trajectories of the normal modes in the configurational spaces without seeking any connection to either superposition principle or time history of such motions. Considering the many degrees of freedom homogeneous from such viewpoint, R. Rosenberg [7, 27] developed essentially nonlinear formulation of the normal modes theory. More historical data, references and details concerning the NNMs theory and applications can be found in [11] and recent monograph [29].

Now basic ideas of NNMs, including stability problems, will be illustrated on the symmetric case of two degree of freedom homogeneous system. The differential equation of motion can be written as $(m = 2n - 1)$

$$\ddot{x}_1 + x_1^m + c(x_1 - x_2)^m = 0,$$
$$\ddot{x}_2 + x_2^m + c(x_2 - x_1)^m = 0. \qquad (20)$$

Differential equations (20) describe two identical homogeneous oscillators (1) coupled by the nonlinear spring of the homogeneous characteristic. The characteristic (nonlinear rigidity) of the connecting spring differs from the characteristic of oscillators by the constant coefficient c. Note that system (20) is conservative with homogeneous potential function of $m + 1$ degree. When $m = 1$ the differential equations become linear and the system possesses the special periodic regimes, such as in-phase and out-of-phase normal modes, on which at any time t one has respectively $x_2(t) = x_1(t)$ and $x_2(t) = -x_1(t)$. Geometrically this is two perpendicular straight lines on the plane of configurations $x_1 x_2$. A unique property of the linear system is that an appropriate combination of the two normal modes completely define any possible motion of the system (linear superposition principle). When $m > 1$, the system is not linearizable, i.e. elimination of the nonlinear terms from the system takes it outside the class of vibrating systems. The above mentioned superposition principle does not work as well. Surprisingly enough, the system still possesses the two periodic solutions on which the equalities $x_2(t) = x_1(t)$ and $x_2(t) = -x_1(t)$ hold for any t (the in-phase and out-of-phase NNMs respectively; in case of n degrees of freedom, at least n NNMs exist [28]). Moreover the system may possess another periodic solutions with straight line trajectories on the plane $x_1 x_2$. Indeed, suppose that

$$x_2(t) = k x_1(t) \quad \text{for any} \quad t, \qquad (21)$$

where k is a slope of the trajectory to be defined (obviously, k can be equal to zero if only c is zero, i.e. the system is decoupled).

Substituting (21) into (20), and multiplying the first equation by k gives

$$k\ddot{x}_1 + k[1 - c(k-1)^m]x_1^m = 0,$$
$$k\ddot{x}_1 + [k^m + c(k-1)^m]x_1^m = 0. \qquad (22)$$

Equating the coefficients followed by x_1^m in the two equations (compatibility condition), one obtain the algebraic equation with respect to the slope k of the form

$$c(k-1)^m(k+1) + k^m - k = 0. \qquad (23)$$

To simplify the illustration, let us $m = 3$. In this case equation (23) admits exact solutions

$$k_{1,2} = \pm 1, \qquad (24)$$

$$k_{3,4} = 1 - \frac{1}{2c}\left(1 \pm \sqrt{1 - 4c}\right). \qquad (25)$$

The first two roots give the in-phase and out-of-phase modes predicted by the physical symmetry of the system. Another two roots are real if the parameter of coupling between the oscillators, c, is sufficiently small: $c < 1/4$. These two new nonlinear modes arise as a result of branching of the aut-of-phase mode at $c = 1/4$. This branching indicates a loosing of stability by the out-phase mode with respect to any small perturbation of the in-phase mode. More complete stability analysis could be done by investigation of the variational equations around the NNMs. An appropriate techniques can be found in [29]. For example, the Ince-algebraization (mentioned at the end of Section 4) was employed for transformation of the variational equations in [30, 31]. Special qualitative method was presented in [32].

In present, the theory of NNMs is not restricted by the homogeneous systems, and not even conservative ones. One can consider the NNMs with curvilinear trajectories in the configurational space as well. However, the first qualitative and quantitative results were originated by the investigation of the homogeneous systems.

6 Conclusion

Family of mechanical systems with a homogeneous potential function includes the linear systems as a particular case. Due to the superposition principle, the linear systems have played a uniquely general role in modern period of mathematical and physical sciences. This role was finalized by development of a very powerful Fourier series/transforms method and the linear (and weakly non-linear) theory of oscillations and waves. A role of the nonlinear representatives of the homogeneous system is relatively shadowed. However, it was shown above that a careful examination of all class of the homogeneous systems brings some ideas applicability of which is not restricted by that class only. For example, the limiting case $n \to \infty$ of the one degree of freedom oscillator gives, in certain meaning an alternative to $\{\sin t, \cos t\}$, pair of non-smooth periodic functions, $\{\tau(t), \dot{\tau}(t)\}$. The two degrees of freedom homogeneous system (20) admits a family of the straight line periodic solutions on the configurational plane. This geometrical simplicity allowed to develop the basic ideas of the NNMs. Now, combining the two mentioned results, one can formulate a reasonable question about the limit $m \to \infty$ in the (at least) two-dimensional case (20). In this case, one has a free moving unit-mass point under the constraints conditions: $|x_1| \le 1$, $|x_1 - x_2| \le 1$ and $|x_2| \le 1$. In this case many (not the straight line only) solutions can be constructed by purely geometrical manipulations. What will happen to these solutions when $m < \infty$?

References

[1] Mitropol'sky, Yu.A. and Martynyuk, A.A. (1988). About some results and actual problems in the theory of periodic motions in nonlinear mechanics. *Prikl. Mekh.*, **24**(3), 3–14 (Russian).

[2] Floquet, G. (1883). Sur les equations differentielles. *Ann. Ecole Norm. Sup.*, **12**, 47.

[3] Lyapunov, A.M. (1956). *Collection of Works*, Vol. 2. Publ. USSR Acad. of Sci., Moscow–Leningrad (Russian).

[4] Kamenkov, G.V. (1972). *Stability and Vibrations of Nonlinear Systems*, Vol. 1. Nauka, Moscow (Russian).

[5] Mitropl'sky, Yu.A. and Senik, P.M. (1961). Construction of asymptotic solution of an autonomous system with strong nonlinearity. *Dokl. Akad. Nauk Ukr.SSR*, **6**, 839–844 (Russian).

[6] Atkinson, C.P. (1962). On the superposition method for determining frequencies of nonlinear systems. *ASME Proc. of the 4-th National Congress of Applied Mechanics*, 57–62.

[7] Rosenberg, R.M. (1962). The normal modes of nonlinear n-degree-of-freedom systems. *J. Appl. Mech.*, **29**, 7–14.

[8] Rosenberg, R.M. (1963). The Ateb(h)-functions and their properties. *Quart. Appl. Math.*, **21**(1), 37–47.

[9] Mickens, R. and Oyedeji, K. (1985). Construction of approximate analytical solutions to a new class of nonlinear oscillator equation. *J. Sound and Vibr.*, **102**, 579–582.

[10] Pilipchuk, V.N. (1985). The calculation of strongly nonlinear systems close to vibroimpact systems. *Prikl. Mat. Mekh.*, **49**(5), 744–752 (Russian).

[11] Manevitch, L.I., Mikhlin, Ju.V. and Pilipchuk, V.N. (1989). *Method of Normal Vibrations for Essentially Nonlinear Systems*. Nauka, Moscow (Russian).

[12] Boettcher, S. and Bender, C.M. (1990). Nonperturbative square-well approximation to a quantum theory. *J. Math. Phys.*, **31**(11), 2579–2585.

[13] Andrianov, I.V. (1993). Asymptotic solutions for nonlinear systems with high degrees of nonlinearity. *Prikl.] Mat. Mekh.*, **57**(5), 941–943 (Russian).

[14] Richtmyer, R.D. (1985). *Principles of Advanced Mathematical Physics*, Vol. 1. Springer, Berlin.

[15] Maslov, V.P. and Omel'yanov, G.A. (1981). Asymptotic solution of equations with a small dispersion. *Usp. Mat. Nauk*, **36**(3), 63–126 (Russian).

[16] Haar, A. (1910). Zur Theorie der orthogonalen Funktionensysteme. *Math. Ann.*, **69**, 331–371.

[17] Benedetto, J.J. and Frazier, M.W. (Eds.) (1994). *Wavelets: Mathematics and Applications*. CRC Press, Boca Raton–Ann Arbor–London–Tokyo.

[18] Pilipchuk, V.N. (1988). A transformation of vibrating systems based on a nonsmooth periodic pair of functions. *Dokl. Akad. Nauk Ukr.SSR, Serie A*, **4**, 37–40 (Russian).

[19] Pilipchuk, V.N. (1996). Analytical study of vibrating systems with strong nonlinearities by employing saw-tooth time transformations. *J. Sound and Vibr.*, **192**(1), 43–64.

[20] Pilipchuk, V.N. (1999). Application of special nonsmooth temporal transformations to linear and nonlinear systems under discontinuous and impulsive excitation. *Nonlin. Dynamics*, **18**, 203–234.

[21] Pilipchuk, V.N. (1996). Calculation of mechanical systems with pulsed excitation. *Prikl. Mat. Mekh.*, **60**, 223–232 (Russian).
[22] Ince, E.L. (1926). *Ordinary Differential Equations*. Longmans Green, London.
[23] Whittaker, E.T. and Watson, G.N. (1986). *A Course of Modern Analysis*, 4th Ed. Cambridge University Press, Cambridge (UK).
[24] Lyapunov, A.M. (1907). Probleme generale de la stabilite du mouvement. *Ann. Fas. Sci. Toulouse*, **9**, 203–474.
[25] Lyapunov, A.M. (1947). *The General Problem of the Stability of Motion*. Princeton University Press, Princeton, (NJ).
[26] Kauderer, H. (1958). *Nichtlineare Mechanik*. Springer Verlag, Berlin–New York.
[27] Rosenberg, R.M. (1966). On nonlinear vibrations of systems with many degrees of freedom. *Adv. Appl. Mech.*, **9**, 155–242.
[28] Van Groesen, E.W.C. (1983). On normal modes in classical Hamiltonian systems. *Int. J. Nonlin. Mech.*, **18**(1), 55–70.
[29] Vakakis, A.F., Manevitch, L.I., Mikhlin, Yu.V., Pilipchuk, V.N. and Zevin, A.A. (1996). *Normal Modes and Localization in Non-Linear Systems*. Wiley-Interscience, New York.
[30] Zhupiev, A.L. and Mikhlin, Yu.V. (1981). Stability and branching of normal oscillations forms of nonlinear systems. *Prikl. Mat. Mekh.*, **45**(3), 450–455 (Russian).
[31] Zhupiev, A.L. and Mikhlin, Yu.V. (1984). Conditions for finiteness of the number of instability zones in the problem of normal vibrations of nonlinear systems. *Prikl. Mat. Mekh.*, **48**(4), 681–685 (Russian).
[32] Zevin, A.A. (1985). Existence, stability and some properties of the class of periodic motions in nonlinear mechanical systems. *Izv. Akad. Nauk SSSR, Mekh. Tverdogo Tela*, **20**(4), 45–55 (Russian).

3.4 STABILITY CRITERIA FOR PERIODIC SOLUTIONS OF AUTONOMOUS HAMILTONIAN SYSTEMS*

A.A. ZEVIN

Transmag Research Institute, Dnepropetrovsk, Ukraine

1 Introduction and Results

We consider the autonomous Hamiltonian system

$$\dot{x} = JH_x(x), \qquad J = \begin{bmatrix} 0 & I_n \\ -I_n & 0 \end{bmatrix}, \tag{1.1}$$

where $x \in R^{2n}$ and I_n is the identity matrix of order n; the Hamiltonian $H(x)$ is supposed to be twice differentiable.

It is known that linear stability of a T-periodic solution $x(t)$ is determined by the corresponding variational equation

$$\dot{y} = JA(t)y, \qquad A(t) = A(t+T) = H_{xx}(x(t)). \tag{1.2}$$

Let ρ_k, $k = 1, \ldots, 2n$ be the Floquet multipliers of equation (1.2). Since equation (1.1) is autonomous and admits the integral

$$H(x(t)) \equiv h = \text{const}, \tag{1.3}$$

there always exists a double multiplier $\rho = 1$ such that $y_1(t) = \dot{x}(t)$ is one of the related solutions (Poincaré [1]). The solution $x(t)$ is called elliptic [2,3], if other multipliers lie on the unit circle, and the corresponding solutions $y_k(t)$ are bounded on the interval $(0, \infty)$. If these multipliers are Krein definite [4], the solution $x(t)$ is called stable or strongly stable.

---

*Advances in Stability Theory (Ed.: A.A. Martynyuk). Stability and Control: Theory, Methods and Applications, Taylor & Francis, London, **13** (2003) 243-253.

For some fixed h, the expression $H(x) = h$ defines the energy surface $M = H^{-1}(h)$. The known results on stability of the periodic solutions on M relate to a surface which bounds the convex compact region Ω. According to Ekeland theorem [2], such a surface carries at least one elliptic orbit, provided that for $x \in M$ the pinching condition holds,

$$2R^{-2}I_{2n} \leq H_{xx}(x) \leq 2r^{-2}I_{2n}, \quad \frac{R^2}{r^2} < 2, \qquad (H_1)$$

where $A \leq B$ means that $(Ay, y) \leq (By, y)$ for any $y \in R^{2n}$; (a, b) is the scalar product of the vectors a and b.

It is clear that $\alpha_1(x)I_{2n} \leq H_{xx}(x) \leq \alpha_{2n}(x)I_{2n}$, where $\alpha_k(x)$ $(0 < \alpha_k(x) \leq \alpha_{k+1}(x)$, $k = 1, \ldots, 2n)$ are the eigenvalues of the matrix $H_{xx}(x)$. Thus, condition (H_1) is equivalent to the inequality

$$\frac{\alpha_+}{\alpha_-} < 2, \qquad (H_1')$$

where $\alpha_- = \min \alpha_1(x)$ and $\alpha_+ = \max \alpha_{2n}(x)$ for $x \in M$.

Condition (H_1) can be dropped, when the Hamiltonian is symmetric $(H(x) = H(-x))$ (Dell'Antonio, D'Onofrio and Ekeland [5]).

Note that these conditions establish stability of the solution $x(t)$ associated with a minimum of the dual functional introduced by Clarke [6]. The results of the first part of this paper also relate to this solution. It will be shown (Theorem 2.1) that condition (H_1') can be relaxed as follows

$$\frac{\alpha_{2n}(x)}{\alpha_1(x)} < 2 \quad \text{for} \quad x \in M. \qquad (H_2)$$

The second part of the paper deals with stability analysis of Lyapunov families of the periodic solutions emanating from the equilibrium point $x = 0$ of system (1.1) $(H_x(0) = H(0) = 0)$. It is assumed that, for some region Ω in the vicinity of zero, the relation holds,

$$A_- < H_{xx}(x) < A_+, \quad x \in \Omega, \qquad (H_3)$$

where A_- and A_+ are symmetric matrices. In terms of the eigenvalues of the matrices JA_- and JA_+, the corresponding conditions guarantee that Lyapunov family $x_k(t, h)$ $(x_k(t, h) \to 0$ as $h \to 0)$ is uniquely continuable in h and remains stable within Ω (Theorem 3.1). Such approach was utilized first in [7] for convex Hamiltonians $(H_{xx}(x) > 0$ for $x \in \Omega)$; here, it is extended to nonconvex Hamiltonians. (Note that for $n = 2$, some stability criteria feasible also for indefinite Hamiltonians were obtained in [8].)

Using this approach, new stability criteria for the solution $x(t) \in M = H^{-1}(h)$ associated with the minimum of the dual functional are obtained. Namely, it is found that the conditions of stability of the solutions $x_n(t, h)$, corresponding to the largest natural frequency of the linearized system, guarantee also stability of the solution $x(t)$ even though they hold for $x \in M$ only.

Note that conditions (H_1) and (H_2) are satisfied when the surface M is sufficiently close to a sphere. This implies the multipliers $\rho_k \neq 1$ of different Krein types to lie on different open semicircles. The stability criteria obtained under condition (H_3) admit an arbitrary disposition of the multipliers on the unit circle.

2 Stability of a Solution on a Convex Energy Surface

Let M be a convex compact energy surface of the Hamiltonian $H(x)$. In this Section, we consider the solution $x(t) \in M$ associated with the minimum of the dual functional

$$\Psi(v) = \int_0^T \left[\tfrac{1}{2} \left(Iv, Pv \right) + H^*(v) \right] dt, \qquad (2.1)$$

where $v(t)$ is a T-periodic function, Pv is its primitive value of zero mean, and $H^*(v)$ is the Legendre transform of $H(x)$. Clarke [6] showed that, if $v_k(t)$ is a critical point of $\Psi(v)$, then $x_k(t) = JPv_k(t) + \zeta$ is a T-periodic solution of (1.1); conversely, a solution of (1.1), $x_k(t)$, yields a critical point of (2.1), $v_k(t) = -J\dot{x}_k(t)$.

For homogeneous Hamiltonian $H'(x)$ $(H'(sx) = s^\alpha H'(x))$ with $\alpha \in (1, 2)$, functional (2.1) reaches its minimum at $v_*(t) \neq 0$ (Ekeland [2]). This implies the existence of the corresponding T-periodic solution $x'(t)$ and, therefore, existence of the family of $T(s)$-periodic solutions $x(t, s) = sx'(s^{\alpha-2}t)$. If $H'(x)$ has the same energy surface $(H'(M) = h)$, then $x(t, s) \in M$ for some s. It is known that orbits of different Hamiltonians on the same energy surface coincide, whence it follows the existence of the required periodic solution $x(t) \in M$ [2, 6].

Further, we assume that the multiplicity of the unit multiplier of equation (1.2) corresponding to this solution is equal two. (Note that this is a generic property for an individual periodic solution $x(t)$.)

First, let us establish an auxiliary result. Consider the eigenvalue problem

$$\dot{y} = \lambda J A(t) y, \qquad y(0) = y(T). \qquad (2.2)$$

Due to $A(t) > 0$, problem (2.2) is of positive type [9]. It follows that the corresponding eigenvalues λ_k are real; let $\lambda_1, \lambda_2, \ldots$ ($\lambda_k \leq \lambda_{k+1}$) be the positive ones. Since the T-periodic function $\dot{x}(t)$ satisfies equation (1.2), there exists an eigenvalue $\lambda_p = 1$ such that $\dot{x}(t)$ is the related eigenfunction. The following lemma determines the number p of this eigenvalue.

Lemma 2.1 *If functional (2.1) attains its minimum at $v = -J\dot{x}$, then $p = 1$; otherwise, $p = 2$.*

Proof Let $v = -J\dot{x}$ be the minimum of $\Psi(v)$. Then the associated quadratic form

$$Q(w) = \int_0^T [(Jw, Pw) + (w, A(t)^{-1}w)] \, dt \qquad (2.3)$$

is nonnegative [2].

By contradiction, suppose that $p > 1$, then the first positive eigenvalue of problem (2.2) $\lambda_1 < 1$. Let $y_1(t)$ be the corresponding eigenfunction. Setting $w = J\dot{y}_1$ in (2.3) and taking into account that $\dot{y}_1 = \lambda_1 J A(t) y_1$ and $A(t) > 0$, we find

$$Q(w) = (\lambda_1^2 - \lambda_1) \int_0^T (A(t) y_1(t), y_1(t)) \, dt < 0.$$

The contradiction obtained shows that in the case considered $p = 1$.

Now, let us suppose that $v = -J\dot{x}$ is not a minimum of $\Psi(v)$. Consider a family of Hamiltonians $H(x,s)$ such that $H(M,s) = h$ for $s \in [0,1]$, $H(x,1) = H(x)$ and $H(x,0) = H'(x)$, where $H'(x)$ is a homogeneous Hamiltonian of order $\alpha \in (1,2)$. The respective trajectories on M and, therefore, a Poincaré map $G(v)$ do not depend on s. It is known that eigenvalues of the matrix of the partial derivatives $G_v(v_0)$, where v_0 corresponds to the periodic solution considered, coincide with the multipliers $\rho_k \neq 1$ of system (1.2). It follows that the multiplicity of the unit multiplier holds double for $s \in [0,1]$. The related solutions can be taken in the form

$$y_1(t,s) = \dot{x}(t,s), \quad y_2(t,s) = f(t,s) + k(s)t y_1(t,s)), \qquad (2.4)$$

where $f(t,s)) = f(t+T(s),s))$, and $k(s) = 0$ when the respective elementary divisors are simple.

The solutions associated with the rest of the multipliers, ρ_k, can be represented as follows

$$y_k(t,s) = \exp[\nu_k(s) t] f_k(t,s), \quad \nu_k(s) = \frac{\ln \rho_k(s)}{T}, \qquad (2.5)$$

where $f_k(t,s)$ are either periodic functions or, in the case of nonsimple elementary divisors, linear combinations of such functions with polynomial, with respect to t, coefficients [9]. Taking into account expressions (2.4) and (2.5), and also known identity $(Jy_p(t), y_k(t)) \equiv c_{pk} = \text{const}$, we obtain

$$(Jy_1(t,s), y_k(t,s)) = 0, \quad k \neq 2. \qquad (2.6)$$

Since the solutions $y_k(t)$, $k = 1, \ldots, 2n$ are linearly independent, it follows from (2.6) that $(Jy_1, y_2) = (Jy_1, f) \neq 0$. We assume that

$$(Jy_1(t,s), f(t,s)) \equiv 1. \qquad (2.7)$$

Note that this condition determines the value $k(s)$ in (2.4).

For $k(s) \neq 0$, the eigenvalue $\lambda(s) = 1$ is simple; thus, the number $p(s)$ holds. Let us put

$$A_0(t,s) = A(t,s) - k(s)R(t,s), \qquad R(t,s) = Jy_1 y_1^T J, \qquad (2.8)$$

where the index T means transposition. Clearly, the matrix $R(t,s)$ is symmetric ($J^T = -J$).

Due to equality (2.6), we have $R(t,s)y_k(t,s) = 0$ for $k \neq 2$, hence, the solutions $y_k(t,s)$, $k \neq 2$ satisfy also equation (1.2) with $A = A_0(t,s)$. Substituting $y_2(t,s)$ in (1.2) and taking into account that $\dot{y}_1 = JAy_1$, we obtain

$$\dot{f} = JA(t,s)f - k(s)y_1. \qquad (2.9)$$

From (2.7), (2.8) and (2.9), it follows that $f(t,s)$ is a periodic solution of equation (1.2) with $A = A_0(t,s)$. Thus, the multiplicity of the unit eigenvalue of the corresponding problem (2.2) equals two.

Since $R(t,s)y_k(t,s) = 0$ for $k \neq 2$, the matrix $R(t,s)$ has a $(2n-1)$-multiple zero eigenvalue; the last eigenvalue is $\beta_{2n} = -(y_1(t,s), y_1(t,s)) < 0$ with $Jy_1(t,s)$ being the corresponding eigenfunction. Therefore, $R(t,s) \leq 0$, and hence, $A_0(t,s) \leq A(t,s)$ for $k(s) < 0$, whereas $A_0(t,s) \geq A(t,s)$ for $k(s) > 0$. The positive eigenvalues of problem (2.2) with $A > 0$ are decreasing under the increase in A [4]. As a result, $p(s)$ is increased by one when $k(s)$ changes its sign. Having observed that $p(0) = 1$, we obtain $p(1) \leq 2$. Since, by supposition, $v = -J\dot{x}$ is not a minimum of $\Psi(v)$, then $p(1) > 1$, i.e. $p(1) = 2$. The lemma is proved.

The theorem below establishes stability of the solution $x(t)$ under condition (H_2) which is more weak than (H_1).

Theorem 2.1 *Under condition (H_2), the solution $x(t)$ is strongly stable.*

Proof Having set $\tau = t/T$, we consider the eigenvalue problem

$$y' = \lambda JA(\tau)y, \qquad y(0) = y(1), \qquad (2.10)$$

$$A(\tau) = A(\tau + 1) = H_{xx}(x(\tau T)), \qquad ' = \tfrac{d}{d\tau}.$$

Clearly, the corresponding eigenvalues $\lambda'_k = T\lambda_k$, $k = 1, 2, \ldots$, hence, $\lambda'_1 = T$ or $\lambda'_2 = T$ by Lemma 2.1.

Since $\alpha_1(x)I_{2n} \leq H_{xx}(x) \leq \alpha_{2n}(x)I_{2n}$, then

$$A_-(\tau) = \alpha_1(\tau)I_{2n} \leq A(\tau) \leq A_+(\tau) = \alpha_{2n}(\tau)I_{2n}, \qquad (2.11)$$

where $\alpha_1(\tau) = \alpha_1(x(\tau T))$, and $\alpha_{2n}(\tau) = \alpha_{2n}(x(\tau T))$.

For $A = A_-(t)$ and $A = A_+(t)$, the multipliers of the first type of equation (2.10) are [9]

$$\rho_k^-(\lambda) = \exp(i\alpha_1^0 \lambda), \quad \rho_k^+(\lambda) = \exp(i\alpha_{2n}^0 \lambda), \quad k = 1,\ldots,n, \qquad (2.12)$$

$$\alpha_1^0 = \int_0^1 \alpha_1(\tau)\,d\tau, \quad \alpha_{2n}^0 = \int_0^1 \alpha_{2n}(\tau)\,d\tau,$$

whereas the multipliers of the second type are $\rho_{k+n}^- = 1/\rho_k^-$ and $\rho_{k+n}^+ = 1/\rho_k^+$.

By (H_2), $\alpha_{2n}^0/\alpha_1^0 < 2$, so for $\lambda = \lambda^* = \pi/\alpha_1^0 + \epsilon$ under sufficiently small $\epsilon > 0$, the multipliers $\rho_k^-(\lambda)$ and $\rho_k^+(\lambda)$, $k = 1,\ldots,n$ lie on the open lower semicircle. The multipliers of the first type, $\rho_k(\lambda)$, $k = 1,\ldots,n$, of equation (2.10) move along the unit circle counterclockwise with an increase in $A(\tau)$ [4]. Thus, in view of (2.11), the corresponding arguments satisfy the inequality

$$\alpha_1^0 \lambda \leq \arg \rho_k(\lambda) \leq \alpha_{2n}^0 \lambda. \qquad (2.13)$$

Therefore, for $\lambda = \lambda^*$, the multipliers of different types of equation (2.10) lie on different open semicircles. Clearly, $\lambda^* < \lambda_1'$; let us follow the behavior of the multipliers as λ increases. For $\lambda = \lambda_1'$, the vanguard multipliers of different types $\rho_1(\lambda)$ and $\rho_{n+1}(\lambda) = 1/\rho_1(\lambda)$ meet each other at the point $\rho = 1$. If $\lambda_1' = T$, then the multipliers of the first type, $\rho_k(T)$, $1 < k \leq n$, lie on the lower semicircle, so that solution $x(t)$ is strongly stable.

Let $T = \lambda_2' > \lambda_1'$, then the multiplier $\rho(\lambda_1') = 1$ corresponds to a Jordan block of order $m = 2$ (in the case $m > 2$, some of the related multipliers would be located outside the unit circle for $\lambda < \lambda_1'$ [9]). Therefore, further increase of λ shifts the multipliers $\rho_1(\lambda)$ and $\rho_{n+1}(\lambda)$ from the unit circle to the real axis, so that $\rho_1(\lambda) \in (0,1)$, and $\rho_{n+1}(\lambda) \in (1,\infty)$. Let us prove that precisely those multipliers will meet again at the point $\rho = 1$ for $\lambda = T$. Really, otherwise $\rho_1(T) \in (0,1)$, because a multiplier can get either *on* or *off* the interval $(0,1)$ only together with a conjugate multiplier. It is seen from the proof of the Lemma that there exists a homogeneous Hamiltonian with a T'-periodic solution $x'(t) \in M$, for which $\lambda = T'$ is the first eigenvalue; the multipliers for $H(x)$ and $H'(x)$ are identical. Let us follow the behavior of the multipliers as λ decreases from T' to zero. For $A > 0$ and small λ, all the multipliers of the first type lie on the upper semicircle [4]. Therefore, the multiplier $\rho_1(\lambda)$ finds itself at the point $\rho = 1$ for some $\lambda < T'$ what is impossible, because T' is the first eigenvalue of the problem. The contradiction obtained shows that $\rho_1(\lambda) = \rho_{n+1}(\lambda) = 1$ for $\lambda = T = \lambda_2'$. Thus, in this case, the multipliers of different types, $\rho_k(T)$ and $\rho_{k+n}(T)$, $k = 2,\ldots,n$, also lie on different open semicircles. The theorem is completely proved.

Remark 2.1 Taking into account that $\arg \rho_1 = 2\pi$ for $\lambda = T$, we obtain $\alpha_1^0 T \leq 2\pi \leq \alpha_{2n}^0 T$ due to inequality (2.13). Therefore, the period of the solution considered satisfies the following bilateral bounds,

$$\frac{2\pi}{\alpha_+} \leq T \leq \frac{2\pi}{\alpha_-}. \tag{2.14}$$

In particular, if the strong inequality is true in (H_1), one can take $A_- = 2R^{-2}I_{2n}$ and $A_+ = 2r^{-2}I_{2n}$, then $\alpha_- = 2R^{-2}$, $\alpha_+ = 2r^{-2}$, and, therefore, $\pi r^2 < T < \pi R^2$.

3 Stability of Lyapunov Families of Periodic Solutions

Let us suppose that $x = 0$ is an equilibrium point of system (1.1) ($H_x(0) = H(0) = 0$) and inequality (H_3) with some symmetric matrices A_- and A_+ is true for $x \in \Omega$, where Ω is a compact region ($0 \in \Omega$). If the frequency ω_k^0 of the linearized equation $\dot{x} = JH_{xx}(0)x$ satisfies the nonresonance condition $\omega_k^0 \neq \omega_p^0/m$, $p \neq k$, $m = 1, 2, \ldots$, then, according to the Lyapunov theorem [10], for sufficiently small h, there exists a family of the $T_k(h)$-periodic solutions $x_k(t, h)$ such that $x_k(t, h) \to 0$, $T_k(h) \to 2\pi/\omega_k^0$, as $h \to 0$.

It was shown in [11] that, under some condition involving the eigenvalues of the matrices JA_- and JA_+, the family $x_k(t, h)$ is uniquely continuable in h within Ω; so that $x_k(t, h) \in \Omega$ for $0 < h < h_k$ and $x_k(t_k, h_k) \in \partial\Omega$ for some t_k and h_k. Our immediate aim is to find out the additional conditions, which guarantee stability of $x_k(t, h)$ for $0 < h \leq h_k$.

Let us assume that the eigenvalues of the matrix $JA(\lambda)$, where $A(\lambda) = A_- + \lambda(A_+ - A_-)$, are purely imaginary for $\lambda \in [0, 1]$; denote them $\pm i\omega_p(\lambda)$ ($p = 1, \ldots, n$; $\omega_p(\lambda) > 0$); let $x_p(\lambda)$ be the corresponding eigenfunctions ($i\omega_p(\lambda)x_p(\lambda) = JA(\lambda) \times x_p(\lambda)$). Having assigned a period T to the matrix $A(\lambda)$, we denote the multipliers of the equation $\dot{x} = JA(\lambda)x$ as

$$r_p(\lambda, T) = \exp[i\omega_p(\lambda)T] \quad \text{and} \quad r_{p+n}(\lambda, T) = r_p^*(\lambda) = \exp[-i\omega_p(\lambda)T]. \tag{3.1}$$

According to the classification of Krein [4], $r_p(\lambda, T)$ is either of the first or second type if, respectively, either $l_p = i(x_p, Jx_p) > 0$ or $l_p < 0$. As known [9], a type of the multiplier holds, as the matrix A is changing; the multipliers $r_p(\lambda, T)$ and $r_p^*(\lambda, T)$ are of different types. If A_- and, therefore, A_+ are positive definite, then $r_p(\lambda, T)$, $p = 1, \ldots, n$ are of the first type.

Since $A(\lambda)$ increases with λ, the multipliers of the first (second) type are moving counterclockwise (clockwise) along the unit circle [4], i.e., the values $\omega_p(\lambda)$ are monotonically changing. We denote $\omega_p^- = \min[\omega_p(0), \omega_p(1)]$ and $\omega_p^+ = \max[\omega_p(0), \omega_p(1)]$. With no loss of generality, we assume that $\omega_p^- \leq \omega_{p+1}^-$.

For $\lambda \in [0, 1]$ and $T \in [2\pi/\omega_k^+, 2\pi/\omega_k^-]$, the multipliers $r_p(\lambda, T)$ and $r_{p+n}(\lambda, T) = r_p^*(\lambda, T)$, $p = 1, \ldots, n$ lie on the arcs, $G_p = [\exp(i\phi_p^-), \exp(i\phi_p^+)]$ and $G_{p+n} = G_p^*$, of the unit circle, where $\phi_p^- = 2\pi\omega_p^-/\omega_k^+$ and $\phi_p^+ = 2\pi\omega_p^+/\omega_k^-$.

Let us suppose that

$$G_p \cap G_q = \emptyset \quad \text{for} \quad p, q \neq k, k+n, \qquad (H_4)$$

when the corresponding multipliers $r_p(\lambda, T)$ and $r_q(\lambda, T)$ are of different types.

If $k < n$, condition (H_4) implies that $\omega_p^+/\omega_p^- < 2$ for $p \geq k$; otherwise, $1 \in G_p \cap G_p^*$ for $p > k$. If $k = n$, we assume *a priori* that

$$\frac{\omega_n^+}{\omega_n^-} < 2. \qquad (H_5)$$

Moreover, we assume that, on the upper semicircle, the arc G_k may have common points only with arcs G_p associated with the multipliers of one type only.

Note that condition (H_4) implies that the arcs G_p and G_q, $p, q \neq k, k+n$ corresponding to multipliers of different types, have no common points (to check it, it is sufficient to consider the upper semicircle only). The rest conditions guarantee that the interval between such arcs is not completely overlapped by the arc G_k.

From condition (H_4) it, in particular, follows that $1 \notin G_p$, $p \neq k$, so

$$\omega_k(\lambda_1) \neq \frac{\omega_p(\lambda_2)}{m} \quad \text{for} \quad \lambda_1, \lambda_2 \in [0, 1], \quad m = 1, 2 \ldots . \qquad (3.2)$$

As shown in [11], under condition (3.2), the family $x_k(t, h)$ is uniquely continuable in h within Ω and the corresponding period $T_k(h) \in (2\pi/\omega_k^+, 2\pi/\omega_k^-)$.

Theorem 3.1 *The solutions $x_k(t, h) \in \Omega$ are strongly stable.*

Proof The multipliers of the linearized equation $\rho_p^0 = \exp(2\pi i \omega_p^0/\omega_k^0) \in G_p$, hence, $x_k(t, h)$ is stable for small h. Suppose that $x_k(t, h_1) \in \Omega$ is unstable for some h_1. Under condition (3.2), the multiplicity of the unit multiplier equals two [11], so that no multiplier can leave the unit circle at the point $\rho = 1$. Therefore, for some $h_2 \leq h_1$, the multipliers of the different types, ρ_p and ρ_q, $p, q \neq k, k+n$, of equation (1.2) meet each other at some point of the unit circle. As was mentioned above, there are points between the arcs, G_p and G_q, where no multiplier, $r_p(\lambda, T)$, $p = 1, \ldots, 2n$, finds itself for $\lambda \in [0, 1]$ and $T \in [2\pi/\omega_k^+, 2\pi/\omega_k^-]$. Therefore, one of the multipliers, either ρ_p or ρ_k, lies at such a point ρ_* for some $h_* \leq h_2$; let $y_*(t, h_*)$ be the corresponding solution $(y_*(t + T_*, h_*) = \rho_* y_*(t, h_*))$, where $T_* = T_k(h_*)$.

Consider the eigenvalue problem

$$\dot{y} = J[A_- + \lambda(R(t) - A_-)]y, \quad y(T_*) = \rho_* y(0), \qquad (3.3)$$

where $R(t) > A_-$. Due to this inequality, the problem is of positive type [9]; let $\lambda_1 \leq \lambda_2 \leq \ldots$ be the positive eigenvalues. By supposition, for $\lambda = 1$ and $R = A_*(t) = A(x_k(t, h_*))$, equation (1.2) has the solution $y_*(t, h_*)$ satisfying boundary condition (3.3); hence, $\lambda = 1$ is the corresponding eigenvalue.

The positive eigenvalues are decreasing with an increase in $R(t)$ [4]; therefore, for $R(t) = A_+ > A_*$, problem (3.3) has an eigenvalue $\lambda_p \in (0,1)$, hence, the equation $\dot{y} = JA(\lambda_p)y$ has the multiplier ρ_* for $T = T_*$. Meanwhile, as shown above, no multiplier of this equation finds itself at the point ρ_* for $\lambda \in [0,1]$ and $T \in [2\pi/\omega_k^+, 2\pi/\omega_k^-]$. The contradiction obtained shows that the multipliers of different types cannot meet on the unit circle, when $x_k(t,h) \in \Omega$. The theorem is proved.

4 Discussion

The conditions of Theorem 3.1 involve the eigenvalues of the matrices A_- and A_+. In the general case, A_- and A_+ can be obtained as follows. Let $(H_2 x, x)$ be the quadratic part of $H(x)$, then $H_2 + \beta_- I_{2n} \leq H_{xx}(x) \leq H_2 + \beta_+ I_{2n}$, where β_- and β_+ are, respectively, the smallest and largest eigenvalues of the matrix $H_{xx}(x) - H_2$ for $x \in \Omega$. So one can set $A_- = H_2 + (\beta_- - \epsilon)I_{2n}$ and $A_+ = H_2 + (\beta_+ + \epsilon)I_{2n}$, where $\epsilon > 0$ is an arbitrary small value. Note that $\beta_-(\Omega), \beta_+(\Omega) \to 0$ as Ω goes to 0, so that ω_k^- and ω_k^+ become close to ω_k^0. Therefore, if the linearized system $\dot{x} = JH_{xx}(0)x$ is strongly stable, one can find a finite region, Ω_k, such that the solutions $x_k(t,h)$ is also strongly stable as long as $x_k(t,h) \in \Omega_k$.

To illustrate the use of Theorem 3.1, let us obtain some stability conditions for the family $x_n(t,h)$ corresponding to the largest natural frequency ω_n^0 of the linearized system. For the sake of certainty, we assume that the Hamiltonian is convex; then the matrix A_- can be taken as positive definite in (H_3), so that the multipliers $r_p(\lambda, T)$ for $p \leq n$ are of the first type, and $\omega_k(0) = \omega_k^-$ and $\omega_k(1) = \omega_k^+$.

Let us suppose that $\arg r_p(0, 2\pi/\omega_n^+) = \phi_p^- = 2\pi\omega_p^-/\omega_n^+ > \pi$ for $p < n$. Clearly, if $\phi_p^+ = 2\pi\omega_p^+/\omega_n^- < 2\pi$, then the arcs, G_p, $p < n$ and G_{p+n}, corresponding to the multipliers of the first and second type lie, respectively, on the lower and upper semicircles, so that condition (H_4) holds. Thus, under condition (H_5), the solutions $x_n(t,h) \in \Omega$ are certainly stable, provided that

$$\omega_p^- > \frac{\omega_n^+}{2}, \quad \omega_p^+ < \omega_n^-, \tag{H_6}$$

Analogously, if $\phi_p^+ < \pi$ for $p < n$, then the arcs G_p, $p < n$ and G_{p+n} lie, respectively, on the upper and lower semicircles, so the stability of $x_n(t,h) \in \Omega$ is guaranteed by the inequality

$$\omega_p^+ < \frac{\omega_n^-}{2}. \tag{H_7}$$

For $n = 2$, the stability conditions are reduced to the above inequalities.

For $n = 3$, stability is also possible, when there are multipliers of different types on each of the semicircles. Let $\phi_1^- < 2\pi - \phi_2^- < \pi$, then G_1 and G_4 lie on the upper semicircle. The condition $G_1 \cap G_4 = \emptyset$ is satisfied, if $\phi_1^+ < 2\pi - \phi_2^+$. The interval between G_1 and G_4 is not overlapped by G_3, if $2\pi - \phi_2^+ > \phi_3^+ = 2\pi\omega_3^+/\omega_3^- - 2\pi$. Therefore, the solutions $x_n(t,h) \in \Omega$ are stable if

$$\omega_1^+ + \omega_2^+ < \omega_3^-, \quad \omega_2^+ + \omega_3^+ < 2\omega_3^-. \tag{H_8}$$

If $2\pi - \phi_2^- < \phi_1^- < \pi$, the required conditions hold when $\phi_1^+ < \pi$, $\phi_3^+ < \phi_1^-$, and $\phi_2^+ < 2\pi$, that implies the following stability conditions

$$\omega_1^- + \omega_2^- > \omega_3^+, \quad \omega_1^+ < \frac{\omega_3^-}{2}, \quad \frac{\omega_3^+}{\omega_3^-} - \frac{\omega_1^-}{\omega_3^+} < 1, \quad \omega_2^+ < \omega_3^-. \tag{H_9}$$

In both cases, the multipliers of different types, ρ_1 and ρ_4, of equation (1.2) lie on the upper semicircles; however, $\arg \rho_1 < \arg \rho_4$ and $\arg \rho_1 > \arg \rho_4$ under conditions (H_8) and (H_9), respectively.

To this end, let us consider again stability of a solution lying on the convex compact energy surface M bounding the region Ω. If inequality (H_3) is true for $x \in \Omega$, then the above conditions guarantee the stability of the solutions $x_n(t,h) \in \Omega$ and, thereby, existence of the stable solution $x(t,h_n) \in M$. Suppose now that (H_3) is true for $x \in M$ only; let us show that the same conditions guarantee stability of the periodic solution $x(t) \in M$ associated with the minimum of dual functional (2.1).

Clearly, in the corresponding eigenvalue problem (2.10), $A_- < A(\tau,s) < A_+$, so that we have

$$\lambda_1^+ = \lambda_2^+ = \frac{2\pi}{\omega_n^+} < \lambda_1, \quad \lambda_2 < \lambda_1^- = \lambda_2^- = \frac{2\pi}{\omega_n^-}.$$

Taking into account that the period of the solution considered $T = \lambda_1$ or $T = \lambda_2$, we find that $T \in (2\pi/\omega_n^+, 2\pi/\omega_n^-)$.

Let $\lambda_k(s)$, $k = 1, 2, \ldots$, be the positive eigenvalues of problem (3.3) with $T_* = T$, $\rho_* = 1$ and $R = R(t,s) = A_+ + s[A(t) - A_+]$; by Lemma 2.1, $\lambda_1(1) = 1$ or $\lambda_2(1) = 1$ $(R(t,1) = A(t))$. Setting $\lambda \equiv \lambda_1(s)$ in equation (3.3), let us follow the behavior of the corresponding multipliers, as s increases on $[0,1]$.

Clearly, $\rho_p(0) \in G_p$, $\rho_n(s) = \rho_{2n}(s) = 1$ $(\lambda_1(s) = 1)$; as it is seen from the proof of Theorem 3.1, for $s \in [0,1]$, the multiplier ρ_p cannot leave the arc G_p. Therefore, for $s = 1$, equation (3.3) is stable that concludes the proof in the case $\lambda_1(1) = 1$. Otherwise, we put $s \equiv 1$ and increase λ from λ_1 to 1. As it is seen from the proof of Lemma 2.1, the multipliers ρ_n and ρ_{2n} are shifting along the real axis and meeting again at the point $\rho = 1$ for $\lambda = 1$.

Note that (H_4) requires, in particular, that $1 \notin G_p$, $p \neq n, 2n$ (otherwise $1 \in G_p \cap G_{p+n}$). It appears that this condition is no longer necessary for stability of the solution considered.

Really, suppose that $\rho_p(s_1) = \rho_{p+n}(s_1) = 1$, $p \neq n$ for some $s_1 < 1$; generically, the multiplicity of the multiplier $\rho(s_1) = 1$, $m = 4$. As it is seen from the proof of Lemma 2.1, in view of $\lambda \equiv \lambda_1(s)$, the multipliers $\rho_p(s)$ and $\rho_{p+n}(s)$ cannot be shifted on the real axis as s increases. Therefore, for $s = 1$, they lie on the unit circle, and $\rho_p(1)$, $\rho_{p+n}(1) \neq 1$; by supposition, $m(1) = 2$.

In particular, the last inequality could be dropped in (H_9) provided that $1 \notin G_2$.

References

[1] Poincaré, H. (1892). *Les Méthodes Nouvelles de la Mécanique Céleste*, Vol. 1. Gauthier-Villars, Paris.
[2] Ekeland, I. (1990). *Convexity Methods in Hamiltonian Mechanics*. Springer, Berlin.
[3] Dell'Antonio, G.F. (1994). Variational calculus and stability of periodic solutions of a class of Hamiltonian systems. *Rev. Math. Phys.*, **6**(5a), 1187–1232.
[4] Krein, M.G. (1955). Principles of the theory of λ-zones of stability of a canonical system of linear differential equations with periodic coefficients. *In Memory of A.A. Andronov*. Izd. Akad. Nauk SSSR, Moscow, 413–498 (Russian).
[5] Dell'Antonio, G.F., D'Onofrio, B. and Ekeland, I. (1992). Les systemes convexes et paires ne sont pas ergodiques en general. *Comptes Rendues Acad. Sci.*, Paris, **315**, 1123–1132.
[6] Clarke, F. (1979). A classical variational principle for periodic Hamiltonian trajectories. *Proc. Amer. Math. Soc.*, **76**, 186–189.
[7] Zevin, A.A. (1986). Non-local criteria for the existence and stability of periodic oscillations in autonomous Hamiltonian systems. *Prikl. Mat. Mekh.*, **50**(1), 45–51 (Russian).
[8] Zevin, A.A. (1998). Stability criteria for periodic solutions of autonomous Hamiltonian systems with two degrees of freedom. *Nonlinearity*, **11**, 19–26.
[9] Yakubovitch, V.A. and Starzhinsky, V.M. (1980). *Linear Differential Equations with Periodic Coefficients*. Wiley, New York.
[10] Lyapunov, A.M. (1966). *Stability of Motion*. Academic Press, New York.
[11] Zevin, A.A. (1997). Nonlocal generalization of the Lyapunov theorem. *Nonlin. Anal. TMA*, **28**(9), 1499–1507.

Part 4
SELECTED APPLICATIONS

4.1 STABILITY IN MODELS OF AGRICULTURE–INDUSTRY–ENVIRONMENT INTERACTIONS*

H.I. FREEDMAN[1†], M. SOLOMONOVICH[1‡#], L.P. APEDAILE[2§] and A. HAILU[3]

[1] *Department of Mathematical Sciences, University of Alberta, Edmonton, Alberta, Canada*
[2] *Peer Diagnostics Ltd., Edmonton, Alberta, Canada*
[3] *Department of Rural Economy, University of Alberta, Edmonton, Alberta, Canada*

1 Introduction

In previous work, the interactions between agricultural wealth, industrial wealth, and the environment have been modelled and analyzed. In Apedaile, *et al.* [1], the environment was held constant and only direct interactions between agriculture and industry were considered. In Solomonovich, *et al.* [2] the environment was introduced with a minimum threshold. In Solomonovich, *et al.* [3], the environment was allowed to degenerate or recover, either through farming, adding of nutrient, or through natural causes.

Previously, wealth was measured in dollars, where a dollar value to environmental quality was assigned. However, it is believed that total assets, rather than dollars, would be a better measure of agricultural and industrial wealth. As to the environment, a better measure of quality would be a pure number between zero and one given by (yield level – catastrophic level)/(maximum possible yield level – catastrophic level). This causes several fundamental changes in the model.

---

*Advances in Stability Theory (Ed.: A.A. Martynyuk). Stability and Control: Theory, Methods and Applications, Taylor & Francis, London, **13** (2003) 255–265.
†Research partially supported by the Natural Sciences and Engineering Research Council of Canada, Grant No. NSERC OGP 4823
‡Research partially supported by the Pacific Institute of Mathematical Sciences and the Pacific Institute for the Mathematical Sciences
#Research partially supported by the Canadian Rural Restructuring Foundation
§Research partially supported by the Alberta Agriculture Initiatives Program

The first change is that, whereas in the previous models agricultural and industrial wealth influence on their growth must of necessity saturate, here that is no longer true. This is because now the land is part of the agricultural assets, and for all practical purposes, through the purchase of more land, agricultural wealth may increase significantly. As to industry, it may acquire machinery, property, etc. which increases its assets.

The importance of the environment in economic growth is given in Smulders [4]. Since now, under the assumption that a catastrophic environment cannot recover on its own, the equation for the change in environment no longer will have either thresholds or source terms.

The organization of the paper is as follows. In the next section we develop our modes. In Section 3, we discuss the agriculture-industry system (constant environment), followed by the agriculture environment submodel in Section 4. The last section will contain a brief discussion of our results.

We leave the analysis of the full three dimensional model to a future paper.

2 The Models

2.1 Two dimensional agriculture-industry model

It is assumed throughout that agricultural assets may increase and decrease both in the absence of the particular industry and due to interaction with industry. In the absence of agricultural interactions, the industrial assets will decline, but interaction with agriculture would have a positive feedback on the industrial assets. This leads to a model of the form

$$\dot{A} = \alpha A - \beta A^2 + \gamma AI, \quad A(0) = A_0 \geq 0 \quad (1a)$$

$$\dot{I} = -\xi I - \eta I^2 + \delta AI, \quad I(0) = I_0 \geq 0, \quad (1b)$$

$$\alpha, \beta, \xi, \eta, \delta > 0.$$

Here A represents the (scaled) assets of agriculture and I the (scaled) assets of industry.

The terms $\alpha A - \beta A^2$ represent the growth of agricultural wealth in the absence of industrial interaction, and, of course, is limited. The term γAI represents the net rate of change of agricultural wealth due to interaction of A with I. γ can be < 0, 0 or > 0.

In the case that $\gamma < 0$, then industrial influence causes a net decrease in agricultural assets. Unfortunately, this is all too often the case [1]. If $\gamma = 0$, then the cost of dealing with industry exactly equals any benefits obtained. However, if $\gamma > 0$, then there is a net benefit to agriculture. Of course soil, etc; may recover on its own [5].

For the I equation, the terms $-\xi I - \eta I^2$ represent the reduction rate of industrial assets in the absence of agricultural interactions. However interacting with agriculture will give rise to a positive change in industrial wealth as given by the term δAI. For an interpretation of the constants (with $\gamma - \delta$ replaced by γ here), we refer the reader to [1].

The above model (1) is a Lotka-Volterra system (see Freedman [6]). If $\gamma < 0$, it is of predator-prey type and if $\gamma > 0$ is a cooperative system.

2.2 Three dimensional model

We now modify system (1) so as to include environmental influence on agricultural wealth, both without and with industrial interactions. We also include an equation for the change of environmental quality.

This leads to a model of the form

$$\dot{A} = \alpha EA - \beta A^2 + \gamma EAI - \theta_1(1 - E)A \tag{2a}$$

$$\dot{I} = -\xi I - \eta I^2 + \delta AI, \tag{2b}$$

$$\dot{E} = \varepsilon E(e - E) - \nu EA + \theta_2(1 - E)A, \tag{2c}$$

$$A(0) = A_0 \geq 0, \quad I(0) = I_0 \geq 0, \quad E(0) = E_0, \quad 0 \leq E_0 \leq 1,$$

$$\alpha, \beta, \delta, \xi, \eta > 0, \quad \varepsilon, \theta_1, \theta_2 \geq 0.$$

The term $\theta_1(1 - E)A$ represents the cost rate of agricultural wealth to improve the environment, whereas $\theta_2(1 - E)A$ represents the corresponding rate of quality increase of the environment. The term $\varepsilon E(e - E)$ represents the ability of the environment to restore itself, in the absence of agricultural activity, to its natural wild state (see [6]), e, where $0 \leq e \leq 1$.

We note that since $\dot{E}\|_{E=0} = \theta_2 A \geq 0$ and $\dot{E}\|_{E=1} = \varepsilon(e - 1) - \nu A \leq 0$, then $0 \leq E_0 \leq 1$ implies that $0 \leq E(t) \leq 1$, $t \geq 0$.

2.3 Agriculture-environment submodel

If interaction with industry is not present, then model (2) becomes

$$\dot{A} = \alpha EA - \beta A^2 - \theta_1(1 - E)A \tag{3a}$$

$$\dot{E} = \varepsilon E(e - E) - \nu EA + \theta_2(1 - E)A. \tag{3b}$$

This would represent interactions between self-sustaining agricultural groups with the environment.

3 Agriculture-Industry Model

In this section we analyze model (1). There are three possible equilibria which are of the forms $F_0(0,0)$, $F_1(\frac{\alpha}{\beta},0)$ and $\widehat{F}(\widehat{A},\widehat{I})$, where $\widehat{A}, \widehat{I} > 0$. Clearly F_0 and F_1 exist. Setting the right hand sides of (1) equal to zero and solving, gives

$$\widehat{A} = \frac{\alpha\eta - \gamma\xi}{\beta\eta - \gamma\delta}, \quad \widehat{I} = \frac{\alpha\delta - \beta\xi}{\beta\eta - \gamma\delta}.$$

Hence in order for $\widehat{A}, \widehat{I} > 0$, we need the inequalities

$$(\alpha\eta - \gamma\xi)(\beta\eta - \gamma\delta) > 0, \quad (\alpha\delta - \beta\xi)(\beta\eta - \gamma\delta) > 0 \qquad (4)$$

to be satisfied. We consider the cases $\gamma \leq 0$, and $\gamma > 0$.

(i) $\gamma \leq 0$. In this case, $\beta\eta - \gamma\delta > 0$ automatically holds. Then $\alpha\eta - \gamma\xi > 0$ also. Hence inequalities (4) reduce to

$$\alpha\delta > \beta\xi. \qquad (5)$$

(ii) $\gamma > 0$. There are two subcases.

(iia): Inequality (5) holds. Then we require

$$0 < \gamma < \min\left\{\frac{\beta\eta}{\delta}, \frac{\alpha\eta}{\xi}\right\}. \qquad (6)$$

(iib): $$\alpha\delta < \beta\xi. \qquad (7)$$
Then we require

$$\gamma > \max\left\{\frac{\beta\eta}{\delta}, \frac{\alpha\eta}{\xi}\right\}. \qquad (8)$$

As a consequence, we have the following theorem.

Theorem 3.1 *System (1) has an equilibrium in the positive $A - I$ plane provided one of the following hold: (Figure 3.1).*

(i) $\gamma \leq 0$, $\alpha\delta > \beta\xi$;
(ii) $\alpha\delta > \beta\xi$, $0 < \gamma < \min\{\frac{\beta\eta}{\delta}, \frac{\alpha\eta}{\xi}\}$;
(iii) $\alpha\delta < \beta\xi$, $\gamma > \max\{\frac{\beta\eta}{\delta}, \frac{\alpha\eta}{\xi}\}$.

We now obtain the stability of the three equilibria when they exist.
Let M be the variational matrix of system (1) about an equilibrium. Then

$$M = \begin{bmatrix} \alpha - 2\beta A + \gamma I & \gamma A \\ \delta I & -\xi - 2\eta I + \delta A \end{bmatrix}.$$

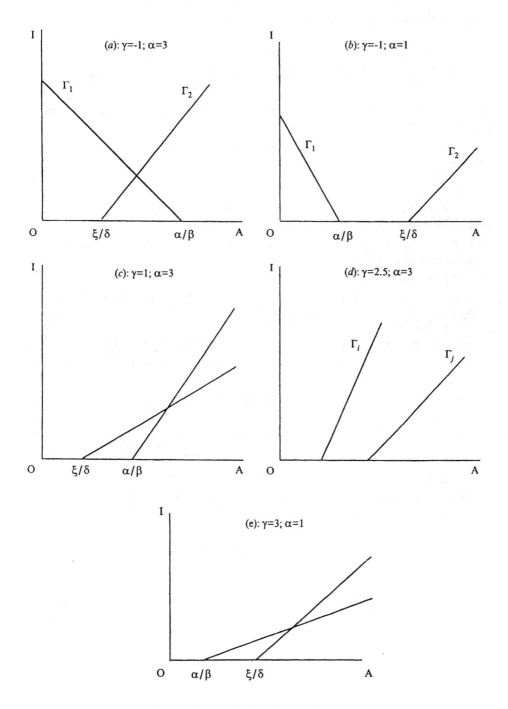

Figure 3.1. Modes of a steady-state.

Using similar notation as for the equilibria, we get

$$M_0 = \begin{bmatrix} \alpha & 0 \\ 0 & -\xi \end{bmatrix}, \quad M_1 = \begin{bmatrix} -\alpha & \frac{\alpha\gamma}{\beta} \\ 0 & \frac{\alpha\delta}{\beta} - \xi \end{bmatrix}, \quad \widehat{M} = \begin{bmatrix} -\beta\widehat{A} & \gamma\widehat{A} \\ \delta\widehat{I} & -\eta\widehat{I} \end{bmatrix}.$$

Clearly F_0 is a saddle point, whereas F_1 is a saddle point if (5) holds and asymptotically stable if (7) holds.

To determine the eigenvalues of $\widehat{M}$, we consider its characteristic equation which reduces to

$$\lambda^2 + (\beta\widehat{A} + \eta\widehat{I})\lambda + (\beta\eta - \gamma\delta)\widehat{A}\widehat{I} = 0. \tag{9}$$

In the most likely case that $\beta\eta - \gamma\delta > 0$, there are no eigenvalues with positive real parts, so that $\widehat{F}$ is locally asymptotically stable.

If, however, $\beta\eta - \gamma\delta < 0$, there is exactly one positive and one negative eigenvalue, and $\widehat{F}$ is a saddle point.

The following are clear.

(i) If $\gamma > 0$, $\frac{\alpha}{\beta} < \frac{\xi}{\delta}$ and $\frac{\beta}{\gamma} \leq \frac{\delta}{\eta}$, $A_0 > 0$, then $\lim_{t\to\infty} (A(t), I(t)) = (\frac{\alpha}{\beta}, 0)$.

(ii) If $\gamma > 0$, $\frac{\alpha}{\beta} < \frac{\xi}{\delta}$ and $\frac{\beta}{\gamma} > \frac{\delta}{\eta}$, then $\widehat{F}$ exists.

Further, there is a separatrix curve such that all solutions initiating beneath the curve with $A_0 > 0$ approach $(\frac{\alpha}{\beta}, 0)$ and all solutions initiating above the curve with $A_0, I_0 > 0$ become unbounded.

(iii) If $\gamma > 0$, $\frac{\alpha}{\beta} > \frac{\xi}{\delta}$ and $\frac{\beta}{\gamma} \leq \frac{\delta}{\eta}$, then all solutions with $A_0 > 0$, $I_0 > 0$ become unbounded.

(iv) If $\gamma > 0$, $\frac{\alpha}{\beta} > \frac{\xi}{\delta}$ and $\frac{\beta}{\gamma} > \frac{\delta}{\eta}$, then if $A_0, I_0 > 0$, $\lim_{t\to\infty} (A(t), I(t)) = (\widehat{A}, \widehat{I})$.

Finally, we consider the case that $\gamma \leq 0$. From Figure 3.1a, b, clearly if $\gamma \leq 0$ and $\frac{\xi}{\delta} \geq \frac{\alpha}{\beta}$, all positive solutions approach $(\frac{\alpha}{\beta}, 0)$.

Hence we consider the final possibility, namely $\gamma \leq 0$, $\frac{\xi}{\delta} < \frac{\alpha}{\beta}$. Then the unique equilibrium $\widehat{F}$ exists. If we choose

$$V(A, I) = u\left(A - \widehat{A} - \widehat{A}\ln\frac{A}{\widehat{A}}\right) + v\left(I - \widehat{I} - \widehat{I}\ln\frac{I}{\widehat{I}}\right),$$

where $u, v > 0$, then $V(A, I)$ is a positive definite function about $\widehat{F}$.

If we compute the derivative of V along solutions of (1) and simplify, we get

$$\dot{V}(A, I) = -\beta u(A - \widehat{A})^2 + (\gamma u + \delta v)(A - \widehat{A})(I - \widehat{I}) - \eta v(I - \widehat{I})^2.$$

If $\gamma < 0$, choose $u = \delta$ and $v = -\gamma$ and get

$$\det V(A, I) = -\beta\delta(A - \widehat{A})^2 + \gamma\eta(I - \widehat{I})^2 < 0 \quad \text{for all} \quad A, I > 0.$$

If $\gamma = 0$, then $\dot{V}(A, I) = -\beta u(A - \widehat{A})^2 + \delta v(A - \widehat{A})(I - \widehat{I}) - \eta v(I - \widehat{I})$. If we choose $v = 1$ and $u > \frac{\delta^2}{4\beta\eta}$, then $\dot{V}$ is again negative definite.

In either case the Liapunov Theory, tells us that $\widehat{\Gamma}$ is globally stable.

4 Agriculture-Environment Submodel

In this section we consider the agriculture-environment submodel obtained from system (2) by setting $\dot{I} = I = 0$. Such a submodel could be construed as helping to understand how self-sufficient farm communities (e.g. Amish, Hutterites, etc.) interact with their environments.

This model takes the form

$$\dot{A} = \alpha E A - \beta A^2 - \theta_1(1 - E)A \tag{10a}$$

$$\dot{E} = \varepsilon E(e - E) - \nu E A + \theta_2(1 - E)A, \tag{10b}$$

$$A(0) = A_0 \geq 0, \quad E(0) = E_0 \geq 0.$$

The equilibria for this system are obtained by solving the equations

$$A[\alpha E - \beta A - \theta_1(1 - E)] = 0 \tag{11a}$$

$$\varepsilon E(e - E) - \nu E A + \theta_2(1 - E)A = 0. \tag{11b}$$

Clearly from (11), $G_0(0,0)$ and $G_1(0,e)$ are equilibria.

To examine the feasibility of a positive equilibrium of the form $\widetilde{G}(\widetilde{A}, \widetilde{E})$, where $\widetilde{A} > 0$, $\widetilde{E} > 0$, write (11a) as

$$A = \beta^{-1}[\alpha E - \theta_1(1 - E)] \tag{12}$$

and substitute into (11b). After simplifying, we get that E must satisfy the quadratic equation

$$[\beta\varepsilon + (\alpha + \theta_1)(\nu + \theta_2)]E^2 - [\beta\varepsilon e + (\alpha + \theta_1)\theta_2 + \theta_1(\nu + \theta_2)]E + \theta_1\theta_2 = 0. \tag{13}$$

Equation (13) has real roots (which must be nonnegative) if and only if the discriminant is nonnegative, or on simplifying, if and only if

$$[\beta\varepsilon e - (\nu\theta_1 + \alpha\theta_2 + 2\theta_1\theta_2)]^2 - 4\beta\varepsilon\theta_1\theta_2 \geq 0. \tag{14}$$

Then one must solve for the $\widetilde{E}$ values, substitute into (12) and check to see if $\widetilde{A} > 0$.

Condition (14), though straightforward to check, does not lead to any insight as to when $\widetilde{G}$ exists. It is, however, possible to obtain a sufficient condition for $\widetilde{G}$ geometrically.

Let Δ_1 be the graph of

$$A = \beta^{-1}[(\alpha + \theta_1)E - \theta_1] \tag{15}$$

obtained by setting (11a) equal to zero. Let Δ_2 be the graph of

$$A = \frac{\varepsilon E(e - E)}{(\nu + \theta_2)E - \theta_2} \tag{16}$$

obtained by setting (11b) equal to zero.

Δ_1 is a straight line with E intercept at $\frac{\theta_1}{\alpha+\theta_1}$ and slope $\frac{\alpha+\theta_1}{\beta}$ (see Figure 4.1a). To discuss Δ_2, we first note that equation (16) can be written as

$$A = -\frac{\varepsilon E}{\nu + \theta_2} + \frac{\varepsilon[\nu e + \theta_2(e - 1)]}{(\nu + \theta_2)^2} - \frac{\varepsilon\theta_2[\nu e + \theta_2(e - 1)]}{(\nu + \theta_2)^2[(\nu + \theta_2)E - \theta_2]} \tag{17a}$$

when $\frac{\theta_2}{\nu + \theta_2} \neq e$,

$$A = \frac{-\varepsilon E}{\nu + \theta_2}, \quad \text{when} \quad \frac{\theta_2}{\nu + \theta_2} = e. \tag{17b}$$

From (17) it is easy to see that in either case, the graph of Δ_2 asymptotically approaches a straight line with negative slope as $E \to +\infty$. Of course, we are only interested in the case $0 \leq E \leq 1$. Note that $\nu e + \theta_2(e - 1) = 0$ if $e \lesseqgtr \frac{\theta_2}{\nu+\theta_2}$, respectively.

We now construct Δ_2. There are three cases to consider.

(i) $e < \frac{\theta_2}{\nu+\theta_2}$. In this case Δ_2 is given in Figure 4.1b. Clearly from Figures 4.1a and 4.1b, $\widetilde{G}$ clearly exists in this case if $\frac{\theta_1}{\alpha+\theta_1} < e$.

(ii) $e = \frac{\theta_2}{\nu+\theta_2}$. In this case Δ_2 is given in Figure 4.1c (when $E \neq e$). Clearly $\widetilde{G}$ does not exist in this case.

(iii) $e > \frac{\theta_2}{\nu+\theta_2}$. In this case Δ_2 is given as in Figure 4.1d. Here one can easily see that $\widetilde{G}$ exists if and only if $\frac{\theta_1}{\alpha+\theta_1} < e$.

The above is summarized in the following theorem.

Theorem 4.1 *System (10) has a unique positive equilibrium in the $E - A$ plane provided $e \neq \frac{\theta_2}{\nu+\theta_2}$ and*

$$\frac{\theta_1}{\alpha + \theta_1} < e. \tag{18}$$

In the case that $e > \frac{\theta_2}{\nu+\theta_2}$, condition (18) is also necessary.

We now investigate the stability of the equilibria, and this leads to yet another set of criteria for $\widetilde{G}$ to exist.

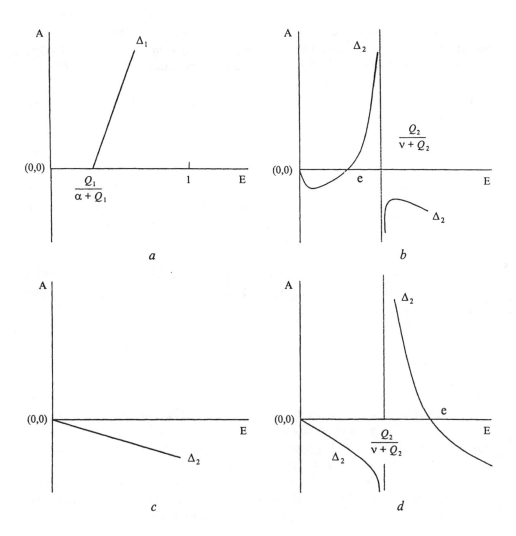

Figure 4.1. Modes of a steady-state.

The variational matrix, M, about a general point of system 10 is

$$M = \begin{bmatrix} \alpha E - 2\beta A - \theta_1(1-E) & (\alpha+\theta_1)A \\ -\nu E + \theta_2(1-E) & \varepsilon e - 2\varepsilon E - (\nu+\theta_2)A \end{bmatrix}.$$

Hence the variational matrices about G_0 and G_1 respectively are

$$M_0 = \begin{bmatrix} -\theta_1 & 0 \\ \theta_2 & \varepsilon e \end{bmatrix}, \quad M_1 = \begin{bmatrix} (\alpha+\theta_1)e - \theta_1 & 0 \\ \theta_2 - (\nu+\theta_2)e & -\varepsilon e \end{bmatrix}.$$

Clearly the eigenvalue of M_0 are $-\theta_1$ and εe, showing that G_0 is a hyperbolic saddle point, unstable locally in the E direction, provided $e > 0$. Similarly the eigenvalues of G_1 are $(\alpha+\theta_1)e-\theta_1$ and $-\varepsilon e$. If $(\alpha+\theta_1)e-\theta_1<0$, then G_1 is locally asymptotically stable, whereas if the reverse inequality holds, M_1 is unstable locally in the A direction (except in case (ii) where Δ_2 is undefined). Hence using persistence theory (see [7–9]) in cases (i) and (iii), one may conclude the existence of $\widetilde{G}$ provided inequality (18) holds.

We now consider the interior of the $E-A$ plane. System (10) can be written in the form
$$\dot{A}=f(A,E)$$
$$\dot{E}=g(A,E), \quad (19)$$

where $f(A,E)=(\alpha+\theta_1)EA-\beta A^2+\theta_1 A$ and $g(A,E)=\varepsilon E(e-E)-(\nu+\theta_2)EA+\theta_2 A$. Let $D(A,E)=\frac{\partial}{\partial A}[A^{-1}E^{-1}f(A,E)]+\frac{\partial}{\partial E}[A^{-1}E^{-1}g(A,E)]$. Then we obtain that
$$D(A,E)=-\frac{\beta}{E}-\frac{\varepsilon}{A}-\frac{\theta_2}{E^2}<0$$

for $E, A > 0$. Hence by Dulac's theorem [10, p. 137] there are no periodic solutions in the interior of the first quadrant. Hence if $\widetilde{G}$ exists uniquely, it must be globally stable.

5. Discussion

In this paper we have developed a new model for the interaction of agriculture with industry and the environment, where agriculture and industry are measured in terms of their assets and environment in terms of its quality. We then analyzed the agriculture-industry and agriculture-environment submodels. Due to limitation of size, we leave the analysis of the full model to a future paper.

In both cases, we have established criteria for the existence of a positive equilibrium, and have shown that in these models, periodic solutions cannot occur. We suspect that in the full model, there may very well be limit cycles.

References

[1] Apedaile, L. P., Freedman, H. I., Schilizzi, S. G. M. and Solomonovich, M. (1994). Equilibria and dynamics in an economic predator-prey model of agriculture. *Math. Comput. Modeling*, **19**, 1–15.

[2] Solomonovich, M., Freedman, H. I., Gebremedihan, A. H., Schilizzi, S. G. M. and Belostotski, L. (1997). A dynamic model of sustainable agriculture and the ecosphere. *Appl. Math. Comp.*, **84**, 221–246.

[3] Solomonovich, M., Freedman, H. I., Apedaile, L. P., Schilizzi, S. G. M. and Belostotski, L. (1998). Stability and bifurcations in an environmental recovery model of economic agriculture-industry interactions. *Nat. Resource Model*, **11**, 35–79.

[4] Smulders, S. (1995). Environmental policy and sustainable economic growth. *Economist*, **143**, 163–195.

[5] Stahlgren, T. J. and Parsons, D. J. (1986). Vegetation and soil recovery in wilderness campsites closed to visitor use. it Environ. Manag., bf 10, 375–380.

[6] Freedman, H. I. (1980). *Deterministic Mathematical Models in Population Ecology*. Marcel Dekker, New York.

[7] Butler, G. J., Freedman, H. I. and Waltman, P. (1986). Uniformly persistent systems. *Proc. Amer. Math. Soc.*, **96**, 425–450.

[8] Freedman, H. I. and Moson, P. (1990). Persistence definitions and their connections. *Proc. Amer. Math. Soc.*, **109**, 1025–1033.

[9] Freedman, H. I. and Waltman, P. (1984). Persistence in models of three interacting predator-prey populations. *Math. Biosci.*, **68**, 213–231.

[10] Farkas, M. (1994). *Periodic Motions*. Springer-Verlag, Heidelberg.

4.2 BIFURCATIONS OF PERIODIC SOLUTIONS OF THE THREE BODY PROBLEM*

V.I. GOULIAEV

Ukrainian Transport University, Kiev, Ukraine

1 Introduction

The problem of three bodies occupies the central place in analytical dynamics, celestial mechanics and cosmodynamics. Notwithstanding the great interest paid to it by many scientists over a protracted period of time, up till now it has not only remained unsolved but even the quantitative behaviour of this system has not been understood. In the general form the three body problem is formulated as follows in [3, 6, 12, 13]. Three particles of arbitrary masses are attracted to one another according to Newton's gravity law. Their initial motion is predetermined and they can occupy any position in space. It is necessary to find their motion.

In this form the three body problem is very difficult. Numerous attempts to simplify it through the use of the ten first integrals stemming from the theorems on a mass center motion and rotation around it, as well as the law of conservation of total mechanical energy turned out to be unsuccessful. Moreover, the established theorems on non-existence of any additional analytical integrals not connected with the ten classical ones lent evidence to the view that this method of construction of particular solutions of the problem is unpromising.

The restricted problem of three bodies is the most important modification of the classical problem of three bodies moving under the action of their mutual gravity. Interest in it has increased significantly owing to intensive study and mastering of outer space. Inasmuch as the equations of the theory of three bodies are not integrable in the general form, the major part of the investigations are dedicated to the study of the particular solutions, corresponding to the points of libration of the three body problem.

---

*Advances in Stability Theory (Ed.: A.A. Martynyuk). Stability and Control: Theory, Methods and Applications, Taylor & Francis, London, **13** (2003) 267–287.

During investigation of periodic orbits within the framework of the three body problem, different kinds of simplifying assumptions are introduced. For example, the body of infinitesimal mass is located in the plane of orbits of the principal gravitating masses always during the motion (the plane restricted problem of three bodies), the body of less mass is moving in circular orbit around the body of greater mass (the circular restricted problem of three bodies). Below the questions of computer simulation of bifurcation states of the considered system and transformation of its periodic orbits are considered.

2 Techniques for Construction of Periodic Orbits within the Framework of the Restricted Problem of Three Bodies

Bring forward the techniques for construction of periodic orbits of the spatial elliptical restricted problem of three bodies and investigation of their stability. Consider a problem on the motion of a system, consisting of three particles M_0, M_1, M_2 mutually gravitating according to Newton's law. Assume that the particles M_0 and M_1 are of finite masses m_0, m_1 and the particle M_2 has an infinitesimal ("zero") mass and does not practically influence the motion of the two finite masses m_0, m_1. The problem named the restricted problem of three bodies comprises investigation of the motion of the infinitesimal mass body M_2 under the action of gravity of the bodies M_0, M_1 of finite masses.

Motion of the body relative the body M_1 is determined stemming from the problem of two bodies. In doing so the particle M_1 orbit is described by the equation

$$r = \frac{p}{1 + e \cos v}, \qquad (2.1)$$

where r is the distance between the particles M_0 and M_1; p is the parameter; e is the eccentricity of its Kepler's orbit; v is the apparent anomaly. The body M_1's orbit may be a circle ($e = 0$), ellipse ($0 < e < 1$), parabola ($e = 1$) or hyperbola ($e > 0$) depending upon the values of the initial velocity of the particle M_1 with respect to the particle M_0.

In celestial mechanics the following three cases are separated: the circular restricted problem, when the particle M_1's orbit is a circle with its center at the point M_0; the elliptical restricted problem, when the particle M_1's orbit in an ellipse with its focus located at the point M_0; the hyperbolic restricted problem, when the particle M_1's trajectory is a hyperbola with its focus located at the point M_0.

If the "zero mass" body M_2 is located during its motion in the plane containing the principal gravitating masses M_0, M_1, the problem is termed a plane one. In the case when the particle M_2 can leave the plane of the particles M_0, M_1's orbits, one has to deal with the spatial restricted problem of three bodies.

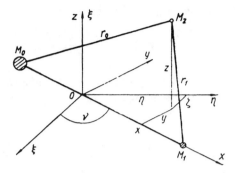

Figure 2.1. Geometrical scheme of the three body problem

Let us construct the equations of motion of the "zero mass" particle M_2 within the framework of the spatial elliptical restricted problem of three bodies. To accomplish this we introduce the coordinate system $Oxyz$ (Figure 2.1) rotating around the axis $O\zeta$ of the immovable coordinate system $O\xi\eta\zeta$ with its origin at the bodies M_0, M_1's mass center's in such a manner that the axis Ox passes through the points M_0 and M_1.

The body M_2 coordinates in the rotating reference frame are determined by the following equations [3]

$$\frac{d^2x}{dt^2} - 2\frac{dv}{dt}\frac{dy}{dt} - \left(\frac{dv}{dt}\right)^2 x - \frac{d^2v}{dt}y = \frac{\partial W}{\partial x},$$

$$\frac{d^2y}{dt^2} + 2\frac{dv}{dt}\frac{dx}{dt} - \left(\frac{dv}{dt}\right)^2 y + \frac{d^2v}{dt}x = \frac{\partial W}{\partial y}, \qquad (2.2)$$

$$\frac{d^2z}{dt^2} = \frac{\partial W}{\partial z},$$

where W is the force function

$$W = f\left(\frac{m_0}{r_0} + \frac{m_1}{r_1}\right), \qquad (2.3)$$

r_0, r_1 are the distances from the "zero mass" M_2 till the basical gravitating bodies M_0, M_1;

$$r_0 = \left[(x-x_0)^2 + y^2 + z^2\right]^{1/2}, \qquad r_1 = \left[(x-x_1)^2 + y^2 + z^2\right]^{1/2};$$

the coordinates x_0, x_1 are specified by the formulae

$$x_0 = -\frac{m_1 r}{m_0 + m_1}, \qquad x_1 = \frac{m_0 r}{m_0 + m_1};$$

the position-vector r is expressed through the parameter p and eccentricity e of the elliptical orbits of the bodies M_0, M_1 by the equality

$$r = \frac{p}{1 + e \cos v};$$

v is the apparent anomaly of the Kepler orbit of the bodies M_0 and M_1 (Figure 2.1).

Taking into account that the derivatives of the apparent anomaly are calculated through the expressions

$$\frac{dv}{dt} = \frac{c}{p^2}(1 + e \cos v)^2, \quad \frac{d^2v}{dt^2} = -\frac{2c^2}{p^4} \sin v (1 + e \cos v)^3,$$

the equations (2.2) are reduced to the form

$$\frac{d^2x}{dv^2} - \frac{2e \sin v}{1 + e \cos v}\frac{dx}{dv} - 2\frac{dy}{dv} - x + \frac{2e \sin v}{1 + e \cos v} y = \frac{p^4}{c^2}(1 + e \cos v)^{-4} \frac{\partial W}{\partial x},$$

$$\frac{d^2y}{dv^2} + 2\frac{dx}{dv} - \frac{2e \sin v}{1 + e \cos v}\frac{dy}{dv} - \frac{2e \sin v}{1 + e \cos v} x - y = \frac{p^4}{c^2}(1 + e \cos v)^{-4} \frac{\partial W}{\partial y},$$

$$\frac{d^2z}{dv^2} - \frac{2e \sin v}{1 + e \cos v}\frac{dz}{dv} = \frac{p^4}{c^2}(1 + e \cos v)^{-4} \frac{\partial W}{\partial z}. \qquad (2.4)$$

In construction of the periodic solutions to the nonlinear differential equations (2.4) for the spatial elliptical problem of three bodies, we assume that a 2π-periodic solution $x^*(v)$, $y^*(v)$, $z^*(v)$ to the system (2.4) for some value of the eccentricity e^* is predetermined. Then in line with the approach proposed in [7], the 2π-periodic solution to the motion equations (2.4) for the parameter value $e + \Delta e$ may be represented in the form

$$x(v) = x^*(v) + \Delta x(v), \quad y(v) = y^*(v) + \Delta y(v), \quad z(v) = z^*(v) + \Delta z(v), \qquad (2.5)$$

where Δx, Δy, Δz are the components of the 2π-periodic solution to the equation system

$$\frac{d^2}{dv^2}\Delta x - \frac{2e^* \sin v}{1 + e^* \cos v}\frac{d}{dv}\Delta x - 2\frac{d}{dv}\Delta y - \Delta x + \frac{2e^* \sin v}{1 + e^* \cos v}\Delta y$$

$$= \frac{p^4}{c^2}(1 + e^* \cos v)^{-4} \frac{\partial}{\partial x}\Delta W + (1 + e^* \cos v)^{-2}\Big\{[2 \sin v(1 + e^* \cos v)$$

$$- e^* \sin 2v]\left(\frac{dx^*}{dv} - y^*\right) - \frac{4p^4}{c^2} \cos v(1 + e^* \cos v)^{-3} \frac{\partial W^*}{\partial x}\Big\}\Delta e,$$

$$\frac{d^2}{dv^2}\Delta y + 2\frac{d}{dv}\Delta x - \frac{2e^* \sin v}{1 + e^* \cos v}\frac{d}{dv}\Delta y - \frac{2e^* \sin v}{1 + e^* \cos v}\Delta x - \Delta y \qquad (2.6)$$

$$= \frac{p^4}{c^2}(1+e^*\cos v)^{-4}\frac{\partial}{\partial y}\Delta W + (1+e^*\cos v)^{-2}\Big\{[2\sin v(1+e^*\cos v)$$

$$- e^*\sin 2v]\left(\frac{dy^*}{dv}+x^*\right) - \frac{4p^4}{c^2}\cos v(1+e^*\cos v)^{-3}\frac{\partial W^*}{\partial y}\Big\}\Delta e,$$

$$\frac{d^2}{dv^2}\Delta z - \frac{2e^*\sin v}{1+e^*\cos v}\frac{d}{dv}\Delta z = \frac{p^4}{c^2}(1+e^*\cos v)^{-4}\frac{\partial}{\partial z}\Delta W + (1+e^*\cos v)^{-2}$$

$$\times\Big\{[2\sin v(1+e^*\cos v)-e^*\sin 2v]\frac{dz^*}{dv} - \frac{4p^4}{c^2}\cos v(1+e^*\cos v)^{-3}\frac{\partial W^*}{\partial z}\Big\}\Delta e$$

linearized in the vicinity of the parameter value e^*, corresponding to the increment Δe.

In equations (2.6) the derivatives of the force function W with respect the coordinates x, y, z are represented in the form

$$\frac{\partial W^*}{\partial x} = -f\Big\{m_0(x^*-x_0^*)[(x^*-x_0^*)^2+(y^*)^2+(z^*)^2]^{-3/2}$$

$$+ m_1(x^*-x_1^*)[(x^*-x_1^*)^2+(y^*)^2+(z^*)^2]^{-3/2}\Big\}x^*,$$

$$\frac{\partial W^*}{\partial y} = -f\Big\{m_0[(x^*-x_0^*)^2+(y^*)^2+(z^*)^2]^{-3/2}$$

$$+ m_1[(x^*-x_1^*)^2+(y^*)^2+(z^*)^2]^{-3/2}\Big\}y^*, \qquad (2.7)$$

$$\frac{\partial W^*}{\partial z} = -f\Big\{m_0[(x^*-x_0^*)^2+(y^*)^2+(z^*)^2]^{-3/2}$$

$$+ m_1[(x^*-x_1^*)^2+(y^*)^2+(z^*)^2]^{-3/2}\Big\}z^*,$$

where the coordinates x_0^*, x_1^* and the position-vector r^* corresponding to the elliptical orbit parameter value e^* equal

$$x_0^* = -\frac{m_1 r^*}{m_0+m_1}, \qquad x_1^* = \frac{m_0 r^*}{m_0+m_1}, \qquad (2.8)$$

$$r^* = \frac{p}{1+e^*\cos v}. \qquad (2.9)$$

The force function derivatives increments are expressed through the "zero mass" M_2 coordinates Δx, Δy, Δz increments with the use of the correlations

$$\frac{\partial}{\partial x}\Delta W = -f\Big\{m_0[(x^*-x_0^*)^2+(y^*)^2+(z^*)^2]^{-3/2}(\Delta x-\Delta x_0)$$

$$- 3m_0(x^*-x_0^*)[(x^*-x_0^*)^2+(y^*)^2+(z^*)^2]^{-5/2}[(x^*-x_0^*)(\Delta x-\Delta x_0)$$

$$+ y^*\Delta y + z^*\Delta z] + m_1\big[(x^* - x_1^*)^2 + (y^*)^2 + (z^*)^2\big]^{-3/2}(\Delta x - \Delta x_1)$$
$$- 3m_1(x^* - x_1^*)\big[(x^* - x_1^*)^2 + (y^*)^2 + (z^*)^2\big]^{-5/2}$$
$$\times \big[(x^* - x_1^*)(\Delta x - \Delta x_1) + y^*\Delta y + z^*\Delta z\big]\Big\}\Delta x, \qquad (2.10)$$

$$\frac{\partial}{\partial y}\Delta W = 3f\Big\{m_0\big[(x^* - x_0^*)^2 + (y^*)^2 + (z^*)^2\big]^{-5/2}\big[(x^* - x_0^*)(\Delta x - \Delta x_0)$$
$$+ y^*\Delta y + z^*\Delta z\big] + m_1\big[(x^* - x_1^*)^2 + (y^*)^2 + (z^*)^2\big]^{-5/2}\big[(x^* - x_1^*)(\Delta x - \Delta x_1)$$
$$+ y^*\Delta y + z^*\Delta z\big]\Big\} - f\Big\{m_0\big[(x^* - x_0^*)^2 + (y^*)^2 + (z^*)^2\big]^{-3/2}$$
$$+ m_1\big[(x^* - x_1^*)^2 + (y^*)^2 + (z^*)^2\big]^{-3/2}\Big\}\Delta y,$$

$$\frac{\partial}{\partial z}\Delta W = 3f\Big\{m_0\big[(x^* - x_0^*)^2 + (y^*)^2 + (z^*)^2\big]^{-5/2}\big[(x^* - x_0^*)(\Delta x - \Delta x_0)$$
$$+ y^*\Delta y + z^*\Delta z\big] + m_1\big[(x^* - x_1^*)^2 + (y^*)^2 + (z^*)^2\big]^{-5/2}\big[(x^* - x_1^*)(\Delta x - \Delta x_1)$$
$$+ y^*\Delta y + z^*\Delta z\big]\Big\} - f\Big\{m_0\big[(x^* - x_0^*)^2 + (y^*)^2 + (z^*)^2\big]^{-3/2}$$
$$+ m_1\big[(x^* - x_1^*)^2 + (y^*)^2 + (z^*)^2\big]^{-3/2}\Big\}\Delta z.$$

Here the increments of the coordinates Δx_0, Δy_0 and the position-vector Δr are represented with the help of the formulae

$$\Delta x_0 = -\frac{m_1\Delta r}{m_0 + m_1}, \qquad \Delta x_1 = \frac{m_0\Delta r}{m_0 + m_1},$$

$$\Delta r = -\frac{p\cos v\Delta e}{(1 + e^*\cos v)^2}.$$

The 2π-periodic solution Δx, Δy, Δz to the linearized equations (2.6) of the elliptical restricted problem of three bodies is looked for in the form

$$\Delta x = \sum_{i=1}^{6}\Delta x_i c_i + \Delta x_e,$$

$$\Delta y = \sum_{i=1}^{6}\Delta y_i c_i + \Delta y_e, \qquad (2.11)$$

$$\Delta z = \sum_{i=1}^{6}\Delta z_i c_i + \Delta z_e,$$

where Δx_i, Δy_i, Δz_i are the elements of the normalized fundamental matrix of solutions to the equations system (2.6) transformed to the homogeneous form;

Δx_e, Δy_e, Δz_e are the Cauchy problem solution for the linearized system (2.6) under the nullified initial conditions; c_i are the constants determined from the periodicity conditions

$$\Delta x(0) = \Delta x(2\pi), \quad \Delta y(0) = \Delta y(2\pi), \quad \Delta z(0) = \Delta z(2\pi), \quad (2.12)$$

$$\frac{d}{dv}\Delta x(0) = \frac{d}{dv}\Delta x(2\pi), \quad \frac{d}{dv}\Delta y(0) = \frac{d}{dv}\Delta y(2\pi), \quad \frac{d}{dv}\Delta z(0) = \frac{d}{dv}\Delta z(2\pi).$$

At realization of the calculation, the solutions $\Delta x_i(v)$, $\Delta y_i(v)$, $\Delta z_i(v)$, $\Delta x_e(v)$, $\Delta y_e(v)$, $\Delta z_e(v)$ are constructed with the use of numerical methods.

Investigation of stability of the motion equations solutions of the elliptic restricted problem of three bodies (2.4) is carried out on the basis of analysis of the eigen-values ρ_i of the monodromy matrix which is deduced as the result of the unit matrix E subtraction from the matrix of coefficients of the left-hand member of the linear algebraic equation system

$$\left[\Delta x_1(2\pi) - 1\right]c_1 + \Delta x_2(2\pi)c_2 + \cdots + \Delta x_6(2\pi)c_6 = -\Delta x_e(2\pi),$$

$$\Delta y_1(2\pi)c_1 + \left[\Delta y_2(2\pi) - 1\right]c_2 + \cdots + \Delta y_6(2\pi)c_6 = -\Delta y_e(2\pi),$$

$$\cdots \cdots \cdots \cdots \cdots \cdots \cdots \cdots \cdots \cdots \cdots \cdots \cdots \cdots \cdots \cdots \cdots \cdots \quad (2.13)$$

$$\frac{d}{dv}\Delta z_1(2\pi)c_1 + \frac{d}{dv}\Delta z_2(2\pi)c_2 + \cdots + \left[\frac{d}{dv}\Delta z_6(2\pi) - 1\right]c_6 = -\frac{d}{dv}\Delta z_e(2\pi),$$

constructed on the basis of the conditions (2.12) with allowance made for the correlations (2.11).

The outlined approach permits us to find critical states of the considered systems, to separate the equation (2.4) stable solutions from the unstable ones, to establish possibilities of the solution bifurcations, to continue the solutions along the bifurcated directions and to construct the "zero mass" motion trajectories corresponding to the precritical and postcritical states [7].

3 Periodic Solutions of the Plane Elliptical Problem of Three Bodies

We perform the numerical construction of periodic trajectories of the plane elliptic restricted problem of three bodies: the Earth, the Moon, a space-craft with the help of the technique described in Section 2. The generating orbits for the considered problem are found through prescription $e = 0$ in equations (2.2). Then (2.2) are transformed into the motion equations of the classical circular restricted problem of three bodies [3]

$$\frac{d^2x}{dt^2} - 2n\frac{dy}{dt} = \frac{\partial \Omega}{\partial x},$$

$$\frac{d^2y}{dt^2} + 2n\frac{dx}{dt} = \frac{\partial \Omega}{\partial y}, \qquad (3.1)$$

$$\frac{d^2z}{dt^2} = \frac{\partial \Omega}{\partial z},$$

where $\Omega = \frac{n^2}{2}(x^2 + y^2) + W$, $n = \frac{c}{a^2} = \frac{dv}{dt}$, a is the radius of the body M_1's circular orbit.

Inasmuch as Ω depends only on the particle M_2 coordinates, the system (3.1) has a first integral, called the Jacobian integral, which looks like

$$\left(\frac{dx}{dt}\right)^2 + \left(\frac{dy}{dt}\right)^2 + \left(\frac{dz}{dt}\right)^2 = n^2(x^2 + y^2) + 2f\left(\frac{m_0}{r_0} + \frac{m_1}{r_1}\right) + 2h,$$

$$\text{or} \quad V^2 = 2P + 2h, \qquad (3.2)$$

where $V = \sqrt{\left(\frac{dx}{dt}\right)^2 + \left(\frac{dy}{dt}\right)^2 + \left(\frac{dz}{dt}\right)^2}$ is the particle M_2's relative velocity; the coordinates x_0, x_1 of the particles M_0, M_1 are the constant values;

$$x_0 = -\frac{m_1 a}{m_0 + m_1}, \qquad x_1 = \frac{m_0 a}{m_0 + m_1}; \qquad (3.3)$$

h is the arbitrary constant, which is entirely determined by an initial position and velocity of the particle M_2.

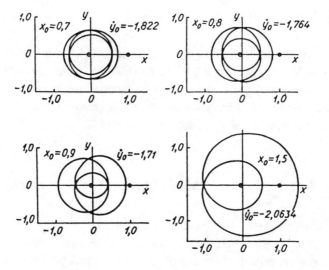

Figure 3.1. Generating trajectories of the "zero mass" body

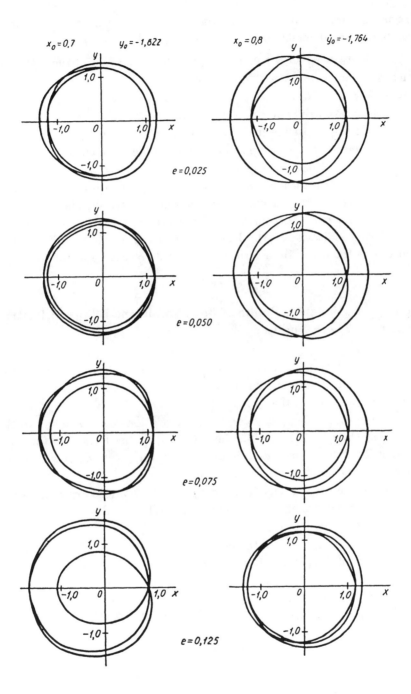

Figure 3.2. Perturbed trajectories of the "zero mass" body

In Figure 3.1 are shown some generating orbits of the plane restricted problem of three bodies for different initial positions x_0 and initial velocities $\dot{y}_0 = \frac{dy_0}{dt}$ $\left(\dot{x} = \frac{dx_0}{dt} = 0,\ y_0 = 0\right)$ of the body M_0 gained as the result of the system (3.1) numerical solution.

The periodic orbits of the body M_2 within the framework of the elliptical restricted problem of three bodies are found with the help of the procedure outlined above (2.4) – (2.13), imparting increments to the eccentricity e of the bodies M_0, M_1's elliptical orbits. Some of the results obtained are represented in Figure 3.2.

Special interest is caused by the orbits with the close passing, when the particle periodically passes near the less (or near the both) of the two gravitating masses. In these cases accuracy of the calculation methods falls drastically and it is necessary to use special numerical procedures in order to achieve adequate satisfaction with the periodicity conditions (2.13). Thus, integration of equations of (2.6) type is performed with the use of the Everhart 11–15 order accuracy method [4].

4 Stability of Triangular Points of Libration in the Elliptical Restricted Problem of Three Bodies

Consider the problem of stability of triangular points of libration in the elliptical restricted problem of three bodies (particles) moving under the action of their mutual gravitational attraction (Figure 2.1).

The three body problem has five particular solutions corresponding to the libration points L_i ($i = 1, 2, \ldots, 5$), where L_1, L_2, L_3 are located in the straight line passing through M_0 and M_1 (Figure 2.1) and the points L_4 and L_5 make up equilateral triangles with the bodies M_0 and M_1.

Inasmuch as the gravitational and centripetal accelerations experienced by the particle M_2 located at the libration point are counterbalanced it remains immovable (in the rotating reference frame).

The most important issues of the problem of the libration points are the questions of stability of the points L_i and periodic orbits in their vicinity [1, 3, 12, 13].

It is known that in the circular restricted problem of three bodies the "rectilinear" points of libration are unstable, but the triangular ones are stable, if the ratio of the masses $\mu = m_1/(m_0 + m_1)$ is sufficiently small [1, 3], that is the inequality

$$0 < 27\mu(1 - \mu) < 1 \qquad (4.1)$$

is satisfied.

On strict investigation of stability it is proved [12] that the triangular points of libration of the plane circular restricted problem of three bodies are stable in

Liapunov's sense in the region of the linear approximation (4.1) for all values of μ, except for two

$$\mu_1 = \frac{45 - \sqrt{1833}}{90} = 0.0242938, \quad \mu_2 = \frac{15 - \sqrt{213}}{30} = 0.013516, \quad (4.2)$$

for which instability takes place.

We outline the numerical technique for stability investigation in linear approximation and construction of the triangular points periodic orbits in the elliptical restricted problem of three bodies. Consider the linearized fourth order system with periodic coefficients. The mathematical problem is set by the system of equations [3]

$$\frac{d^2x}{dv^2} - 2\frac{dy}{dv} - \frac{h_2 x}{1 + e \cos v} = 0,$$

$$\frac{d^2y}{dv^2} + 2\frac{dx}{dv} - \frac{h_1 x}{1 + e \cos v} = 0,$$
(4.3)

where $h_{1,2} = \frac{3}{2}\left[1 \pm \sqrt{1 - 3\mu(1-\mu)}\right]$; e is the eccentricity of the orbit of the two masses m_0, m_1; μ is the ratio between the smaller mass and sum of both masses.

The linear equation (4.3) describes the motion of a particle near a triangular point in the restricted problem of three bodies. Consider the possibility of motion stability loss with the periods 2π, 3π, 4π, etc. We solve the equation system in the segments $0 \leq v \leq 2\pi$, $0 \leq v \leq 4\pi$, $0 \leq v \leq 6\pi$, etc. under the following initial conditions

$$\begin{aligned} x_1(0) &= 1, & \dot{x}_1(0) &= 0, & y_1(0) &= 0, & \dot{y}_1(0) &= 0, \\ x_2(0) &= 0, & \dot{x}_2(0) &= 1, & y_2(0) &= 0, & \dot{y}_2(0) &= 0, \\ x_3(0) &= 0, & \dot{x}_3(0) &= 0, & y_3(0) &= 1, & \dot{y}_3(0) &= 0, \\ x_4(0) &= 0, & \dot{x}_4(0) &= 0, & y_4(0) &= 0, & \dot{y}_4(0) &= 1. \end{aligned} \quad (4.4)$$

The system (4.3) has a T-periodical solution $x(v)$, $y(v)$ for some values of e and μ according to Floquet's theory if and only if the equality

$$I = \det \begin{vmatrix} x_1(T) - 1 & x_2(T) & x_3(T) & x_4(T) \\ \dot{x}_1(T) & \dot{x}_2(T) - 1 & \dot{x}_3(T) & \dot{x}_4(T) \\ y_1(T) & y_2(T) & y_3(T) - 1 & y_4(T) \\ \dot{y}_1(T) & \dot{y}_2(T) & \dot{y}_3(T) & \dot{y}_4(T) - 1 \end{vmatrix} = 0 \quad (4.5)$$

occurs.

At $e = 0$ the points are found, where $I = 0$. Proceeding from these points the curves, representing boundaries between the stable and unstable regions and resonant lines are constructed. To accomplish this the parameters e and μ are imparted such small increments Δe, $\Delta \mu$ that the equality (4.5) remains true. In this case the correlation

$$\Delta I = \frac{\partial I}{\partial e} \Delta e + \frac{\partial I}{\partial \mu} \Delta \mu \approx 0 \tag{4.6}$$

allows us to establish an approximate relationship between them. Using it, one can find new values of the parameters $e + \Delta e$, $\mu + \Delta \mu$, for which

$$r = I = 0.$$

Continuing further the process of the parameters e and μ variation with allowance made for the residual r in equation (4.5), we find the approximate relation

$$\frac{\partial I}{\partial e} \Delta e + \frac{\partial I}{\partial \mu} \Delta \mu - r = 0, \tag{4.7}$$

on the basis of which it is possible to construct the curves approximately satisfying the periodic solutions to equations (4.3) in the plane $e - \mu$. Equation (4.7) coefficients are calculated using the formulae [7]

$$\begin{aligned}\frac{\partial I}{\partial e} &= \sum_{\substack{i=1 \\ j=1}}^{n} \frac{\partial u_i^j(T)}{\partial e} B_j^i, \\ \frac{\partial I}{\partial \mu} &= \sum_{\substack{i=1 \\ j=1}}^{n} \frac{\partial u_i^j(T)}{\partial \mu} B_j^i,\end{aligned} \tag{4.8}$$

where B_j^i is the adjunct of the corresponding element of the matrix (4.5).

The functions $\frac{\partial u_i^j(T)}{\partial e}$ and $\frac{\partial u_i^j(T)}{\partial \mu}$ are found as the solutions to the equations

$$\begin{aligned}\Delta \ddot{x} - 2\Delta \dot{y} - \frac{h_2}{1 + e \cos v} \Delta x + \frac{h_2 x \cos v}{(1 + e \cos v)^2} \Delta e &= 0, \\ \Delta \ddot{y} + 2\Delta \dot{x} - \frac{h_1}{1 + e \cos v} \Delta y + \frac{h_1 y \cos v}{1 + e \cos v} \Delta e &= 0, \\ \Delta \ddot{x} - 2\Delta \dot{y} - \frac{h_2}{1 + e \cos v} \Delta x - \frac{x}{1 + e \cos v} \frac{\partial h_2}{\partial \mu} \Delta \mu &= 0, \\ \Delta \ddot{y} + 2\Delta \dot{x} - \frac{h_1}{1 + e \cos v} \Delta y - \frac{y}{1 + e \cos v} \frac{\partial h_1}{\partial \mu} \Delta \mu &= 0, \\ \frac{\partial h_1}{\partial \mu} = \frac{g}{2} \frac{2\mu - 1}{\sqrt{1 - 3\mu(1 - \mu)}}, \quad \frac{\partial h_2}{\partial \mu} = -\frac{g}{4} \frac{2\mu - 1}{\sqrt{1 - 3\mu(1 - \mu)}},\end{aligned} \tag{4.9}$$

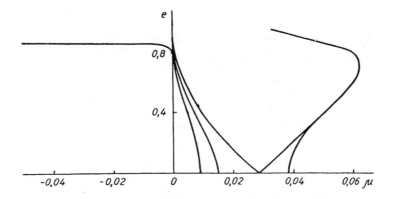

Figure 4.1. Diagram of stability

issuing from the system (4.3) variation with respect to e and μ under the nullified initial conditions.

Using the equality (4.7) and specifying one of the parameters e or μ as leading the increment of the other one finds the following

$$\Delta e = -\left(\frac{\partial I}{\partial \mu}\Delta\mu + r\right)\left(\frac{\partial I}{\partial r}\right)^{-1} \quad \text{or} \quad \Delta\mu = -\left(\frac{\partial I}{\partial e}\Delta e + r\right)\left(\frac{\partial I}{\partial \mu}\right)^{-1}.$$

At construction of resonance curves obtained in dependence on the period multiplicity with the value 2π, the eigen-values of (4.5) should be analysed. If all of them but one corresponding to the specified period are less then unit, the resonant line is in the boundary of stability, in the case of exceeding the unit value at least by one of the matrix eigen-values the considered point is unstable.

Note that the first completed results in the investigation of stability of the libration triangular points in the elliptical problem were obtained by Danby [1]. In the plane $e - \mu$ he built up the stability domain with the use of a numerical technique. It is located inside two curvilinear triangles connected by adjacent vertices of angles at the basical point of principal parametrical resonance.

Markeev [12] investigated this problem by the application of a nonlinear set up. Using analytical and numerical methods he separated the curves inside the stability domain, where the resonant correlations of the third and fourth orders are satisfied.

If is found by us that the stability domain of the considered problem turned out to be wider. It is established that the manifold of these solutions has a continuation in the form of a narrowing belt proceeding from the upper angle of the right triangle and adjoining presumably the vertex $e = 1$, $\mu = 0$ of the left triangle.

Figure 4.1 shows the diagram reflecting the correlation between e and μ. The curves correspond to the periodic solutions to the equation (4.3) system. It should

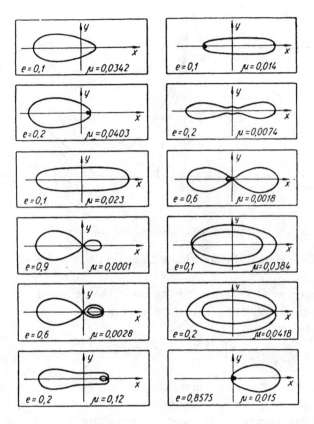

Figure 4.2. Periodic trajectories of the "zero mass" body at critical states

be noted that in the diagram only some resonant lines are presented because most of them are depicted in [12]. Us of the elaborated technique allowed us to continue the right branch of the curve $e - \mu$ from the point $e = 0.3500$, $\mu = 0.04898$ up to the point $e = 0.9500, \mu = 0.03491$. In this segment of the curve, convergence of the numerical process significantly deteriorates. As the straight line $e = 1$ is approached, the calculational procedure begins to diverge.

The region of the parameter μ negative values is also investigated on the basis of the proposed technique. The results of these investigations may be used in quantum mechanics for analysis of the libration points stability under conditions of availability of both attraction and repulsion between interacting particles.

In Figure 4.2 the orbit geometry is shown for different points in the curves of correlation between e and μ (Figure 4.1), where solutions to the equation system (4.3) are periodic. In their construction allowance was made for the fact that the determinant of the periodicity matrix (4.5) equals zero in the specified curves. This fact permits us to remove one of the equations (for example, j-th) expressing the periodicity conditions and to consider the corresponding j-th unknown c_j as being freely variated. Imparting some value to it, the remaining unknown parameters c_j ($i \neq j$) can be found from the intact system of equations. Thereafter the coordinates

$$x = \sum_{i=1}^{4} x_i c_i, \quad y = \sum_{i=1}^{4} y_i c_i$$

of points of the periodic trajectories are calculated.

5 Chaotization of Solutions of the Plane Restricted Problem of Three Bodies through the Sequence of Period-Doubling Bifurcations

It is shown in the papers [5, 10], that in the simplest nonlinear mechanical system with dissipation under the action of periodic time disturbances, generation of chaotic vibrations as the result of an infinite sequence of period-doubling bifurcations is possible. The availability of universal by Feigenbaum cascades of period-doubling bifurcations is found for series of conservative mechanical systems modelled by two-dimensional Hamilton maps [8, 10]. The specified cascades converge geometrically to some limit point. In its vicinity all the $2^n T$-periodic circles, including $n = \infty$, are unstable. Other converging sequences of bifurcations are found for mappings alongside the period-doubling bifurcations.

Note, that in the papers [5, 10] basical objects of investigation are one- and two-dimensional mappings. Below are the results of investigations of evolution of periodic solutions to the nonlinear Hamiltonian system of fourth-order differential equations, describing the "zero mass" body motion for the plane restricted problem of three bodies when the eccentricity of gravitating centres motion trajectory is changing. A regular sequence of period-doubling bifurcations with the velocity distinct from the geometrical one is found. The obtained results [9] allow us to conclude that initiation of chaotic modes through the sequence of period-doubling bifurcations not possessing the universal properties is possible in multidimensional nonlinear conservative systems [5, 8, 10].

The method of attack outlined below is based on consecutive linearization of equations of motion, continuation solutions by parameter and methods of bifurcation theory [7].

We follow the evolution of periodic solutions to the system of nonlinear ordinary differential equations

$$\dot{x} = f(x, \lambda, t), \tag{5.1}$$

when the scalar parameter λ changes. Here f is the n-dimensional T-periodic relative t vector function differentiable the necessary number of times with respect to x and λ.

Assuming that the solution $x(t)$ to the system (5.1) continuously depends on the initial condition $x(0)$ and the parameter λ, it is possible to deduce the following form of the T-periodicity conditions

$$x(0) = x(x(0), \lambda, T). \tag{5.2}$$

Let the system (5.1) have a T-periodic solution $x_0(t)$, satisfying conditions (5.2) at some value of the parameter $\lambda = \lambda_0$. Then both parts of (5.2) being variated in the vicinity of the state $\lambda = \lambda_0$, $x(t) = x_0(t)$ and the value of the increment $\Delta \lambda_0$ being specified, the appropriate increments of the initial conditions $\Delta x_0(0)$ could be found as the solutions to the linear system of equations

$$[Y(T) - E]\Delta x_0(0) = -y_\lambda(T)\Delta \lambda_0. \tag{5.3}$$

Here $Y(T) = \partial x(T)/\partial x(0)$ is the monodromy matrix determined as the solution to the homogeneous matrix equation

$$\dot{Y} = f_x Y \quad (Y(0) = E), \tag{5.4}$$

E is the unit matrix; $y_\lambda(T) = \partial x(T)/\partial \lambda$ is the n-dimensional vector determined from the particular solution to the nonhomogeneous system of variated equations

$$\dot{y}_\lambda = f_x y_\lambda + f_\lambda \quad (y_\lambda(0) = 0).$$

Refinement of the initial conditions $x_1(0) = x_0(0) + \Delta x_0(0)$ at $\lambda_1 = \lambda_0 + \Delta \lambda_0$ is produced via use of the Newton–Kantorovich technique.

Continuing further the parameter λ varying, one can find the system (5.1) approximate T-periodic solutions $x_m(t)$ corresponding to the values λ_m.

The monodromy matrix $Y(T)$ calculated at every step of the continuation procedure characterizes the conditions of existence and uniqueness of the periodic solution to the linearized system in the considered vicinity and permits us to analyse its stability [2, 11]. The state, when the conditions $|\rho_i| = 1$, $\arg \rho_i = 2\pi/n$ are satisfied at least for one of the matrix $Y(T)$ multiplicators, is bifurcational and the appropriate point in the state space is the furcation point. The nT-periodic solution to equation (5.1) furcating from this point can be constructed only through the use of a bifurcation equation, because at the considered point

$$\det(Y(nT) - E) = 0.$$

To construct the approximate bifurcation equation take into account the smaller quantities with higher order in (5.3).

With allowance made for the small quantities with the second order in (5.3), the approximate equations of bifurcation take the form

$$[Y(nT) - E]\Delta x(0) + y_\lambda(nT)\Delta\lambda + \frac{1}{2!}\{Z(nT)\Delta x(0), \Delta x(0)\}$$
$$+\{Z_\lambda(nT)\Delta x(0), \Delta\lambda\} + \frac{1}{2!}\{z_\lambda(nT)\Delta\lambda, \Delta\lambda\} = 0. \tag{5.5}$$

$$Z(nT) = \frac{\partial Y(nT)}{\partial x(0)}; \quad Z_\lambda(nT) = \frac{\partial Y(nT)}{\partial \lambda}; \quad z_\lambda(nT) = \frac{\partial y_\lambda(nT)}{\partial \lambda}. \tag{5.6}$$

For the three-index functional matrix to be constructed, differentiate both the matrix equations (5.4) with respect to $x(0)$, taking into consideration the assumed designations (5.6). In consequence of this we can write

$$\dot{Z} = \{f_{xx}Y, Y\} + f_x Z. \tag{5.7}$$

By this means the matrix $Z(nT)$ may be generated through the solution of the Cauchy problem for equations (5.7) with the initial conditions $Z(0) = 0$.

In a similar manner the equations for $Z_\lambda(t)$, $z_\lambda(t)$ are constructed. They are as follows

$$\dot{Z}_\lambda = \{(f_{xx}y_\lambda + f_{x\lambda}), Y\} + f_x Z_\lambda, \qquad (Z_\lambda(0) = 0),$$
$$\dot{z}_\lambda = \{(f_{xx}y_\lambda + f_{x\lambda}), y_\lambda\} + f_x z_\lambda + f_{\lambda\lambda}, \qquad (z_\lambda(0) = 0).$$

The approximate equations of furcations represent a system of nonlinear algebraic equations. Their solutions are constructed by the Newton method.

If the system (5.5) has no solutions or there are multiple ones among its roots, it is necessary to take into account the successive terms of expansion and to iterate the calculations. Furthermore, the nT-periodic solution may be continued along every found branch by the outlined technique. At every step of the continuation procedure, analysis of the matrix $Y(nT)$ multiplicators allows us to find the next point of furcation with m-tuple increase of the period (if there is any), etc.

To ensure the high accuracy required in the construction of long period solutions and analysis of their furcation possibilities, the 11–15 order method of Everhart [4] is used.

Using the described technique, numerical investigation of the "zero mass" body of the plane restricted problem of three bodies was performed. In the rotating barycentric reference frame, motion of a particle is determined by the following nonlinear equations [3]

$$\xi'' - 2\eta' - \frac{1}{1 + e\cos v}\xi = \frac{1}{1 + e\cos v}\frac{\partial W}{\partial \xi},$$
$$\eta'' + 2\xi' - \frac{1}{1 + e\cos v}\eta = \frac{1}{1 + e\cos v}\frac{\partial W}{\partial \eta},$$

stemming from (2.2). Here, as previously, e is the eccentricity of orbit of the masses m_0 and m_1, v is the apparent anomaly, the prime denotes differentiation with respect to v,

$$W = \frac{1-\mu}{r_0} + \frac{\mu}{r_1},$$

$$r_0 = \left[(\xi+\eta)^2 + \eta^2\right]^{1/2}, \quad r_1 = \left[(\xi+\eta-1)^2 + \eta^2\right]^{1/2}.$$

As an initial 2π-periodic solution assumed at $e_0 = 0$, the stable stationary state of the "zero mass" body at the triangular libration point L_4 ($\xi_0 = (1-2\mu)/2$, $\eta_0 = \sqrt{3}/2$) was selected. It will be recalled, that the "rectilinear" points of libration are unstable and the "triangular" ones are stable for the circular restricted

Table 5.1. Bifurcational values of periodic orbit eccentricity and initial conditions of the "zero mass" particle vibrations

i	$e_{(i)}$	$\delta_{(i)}$	$\xi_{(i)}(0) - \xi_0;\ \xi'_{(i)}(0)$	$\eta_{(i)}(0) - \eta_0;\ \eta'_{(i)}(0)$
1	0.549	12.7684399	0 0	0 0
2	0.580602800415764	7.305849	0.204154462402339 -0.013967041009111	-0.112725297847557 -0.091265331421095
3	0.583077871761819	11.737178	0.204367815046881 -0.011574655593698	-0.110244997700553 -0.092876179936714
4	0.583416651188409	86.20768	0.201337751179594 -0.010073702719872	-0.107219825814351 -0.091866609210255
5	0.583445514976422		0.201150117317916 -0.009973804064603	-0.107024639206067 -0.091817185567852
6	0.583445849792199		0.201136063209813 -0.009971084985379	-0.107014666859876 -0.091809932228831
7	0.583446043767878		0.201133970681293 -0.009969171554483	-0.107011904772506 -0.091809376528804

problem of three bodies if the value $\mu = m_1/(m_0 + m_1)$ is sufficiently small, that is to say, the inequality (4.1) is satisfied.

The investigation of the periodic solutions evolution at the varying of the parameter of eccentricity e was fulfilled for the case $\mu = 0.005$. It is established that the "zero mass" body remains at the stationary state with the parameter e changing from 0 to $e_{(1)}$ (see Table 5.1).

At $e = e_{(1)}$ the stationary state becomes unstable and in response to the Andronov–Hopf bifurcation accompanied by the concurrent period-doubling the stable 4π-periodic vibrations come into being in the vicinity of the triangular point of libration. Further increase of eccentricity causes enlargement of the "zero mass"

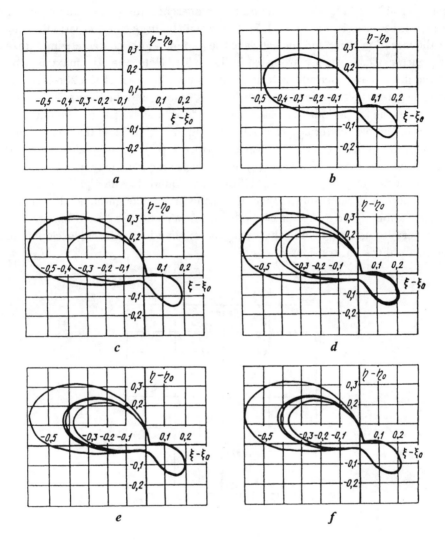

Figure 5.1. Motion trajectories of the "zero mass" body with periods 2π – (a), 4π – (b), 8π – (c), 16π – (d), 32π – (e), 64π – (f) for the eccentricity $e_{(i)}$ ($i = \overline{1,6}$) bifurcational values

body vibration amplitude. At $e = e_{(2)}$ the periodic trajectory endures the next bifurcation of period-doubling with generation of a stable 8π-periodic trajectory and so on. The sequence of six period-doubling bifurcations of the particle vibrations at eccentricity change was constructed. At $e = e_{(7)}$ the 128π-periodic trajectory loses its stability and a pair of conjugate complex multiplicators of the monodromy matrix $Y(128\pi)$ falls outside the unit circumference $|\rho_{i,j}| = 1$. The bifurcational values of eccentricity $e_{(i)}$ $(i = \overline{1,7})$, the initial conditions of $2^i\pi$-periodic solutions at the furcation point and the values of the parameter $\delta_{(i)}$ characterizing rate of the sequence $e_{(i)}$ convergence are listed in Table 5.1. Note, that the maximum period of the investigated periodic solutions makes up 128π. All of the $2^i\pi$-periodic trajectories are stable inside the intervals $e_{(i-1)} < e < e_{(i)}$. In Figure 5.1(a) – (f) are presented the 2π-periodic trajectories of the "zero mass" body for the bifurcational values of the parameter $e_{(i)}$ $(i = \overline{1,6})$. Shown in Figure 5.2 is the "furcation tree" representing a section of the subspace $(\xi(v), e, \xi'(v))$ by the plane $\xi'(v) = 0$. It constitutes the extremal values $\xi(v)$ dependence on eccentricity e at $0.5 < e < 0.6$. The sequence of period-doubling bifurcations found does not possess the universal properties [5, 8] established for the second order differential equations, but the complicating of the "zero mass" body trajectory and the "furcation tree" structure permits us to conclude that a regime of chaotic motion is developing in the mechanical system considered.

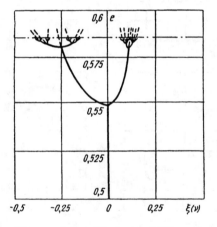

Figure 5.2. The tree of bifurcations

References

[1] Danby, J.M. (1964). Stability of the triangular points in the elliptic restricted problem of three bodies. *Astron. J.*, **69**, 165–174.

[2] Demidovich, B.P. (1967). *Lectures in Mathematical Theory of Stability.* Nauka, Moscow (Russian).

[3] Duboshin, G.N. (1968). *Celestial Mechanics. Principal Problems and Methods.* Nauka, Moscow (Russian).

[4] Everhart, E. (1974). *An Efficient Integrator of Very High Order and Accuracy with Appendix Listing of RADAU.* Univ. of Denver, Denver.

[5] Feigenbaum, M.J. (1980). Universal behaviour in nonlinear systems. *Los Alamos Sci.*, **1**, 4–27.
[6] Gouliaev, V.I. (1998). Celestial mechanics. In: *Development of General Mechanics in Russia and Ukraine in the 20-80th Years of XX Century* (Ed.: A.Yu. Ishlinsky and G.S. Pisarenko). Nauka, Moscow; Phoenix, Kiev, 54–84 (Russian).
[7] Gouliaev, V.I., Bazhenov, V.A., Gotsuliak, E.A., et al. (1983). *Stability of Periodical Processes in Nonlinear Mechanical Systems*. Vyshcha Shkola, Lvov (Russian).
[8] Gouliaev, V.I., Zavrazhina, T.V. and Koshkin, V.L. (1996). Universal regularities of transition to chaos of dissipative and conservative nonlinear vibrational systems. *Mekh. Tver. Tela*, **3**, 12–20 (Russian).
[9] Gouliaev, V.I., Zubritska, A.L. and Koshkin, V.L. (1989). Sequence of period-doubling bifurcations of solutions of the restricted problem of three bodies. *Dokl. Akad. Nauk Ukr. SSR*, **3**, 35–39 (Russian).
[10] Holmes, P.J. and Moon, F.C. (1983). Strange attractors and chaos in nonlinear mechanics. *Trans. ASME /J. Appl. Mech.*, **50**, 1021–1032.
[11] Malkin, I.G. (1966). *Theory of Stability of Motion*. Nauka, Moscow (Russian).
[12] Markeev, A.P. (1978). *Points of Libration in Celestial Mechanics and Cosmodynamics*. Nauka, Moscow (Russian).
[13] Szebehely, V. (1967). *Theory of Orbits. The Restricted Problem of Three Bodies*. Academic Press, New York, London.

4.3 COMPLEX MECHANICAL SYSTEMS: STEADY-STATE MOTIONS, OSCILLATIONS, STABILITY*

A.Yu. ISHLINSKY[1], V.A. STOROZHENKO[2] and M.E. TEMCHENKO[2]

[1] *Institute of Problems of Mechanics of Russian Academy of Sciences, Moscow, Russia*
[2] *Institute of Mathematics of National Academy of Sciences of Ukraine, Kiev, Ukraine*

1 Introduction

The paper provides an analysis of works dealing with the investigation of the dynamics of an absolute solid suspended on an inertialess absolutely elastic nonextendable string (thread). Systematic studies in this direction were undertaken in Kiev in the 1950s and have continued up to this time with few breaks.

Section 2 describes the experimental and theoretical stability investigations of axisymmetric solid rotation on a string and bifurcation processes occurring in this mechanical system.

In Section 3 papers are analysed that deal with the investigation of a gyroscope and suspended body with a cavity completely filled with an ideal incompressible liquid.

The peculiarities of stationary motion of a body suspended on a string at a point out of its axis of symmetry are scrutinized in Section 4. Some motion forms being paradoxical at first sight are established.

Further, in Sections 5 and 6 papers are reviewed that dwell on the performing devices incorporating the selfbalancing of quickly rotating solids suspended on a string.

Advances in Stability Theory (Ed.: A.A. Martynyuk). Stability and Control: Theory, Methods and Applications, Taylor & Francis, London, **13** (2003) 289–320.

Section 7 contains the results of investigations of steady state oscillations of solids that can be used in problems of the experimental determination of the dynamic characteristics of bodies of irregular geometric form.

The final section presents some methods that allow essential simplifications in the investigation of eigen oscillations of complex gyroscopic systems (gyrocompasses, gyrostabilizators, etc.).

2 Steady-State Motions and Bifurcation of Rigid Bodies Rotating on a String

The description of the motion of rigid bodies comprises one of the main branches of theoretical mechanics. The solution of problems in this area involves the integration of a set of nonlinear differential equations, the latter invariably including the equations of rigid-body dynamics established by Euler. The motion of a free rigid body or a body that has a fixed point (or moves in a prescribed manner) is the first subject addressed in the given branch of mechanics. Formidable mathematical obstacles have been encountered in dealing with such problems, and the advances that have been made are due to the efforts of such eminent researchers as L. Euler, J.L. Lagrange, L. Poinsot, S.D. Poisson, W. Hess, S.V. Kovalevskaya, D.B. Bobylev, V.A. Steklov, S.A. Chaplygin.

A new direction was added to theoretical mechanics in the 1940s and 1950s when progress was made in the study of the motion of a rigid body suspended on a weightless, ideally flexible inextensible thread (henceforth referred to as a string). However, the number of degrees of freedom in such a system increases to five or six if the string is replaced by a weighable rod with ideal spherical linkages at its ends. In the best case, the number of first integrals remains the same as in the case of a rigid body with a fixed point. Thus, it is no longer certain that a general analytic solution can be found to the problem in general form. It is also difficult to solve numerically this problem with arbitrary initial conditions, due to the extreme awkwardness of the initial differential equations.

This new direction of theoretical mechanics was founded by Academician M.A. Lavrentjev and a well-known scientist-experimentator S.V. Malashenko in the 1940s at the National Academy of Sciences of Ukraine while solving problems on the detonation of shaped charges during which the rotation motion of a body suspended on a string was first studied [1, 2]. The design under experiment consisted of a vertically installed adjusted motor with a flexible string attached to a rotor axis where a body was fixed. A motor rotation was transmitted to the body through a string as well as an angular velocity changing within 20–25 thousand rotations per minute.

The researches conducted by S.V. Malashenko in the 1950s asserted that while rotating axisymmetric rigid bodies suspended on a string in a vertical mode of a

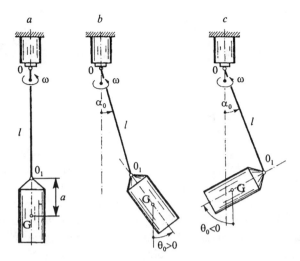

Figure 2.1. Modes of steady-state motion of an axisymmetric rigid body suspended on a string at one point of its dynamic symmetry axis: (a) is the vertical mode of a steady-state motion; (b) is the pendulum mode; (c) is the conical mode with different deviation from a string vertical line and a symmetry axis of a body.

body steady-state motion (Figure 2.1a) under certain angular velocity ω turned into conical mode when a string and a symmetry axis deviated on both sides from a vertical line (Figure 2.1c). Theoretical studies of this initially unexpected effect [3] showed that on a diagram "angular velocity of a steady-state motion ω – angle of string deviation from the vertical" a bifurcation point referred to the mentioned angular velocity value. Moreover these studies predicted the existence of one more mode of steady-state motion appearing at a lower value of angular velocity. In the new mode termed the pendulum mode the string and the symmetry axis of the body had one and the same deviation side (Figure 2.1b). More careful experimental studies [4] confirmed the existence of this mode motion. The equation for determining critical angular velocities of steady-state motion can be preserved in the form [3, 4]

$$\omega^4 - \frac{g}{l}\left[1 + \frac{ma(l+a)}{A-C}\right]\omega^2 + \frac{g}{l}\frac{mga}{A-C} = 0, \qquad (2.1)$$

where A, C are equatorial and axial inertia moments of an axisymmetric rigid body; m, a are correspondingly its mass and a distance from a mass body centre to the point of its attachment to a string; g is acceleration due to gravity; ω is a body angular velocity.

Roots of this equation

$$\omega_{1,2,3,4} = \pm \left\{ \frac{g}{l}\left(\frac{1+\mu+\nu}{2} \pm \left[\left(\frac{1+\mu+\nu}{2}\right)^2 - \mu\right]^{1/2}\right)\right\}^{1/2}, \quad (2.2)$$

$$(\mu = mal/(A-C); \quad \nu = \mu a/l)$$

are valid for $A > C$ (an oblong body). In this case there exist two different absolute magnitude bifurcation values of angular velocity. For $A < C$ (for example, disk) by formulae (2.1) there exists only one bifurcation value of angular velocity and corresponding pendulum mode of steady-state motion, shown on Figure 2.1(b).

In [5] the impact of string length on the stability of a suspended axisymmetric rigid body was studied. While solving the problem it was supposed that the string was weightless, inextensible and ideally flexible, i.e. it was considered to be like a geometrical bond. By means of a linearized motion equation of the considered mechanical systems the necessary stability conditions of a steady-state motion body were obtained with a symmetry axis and a string coinciding with the vertical. It was shown that for the given mode of motion to be steady it is necessary that all roots of the characteristic equation corresponding to a set of differential equations describing the motion of the considered mechanical system

$$\lambda^4 - \frac{C\omega}{A}\lambda^3 - \frac{g}{l}\left[1 + \frac{ma(l+a)}{A}\right]\lambda^2 + \frac{g}{l}\frac{C\omega}{A}\lambda + \frac{mg^2 a}{Al} = 0. \quad (2.3)$$

be real.

When the point of an attaching body to a string is above the centre of mass ($a > 0$) then all roots of eq. (2.3) are real. If the centre of a body mass is above the point of an attaching body to a string ($a < 0$), then in [5] there exists a condition imposed on a body parameter, a length of a string l and angular velocity ω for a body motion to be steady. In [6, 7] a simple method of graphical analysis is proposed permitting as to define the number of real roots of the characteristic eq. (2.3) for $a < 0$ depending on the variation of the parameters ω and l mentioned above. By all these means the necessary conditions of a steady rotation of a rigid body suspended on a string were studied. Note, that at this period of time in [8] the sufficient conditions were obtained and precisely described in [9].

In [8] stability according to Liapunov was proved for a vertical mode of an axisymmetric rigid body rotation for any value of angular velocity in the case when the point of an attaching body to a string is above the centre of mass. The steadiness of the pendulum mode of a steady-state motion shown in Figure 2.1(b) was studied in [10] where using Chetaev's method [11] for corresponding perturbed motion equations Liapunov's function was constructed in a form of linear connection of two first integrals, i.e. energy integral V_1 and squares integral V_2

$$V = V_1 - \omega V_2. \quad (2.4)$$

The pendulum mode of a steady-state motion of a body was determined to be steady for any values of the angular velocity ω.

Thus, for all values $\omega \in [\omega_1, \infty)$ both the principal (vertical) mode and a deviated pendulum mode of a steady-state motion were steady. This fact might seem paradoxical at first sight, not corresponding to the Poincaré law on the change of steady modes of a steady-state motion in bifurcation points [11]. As shown in [12] such distribution of stability of steady-state motion is explained by considering only steady-state motions without eigen rotations of a body around the axis of dynamic symmetry. While considering a set of steady-state motions taking as an example not cylindrical velocities but corresponding generalized impulses [13] distribution of stability will exactly follow the Poincaré law.

Studying the stability of the second deviating in a steady-state motion of the body in Figure 2.1(c) turned out to be hard. So far this issue has not been entirely solved. Some time ago V.N. Skimel showed that such stability of this motion mode (if it holds) was of gyroscopic character [14]. This made research more difficult in comparison with the investigation of steady-state motions with a secular stability. In [15] some analytical calculations were conducted concerning variation equations of a system in which the string length is much more than the linear dimensions of the body. The possibility of instability of the given steady-state motion over a comparatively short range of angular velocity was found. This result had been confirmed in [16] where a sign definite integral of variation. It generalizes Chetaev's gyroscopic integral obtained for a simpler system of differential equations [11] was constructed.

In the above-mentioned studies of the motion of an axisymmetric rigid body suspended on a string the angle of an eigen rotation of the body φ is supposed to be constantly equal to 0. If a body is suspended in such a way that the angle φ is equal to zero not only in steady-state motion of the body but also in its perturbed motion, then the system under consideration will have not 5 degrees of freedom as before but only 4. In this case the three modes of steady-state motion presented in Figure 2.1 are valid. And the first two modes are steady for any value of angular velocity. As for the third mode of motion presented in Figure 2.1(c) there exists a sufficiently big value $\omega = \omega_*$ for all $\omega \in [\omega_*, \infty)$ such that this mode of motion is unsteady [15]. By this fact the motion of an axisymmetric rigid body without its eigen rotation differs from motion of a body suspended in a way that admits eigen rotation. In the last case we deal with the first approximation stability of steady-state motion for sufficiently big ω. This fact is interesting because a decrease in some mechanical system degrees of freedom leads to the appearance of a large number of instability zones. Furthermore we change spherical joints by cylindrical ones. This reduces the number of degrees of freedom to 3. The steady-state motion mode shown in Figure 2.1(c) becomes unsteady for any ω. Next to the vertical mode of motion there appears an instability zone between its first and second bifurcation values. For $\omega > \omega_2$ the vertical form admits gyroscopic stabilization. Thus, the Poincaré law on stability change has been preserved.

It should be noted that in most studies of the motion of a body suspended on a string it was supposed that the length of the string was much larger than the body dimensions. Experimental research has been carried out for such relations between parameters of a system. But it was established that for a "short" string the modes of relative equilibrium can undergo substantial changes. The appearance of new modes is possible. First the problem of studying these modes of steady-state motion was considered in detail by Bugaenko and Veligots'ky [17]. They showed that on the relative equilibrium curve corresponding to the mode represented in Figure 2.1(c) under certain relations between system parameters, i.e.

$$l = a + (\boldsymbol{A} - \boldsymbol{C})/2ma, \qquad (2.5)$$

there exists a point where two new modes of motion branched off as if the formation of one bifurcation upon the other occurred. This process has been described in detail in [15].

3 Stability Study of Vertical Rotations of Rigid Bodies Filled with Fluid

Now let us start to examine the motion of bodies with cavities completely filled with an ideal incompressible fluid. Pioneering experimental studies were conducted in the 1950s by M.A. Lavrent'ev and S.V. Malashenko. Several other researchers theoretically studied the stability of motion of rigid bodies with a liquid filler during that decade. Among them was S.G. Crane, whose findings were contained in his doctoral dissertation. Using Hermitian operators in a Hilbert space, these investigations examined the motion of rigid bodies with a liquid filler about their centre of mass.

Academician S.L. Sobolev was the first to study the stability of motion of gyroscopes with cavities containing a turbulent fluid. In this research, conducted during the war years, Sobolev assumed that a mechanical system comprising a "gyroscope+fluid" rotates as a single rigid body [18] during steady-state motion. During the course of solving this problem, the combined motion of the body and fluid was examined with respect to a nonrotating coordinate system not connected to the body of the rotating gyroscope. This approach complicated the problem greatly. Sobolev employed the methods of functional analysis to find a solution. It was shown in [19] that the solution of the problem addressed in [18] becomes simple and more natural if the motion of the fluid filling the cavity is referred to the coordinate system connected to the body of the hollow gyroscope. It was also determined why gyroscopes have no more than three critical rotational velocities when the cavity is ellipsoidal but have an infinite number of such velocities when the cavity is cylindrical.

The conditions necessary for stable motion of a gyroscope with an ellipsoidal cavity completely filled with an ideal incompressible fluid were reduced to the conditions necessary for all the roots of the third-degree characteristic equation to be real-valued. A detailed analysis of this case was presented in [18, 20]. It was shown in [18, 20] that when the gyroscope contains a cavity in the form of a right circular cylinder filled with an ideal incompressible fluid, there is an infinite (countable) set of partial motions of the fluid in which the moment of force exerted on the wall of the cavity by the fluid is nontrivial. In [21] a technique of starting up a gyroscope is described and a comparison of theoretical and experimental data which turn out to be compatible is presented.

The method of investigation proposed in [19] was also used in [22] to examine the stability of an axisymmetric rigid body rotating on a string and having an ellipsoidal cavity filled with an ideal incompressible fluid. It was assumed that during steady-state motion the body and fluid rotate about the vertical as a single body with a constant angular velocity ω. The string was regarded as noninertial, inextensible, and perfectly flexible, i.e. it was regarded as a holonomic constraint.

The problem was solved within the framework of the theory of linear differential equations. The necessary conditions of stability were reduced to the requirement that the roots of the following characteristic equation be real-valued:

$$(A^0 + k\eta)\lambda^5 - \omega[C + (A^0 + k)\eta]\lambda^4 + \left[-K - (A^* + k\eta)\frac{g}{l} \right.$$
$$\left. + C\omega^2\eta + k\omega^2(1-\eta) \right]\lambda^3 + \left[K\eta + C\frac{g}{l} + \frac{g}{l}(A^* + k)\eta \right]\omega\lambda^2 \qquad (3.1)$$
$$+ \frac{g}{l}[K - k\omega^2(1-\eta) - C\omega^2\eta]\lambda - K\omega\eta\frac{g}{l} = 0.$$

Here,

$$A^0 = A^* - z_c(ml_1 + m_1l_2); \quad A^* = A + ml_1^2 + m_1l_2^2; \quad \eta = \frac{c^2 - a^2}{c^2 + a^2},$$
$$z_c = \frac{ml_1 + m_1l_2}{m + m_1}; \quad k = \frac{4}{15}\pi\varrho a^2 c(c^2 - a^2); \quad K = g(ml_1 + m_1l_2). \qquad (3.2)$$

In the above a and c are the semi-axes of the ellipsoidal cavity; ϱ and m_1 are the specific density and mass of the fluid inside the cavity; l_2 and l_1 are the distances from the point O, where the body is attached to the string to the center of the cavity and the center of mass of the body, respectively; l is the length of the string.

Equation (3.1) was studied by the method of graphical analysis described in [22]. It was shown that the given mechanical system may have one or two regions of angular velocity in which its motion is unstable. The number of regions will depend on the relationship between the parameters of the body and the cavity.

The authors of [23, 24] used the method proposed in [22] to examine the problem of the stability of motion of a rigid body attached to a string and having a cavity in the form of a right circular cylinder filled with an ideal incompressible fluid. M.L. Gorbachuk used the method in [18] to study the complex characteristic equation in the problem as a part of the solution process. It was shown that when certain conditions are imposed on the shape of the cavity, the range $(0, \infty)$ of variation of ω can be divided into two or three intervals $(0, \infty) = (0, \omega_1) \cup (\omega_1, \infty)$ or $(0, \infty) = (0, \omega_1) \cup (\omega_1, \omega_2) \cup (\omega_2, \infty)$ in which the number of negative squares of the indefinite metric is constant and equal to 0,1 or 1,2 respectively. The exact result depends on the coefficients of the system of equation corresponding to the given problem. If $\omega \in (0, \omega_1)$, then the motion of the system will be stable in the first approximation for any type of cavity. For other changes in ω, the stability of the system will depend on the shape of the cavity.

Let us pause to examine one question: Is there a bifurcation in the case of a rigid body filled with a fluid and rotating on a string? To find the answer, one of the authors of this survey examined the problem of finding the branch points of an axisymmetric rigid body with a liquid filler of ellipsoidal form. The following equation was obtained to determine these points

$$\left[(A^* - C + k)l - \frac{Kkz_c}{g}\right]\omega^4 - [g(A^* - C + k) + Kl]\omega^2 + Kg = 0. \tag{3.3}$$

Here, all notations are the same as in equation (3.1).

It follows from analysis of equation (3.3) that there are two angular velocities at which the vertical mode of steady-state motion will give rise to two deflected modes. It was also shown that the solid and liquid will rotate as a single body if the axis of symmetry of the body deviates slightly from the vertical.

Equation (3.3) for the branch points can also be obtained directly if we make $\lambda = \omega$ in the characteristic equation (3.1). This result, seemingly paradoxical at first, was explained in detail in the survey [15] and given the name of the $\lambda - \omega$ theorem. In [25] the proof of this assertion for a system composed of rigid bodies connected one to another by universal joints is presented (see also [26]). The above-mentioned theorem significantly simplifies bifurcation processes in rotating systems enabling an equation to be obtained for determining bifurcation values of angular velocity from the characteristic equations of the system avoiding additional complicated transformations.

4 Steady-State Motions of an Axisymmetric Rigid Body Suspended on a String at a Point off the Axis of Symmetry

Several interesting results have been obtained from the study of an axisymmetric rigid body attached to a string at arbitrary points. It has been found in particular

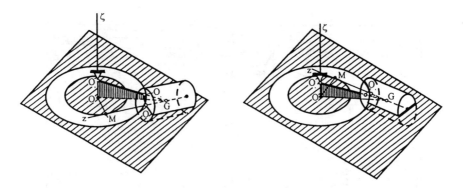

Figure 4.1. Modes of steady-state motion of an axisymmetric rigid body with a dynamic symmetry axis not being in a rotating plane. M is a tangent point of the dynamic symmetry axis of a body z on a circle, a fixed point O_1, a point O_2 of the attaching body to a string and centre of body masses G are on the same line.

that a rotating rigid body whose central ellipsoid of inertia is an ellipsoid of revolution and which is attached to the string at a point not on its axis of symmetry behaves in a surprising manner. As long as the angular velocity is below a certain value ω_0 determined by the equality

$$\omega_0^2 = \frac{g}{(l+a)\sin\tau}, \tag{4.1}$$

(τ being the angle formed by the axis of symmetry of the body and a straight line connecting the centre of mass of the body and the point of its attachment to the string), the axis of symmetry and the string are in the same plane. Stable motion occurs when the body is attached to the string at a point above the symmetry axis. An increase in ω is accompanied by an increase in the angle between the vertical and the axis of the body, this angle finally reaching $\pi/2$ at $\omega = \omega_0$. The symmetry axis turns out to be horizontal in this case, leaving the vertical plane containing the straight line and the string. The axis of symmetry may deviate from this plane either in the direction of rotation or in the opposite direction (Figure 4.1). Such behavior of the body was predicted analytically before being observed experimentally [27].

It was shown in [27–30] that in every mode except the two singular symmetric modes of steady-state motion discussed above, the symmetry axis of the body is located in a revolving vertical plane that contains the string and the centre of mass of the body. All of these other modes of steady-state motion correspond to the

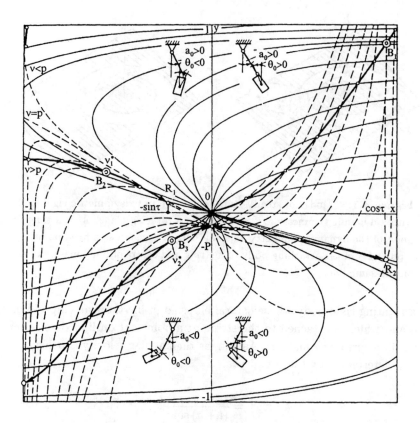

Figure 4.2. Graphic analysis on a plane (x, y) of equations describing the steady-state motion of a prolate axisymmetric rigid body, suspended on a string at a point which is not on the axis of a body's dynamic symmetry.

following conditions of dynamic equilibrium:

$$x = \frac{y}{\kappa}\left(\frac{\nu}{\sqrt{1-y^2}} - 1\right),$$
$$y = \frac{x(\nu + \rho x)}{\sqrt{1-x^2}} - (\kappa + \sigma)x - \rho\sqrt{1-x^2},$$
(4.2)

where we have added the notation

$$x = \sin\theta^0; \quad y = \sin\theta_1^0; \quad \kappa = a/l; \quad \nu = g/(l\omega^2),$$
$$\rho = 2(A - C)\sin\tau\cos\tau/mal; \quad \sigma = 2(A - C)\cos 2\tau/mal,$$
(4.3)

θ_1^0 and θ^0 being the angles between the vertical and, respectively, the string and a

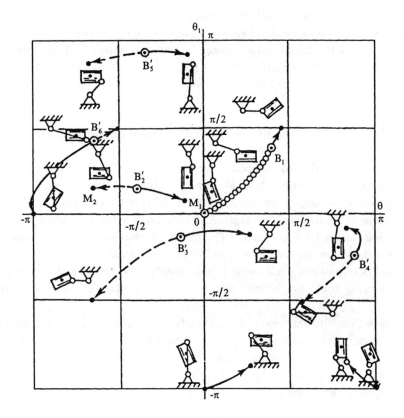

Figure 4.3. The distribution of steady-state motion stability of a prolate rigid body with a dynamic symmetry axis sharing a vertical plane with a string $\left(|\theta_1^0| \leq \frac{\pi}{2}\right)$ or a rigid weightless rod $\left(\frac{\pi}{2} \leq |\theta_1^0| \leq \pi\right)$; steady-state motions with a secular stability are denoted by a line of circles; unstable modes of a motion and modes admitting gyroscopic stabilization are denoted by dotted line and full line correspondingly.

line segment of length a joining the center of mass of the body and the point of its attachment to the string.

Equations (4.2) can be regarded as equations in the unknowns x and y with arbitrary ν (or ω) and fixed values of the other parameters.

By solving (4.2) and using graphical analysis, researchers were able to construct the pattern of steady-state motion in the given case for a prolate body, in particular, a cylinder (Figure 4.2) and an oblate body (disk) [15, 31]. Here, it was assumed that the point of attachment of the body to the string (the suspension point) was located in the plane of the base at a distance from the axis of dynamic symmetry of the body that was short as compared with the radius of the base. It was also assumed that the length l of the string was larger than the distance between the center of mass of the body and the suspension point.

The stability of each mode of steady-state motion of prolate rigid bodies ($A > C$) was studied in [32] by constructing a secular equation and determining the number of positive and negative roots it had. The pattern of steady-state motion shown in Figure 4.3 was obtained as a result.

5 Steady-State Motions of Dynamically Asymmetric Rigid Bodies

In experimental studies of different modes of motion of a rigid body on a string with high angular velocities, S.V. Malashenko observed that the body underwent dynamic self-balancing. This process consists of steady-state motion of the body in which one of its principal central axes of inertia is arbitrarily close to a vertical line passing through a fixed point. The discovery that such a process takes place was the inspiration for the idea (similar to Laval's well-known concept of self-centering) of dynamically balancing rapidly rotating bodies on a string suspension. Such balancing could be theoretically achieved by objects of unusual weight and size.

In connection with the need to theoretically substantiate the idea just referred to, the motion of a rigid body on a string was studied in the most general formulation possible: the body was of arbitrary form and the suspension point was not located on any of the principal central axes of inertia. The equations of relative equilibrium of such a body can be reduced to the form

$$\nu \sin\theta_1 - (\sin\theta_1 + \kappa \sin\theta)\cos\theta_1 = 0,$$

$$\frac{mal \sin(\theta - \theta_1)}{\cos\theta_1} + (A - B)\cos(x^*, \eta_1)\cos(x^*, \zeta_1)$$
$$+ (C - B)\cos(z^*, \eta_1)\cos(z^*, \zeta_1) = 0, \quad (5.1)$$

$$(A - B)\cos(x^*, \xi_1)\cos(x^*, \zeta_1) + (C - B)\cos(z^*, \xi_1)\cos(z^*, \zeta_1) = 0,$$

where x^*, y^*, z^* and A, B, C are respectively the principal central axes and principal central moments of inertia of the body; $\xi_1 \eta_1 \zeta_1$ is the coordinate system rotating with the body; θ and θ_1 are the angles characterizing the position of the body and the string; $\nu = g/l\omega^2$, κ/l are dimensionless parameters.

It was shown in [33] that with an infinite increase in angular velocity, i.e. with $\nu \to 0$, (5.1) are satisfied if one of the principal central axes of inertia of the body coincides with the vertical axis ζ_1 and $l\sin\theta_1 = -a\sin\theta$. The latter equation can be satisfied only if the center of mass of the body is located exactly on a stationary vertical. Thus, with infinitely large values of angular velocity for a body suspended on a string, the limiting mode of steady-state motion (with $\omega \to \infty$) corresponds to rotation of the body about its principal central axis of inertia. This

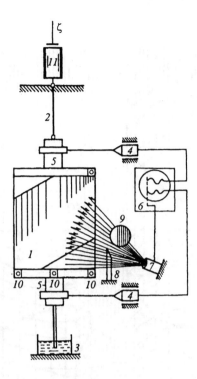

Figure 5.1. The device of a rotor dynamic balancing on a string suspension: 1 is a balanced body; 2 is a string; 3 is a stabilizer of transverse vibrations; 4 is a transmiter of a trunnion displacement; 5 are trunnions of the rotor; 6 is an oscillation registration device; 7 is a stroboscope; 8 is an optical guided line; 9 is a photoelement; 10 are balancing masses; 11 is an electric motor.

result substantiates the notion of the dynamic self-balancing of rigid bodies on a string suspension.

Experimental studies set up to refine the method described above for the dynamic self-balancing of rapidly rotating bodies were conducted on a specially built unit depicted in Figure 5.1. A detailed description of the unit was given in [33]. In the experiments, dynamic balancing was considered to be a process in which, with a specified accuracy, the axes of the journals of model 1 coincided with the model's principal central axis of inertia. Experiments, supported by numerical calculations, showed that for a relatively low angular velocity (of the order of 200 rad/sec) the given method of dynamic balancing makes it possible to determine the location of a 4-gr debalancing weight on the surface of a 20-kg model [33].

In studying the vertical rotation of a dynamically asymmetric body about one of its principal central axes of inertia when suspended on a string, it is found that the distribution of stability along the axis characterizing the changes in angular

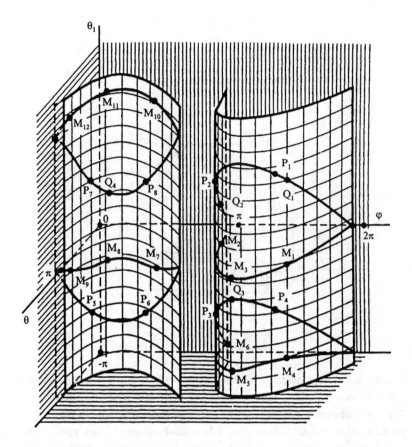

Figure 5.2. Graphico-analytic solution in a space $(\theta, \theta_1, \varphi)$ of equations of relative equilibrium of an arbitrary rigid body suspended on a string or on a weightless rod; P_i, M_i are limit points ($\omega \to \infty$) of the steady-state motion curve; Q_i are points corresponding to steady-state motions with the vertical arrangements of the principal central inertia axis.

velocity is fairly complex [34]. There are almost no effective qualitative methods of studying stability in this case, and researchers usually resort to numerical methods. Nevertheless, the authors of [34] proposed a method of graphical analysis that in a number of cases answers the question: do the gyroscopic terms in the equations of motion lead to gyroscopic stabilization?

The author of [35] used graphical analysis to construct a three-dimensional representation (Figure 5.2) of the solutions of equations (5.1), describing steady-state motions of an arbitrary rigid body suspended on a weightless undeformable rod. The figure shows the modes of motion corresponding to limit points ($\omega \to \infty$) on

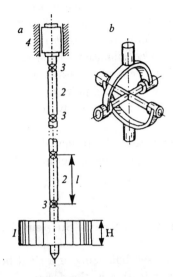

Figure 5.3. Uniform rotation of a rigid body suspended by a system of n weight rigid rods; (a) a general view of the mechanical system: 1 is a tested body; 2 is a rod; 3 is a Kardan–Gook joint; 4 is an electric motor; (b) is the scheme of a Kardan–Gook joint.

the curve of steady-state motion P_i and M_i, as well as those modes corresponding to points representing a static equilibrium position ($\omega = 0$).

Figure 5.3 also shows steady-state motions of the body in which one of its principal central axes of inertia is exactly vertical and angular velocity is finite (these motions correspond to the points designated by the letter Q_i on the curve of steady-state motion). These modes find practical application in experimental determination of the principal axes of inertia of a body. There are many established methods of empirically determining the moments of inertia of a rigid body relative to some fixed axis [36], but less progress has been made in regard to using experiments to determine the direction of the principal central axes of inertia in rigid bodies of arbitrary form. One method that might be used to solve this problem is based on the modes of steady-state motion of a rigid body on a string that were discovered in [37]. In these modes, one of the principal central axes of inertia is exactly vertical. Such modes can be realized at certain finite values of angular velocity of the body. In fact, let $z^* \| \zeta_1$. In this case, equations (5.1) become

$$\nu \sin \theta_1 = (\sin \theta_1 + \kappa \sin \theta) \cos \theta_1; \quad mal \frac{\nu}{\cos \theta_1} \sin(\theta - \theta_1) = 0. \quad (5.2)$$

With allowance for notation (4.3), we find from (5.2) that

$$\theta_1 = \theta; \quad \omega^2 = \omega_*^2 = \frac{g}{(l+a)\cos\theta}. \quad (5.3)$$

Since there are three principal axes, there are also three values of ω at which one of the principal central axes of inertia f the body will be vertical. It follows from the first equality of equations (5.3) that the center of mass of the body is constantly on a straight line serving as a continuation of the line of the string. The inverse is also true. Specifically, if the center of mass of a body undergoing steady-state motion is located on one line with a fixed point and a suspension point, then one of the principal central axes of inertia of the body during such motion will be exactly vertical.

Studies [38, 39] have shown that the principal axes of inertia in a rigid body of arbitrary form can also be determined experimentally using the other modes of steady-state motion of the body.

6 Stability of Centrifuges on a String Suspension

The phenomenon of dynamic self-balancing of a rigid body rotating rapidly on a string has found practical application in the design of centrifugal test stands and centrifuges. A high speed centrifuge was specifically designed to separate heavy isotopes for experimental nuclear research (Figure 6.1). A detailed description of this centrifuge is presented in [40, 41]. Theoretically this suspension resembles a string one, but instead of a string a supporting leg is used. Restoring moments are represented not by gravity force, but by the elastic force produced by magnetic fields of solenoid magnets in 4 and 5. In use this centrifuge could lose its stability for large values of angular velocity (up to 10^5 rot. per min.). In [40] and [41] research into centrifuge stability and bifurcation processes that look place while it was in use is presented. It was shown that the mentioned loss of stability at large angular velocities can be explained by a technological imperfection as well as by some other reasons.

A string suspension is also applicable for the construction of huge centrifugal test stands (some meters in diameter). Formidable technical problems are usually encountered in designing suspensions for such large objects, but the task is made considerably easier when the suspension is made in the form of a multi-link system of weighable rods connected by universal joints. As studies have shown ([6]) such a rod system has many of the same properties as a weighable string. It has therefore been referred to as a "string" suspension.

In contrast to an ideal string, the "string" suspension may have a large intrinsic mass. This must be taken into account when analyzing the motion of the body – especially in the case of motion with high angular velocities. Several investigations have examined this problem. It was first studied in a simplified formulation in [42], where the string suspension was replaced by a single weighable rod hinged (without friction) both to the rigid body being tested (the centrifuge) and to a certain fixed point. It was shown in a first approximation that the fundamental

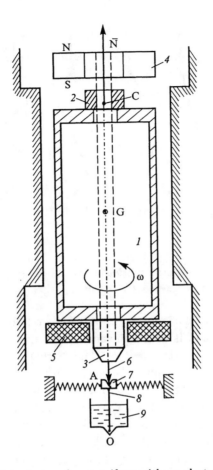

Figure 6.1. The scheme of a centrifuge with an electro-magnetic suspension: 1 is the centrifuge body; 2, 3 are additional ferromagnetic masses; 4, 5 are electromagnets; 6 is a needle supported by a bearing 7; 8 is a leg; 9 is a viscoelastic medium for damping transverse vibrations.

mode of steady-state motion of the given mechanical system is stable. Here, the symmetry axes of both bodies coincide with a vertical passing through the fixed point. As in the case of the motion of an axisymmetric body suspended on a string, the given system also has two critical values of angular velocity ω_1^* and ω_2^* at which two different modes of steady-state motion branch off from the fundamental mode. In the first of these modes, the axes of symmetry of the rod and the body deviate to one side of the vertical. In the second mode, these axes are deflected in opposite directions. The critical values ω_1^* and ω_2^* are located between the analogous values ω_1 and ω_2 corresponding to the case when the weighable rod is replaced by a string. Numerical analysis showed that ω_1^* and ω_2^* differ negligibly from ω_1 and ω_2 even

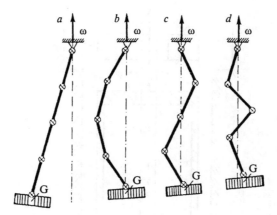

Figure 6.2. The steady-state motion of a rigid body suspended by a system of weight rods joined by Kardan–Gook joints.

when the mass of the rod that replaced the string is relatively large.

A centrifugal test stand is another example of a unit that employs a "string suspension with a large intrinsic mass". The "string" suspension consists of a system of n weighable rods serially connected one to another by universal joints (Figure 5.3). The stability of the vertical rotation of a similar centrifuge was studied in [43], where each weighable rod was represented as a right circular cylinder. It turned out that the vertical mode of dynamic equilibrium of such a system is stable for any value of angular velocity. It was also proved that branch points exit, i.e. that there are values of angular velocity at which new modes branch off from the vertical mode of the dynamic equilibrium. The number of such points is equal to n – the number of rods comprising the "string" suspension. The new branched modes of dynamic equilibrium are shown in Figure 6.2 for $n = 4$.

A more general problem was examined in [44], where a study was made of the stability of the vertical rotation of a statically nonequilibrium system composed of different axisymmetric bodies connected one to another and the rotor of a vertical electric motor by universal joints. The equation of the first approximation of the perturbed motion of the system about the unstable vertical position of equilibrium was obtained in the form

$$\left(m_j a_j + l_j \sum_{k=j+1}^{n} m_k \right) \sum_{p=1}^{j-1} l_p (\ddot{z}_p + 2i\dot{z}_p - z_p) + C_j(-i\dot{z}_j + z_j) \qquad (6.1)$$

$$+ \left(A_j + m_j a_j^2 + l_j^2 \sum_{k=j+1}^{n} m_k \right)(\ddot{z}_j + 2i\dot{z}_j - z_j) + l_j \sum_{k=j+1}^{n} \left(m_k a_k \right.$$

$$+ l_{k+1} \sum_{p=k+2}^{n} m_p \Bigg) (\ddot{z}_k + 2i\dot{z}_k - z_k) - \frac{g}{\omega^2} m_j z_j \Bigg(a_j + \sum_{k=j+1}^{n} l_k \Bigg)$$

$$+ 2\frac{q}{\omega^2} z_j - \frac{q}{\omega^2} z_{j+1} - \frac{q}{\omega^2} z_{j-1} = 0, \quad (j = 1, 2, \ldots, n).$$

In equation (6.1) $z_j = \alpha_j + i\beta_j$ are complex variables characterizing the deviation of the j-th body with mass m_j from the vertical and the moments of inertia $A_j = B_j$, C_j; l_j is the distance between the centers of j-th and $(j+1)$-th universal joint; a_j is the distance between the center of the j-th universal joint and the center of mass of the j-th body; q is a coefficient characterizing the elastic properties of each joint. In the first approximation of the given mechanical system, the condition of stability is the existence of real-valued roots of the characteristic equation

$$\det \| - \boldsymbol{A}\lambda^2 + \boldsymbol{C}\lambda - g\boldsymbol{G}/\omega^2 + q\boldsymbol{Q}/\omega^2 \| = 0, \tag{6.2}$$

where $\boldsymbol{A} = \|a_{ij}\|$; $\boldsymbol{C} = \|c_{ij}\|$; $\boldsymbol{G} = \|g_{ij}\|$; $\boldsymbol{Q} = \|q_{ij}\|$ are matrices whose elements are determined by the formulas

$$a_{ii} = A_i + m_i a_i^2 + l_i^2 \sum_{p=i+1}^{n} m_p,$$

$$a_{ij} = a_{ji} = l_i \Bigg(m_j a_j + l_j \sum_{p=j+1}^{n} m_p \Bigg) \quad (i < j), \tag{6.3}$$

$$c_{ij} = \delta_{ij} C_i, \quad g_{ij} = \delta_{ij} \Bigg(m_j a_j + l_j \sum_{p=j+1}^{n} m_p \Bigg),$$

$$q_{11} = q_{nn} = 1, \quad q_{i,i+1} = q_{i+1,i} = 0,$$

$$q_{kk} = 2, \quad q_{i,i+k} = q_{i+k,i} = 0 \quad (2 < k < n-1),$$

and δ_{ij} is the Kronecker symbol.

Using the properties of the eigenvalues of second-degree matrix bundles, we succeeded in reducing the stability condition to the inequality

$$\omega^2 \geq 4g \max_{r} (\boldsymbol{r} * \boldsymbol{A}\boldsymbol{r})(\boldsymbol{r} * \boldsymbol{G}\boldsymbol{r})(\boldsymbol{r} * \boldsymbol{C}\boldsymbol{r})^{-2}, \tag{6.4}$$

were $(\boldsymbol{r} * \boldsymbol{A}\boldsymbol{r})$, $(\boldsymbol{r} * \boldsymbol{G}\boldsymbol{r})$, $(\boldsymbol{r} * \boldsymbol{C}\boldsymbol{r})$ are Hermitian forms generating the matrices $\boldsymbol{A}$, $\boldsymbol{G}$ and $\boldsymbol{C}$.

Inequality (6.4) has the same form as the well-known Maievsky–Chetaev criterion for the stability of vertical rotation of a single unbalanced Lagrangian gyroscope [11]. Specifically,

$$\omega^2 \geq 4mga(A + ma^2)C^{-2}. \tag{6.5}$$

Condition (6.5) follows naturally from equation (6.4) when $n = 1$. In contrast to the Maievsky-Chetaev condition – which establishes the relationship between angular velocity ω on the one hand and the moments of inertia A and C and potential energy mga, on the other hand – inequality (6.4) establishes the dependence of ω on the Hermitian forms $(r * Ar)$, $(r * Gr)$ and $(r * Cr)$. The coefficients of these forms are the analogous dynamic characteristics, but for a system of coupled rigid bodies. This allows us to regard inequality (6.4) as a generalization of the Maievsky–Chetaev condition to the case of a system of Lagrangian gyroscopes.

The paper [44] proposed the method of finding of the maximum of the functional expressed by right-hand side of (6.4). The authors of this study also presented certain variants of sufficient conditions of stability in a first approximation of the given mechanical system. There conditions were obtained by estimation of the eigenvalues of the matrices on the right side of equation (6.4).

7 Steady Oscillatory Motions of a Rigid Body

In the problem described research into the rotation motions of rigid bodies most of which have been practically applied was conducted. Now consider some results of investigating the properties of oscillatory motions of rigid bodies which can be also applied in practice.

First, consider a pendulum with a horizontal oscillation axis (Figure 7.1) and admit that a pendulum body rotates freely around axis z, connecting the centre of the pendulum mass with an oscillation axis. The motion of this specific pendulum is studied in [45]. It is shown that after a transition such oscillatory motions of the pendulum are set up that in the oscillation plane besides axis z there is another of the principal central inertia axes of the pendulum. The stability of this mode of motion was studied. In the case when additional rotation axis 2 is not principal the equation of pendulum motion can be represented in the form

$$(I_{xx} \cos^2 \gamma + I_{yy} \sin^2 \gamma + I_{xy} \sin 2\gamma)\ddot{\alpha}$$
$$- [(I_{xx} - I_{yy}) \sin \gamma \cos \gamma - I_{xy} \cos 2\gamma]\dot{\alpha}\dot{\gamma}$$
$$- (I_{xz} \cos \gamma - I_{yz} \sin \gamma)\ddot{\gamma} + (I_{xz} \sin \gamma + I_{yz} \cos \gamma)\dot{\gamma}^2 + mgl \sin \alpha = 0, \qquad (7.1)$$
$$I_{zz}\ddot{\gamma} = -\tfrac{1}{2}[(I_{xx} - I_{yy}) \sin 2\gamma - 2I_{xy} \cos 2\gamma]\dot{\alpha}^2 + (I_{xz} \cos \gamma - I_{yz} \sin \gamma)\ddot{\alpha}.$$

In [45] the following approximate solution of the system of equation (7.1):

$$\alpha = \alpha_*(t), \qquad \gamma = \gamma_* + \frac{I_{xz} \cos \gamma_* - I_{yz} \sin \gamma_*}{I_{zz}} \alpha_*(t), \qquad (7.2)$$

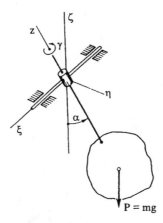

Figure 7.1. Oscillatory motions of a pendulum with an additional rotation axis: α, γ are angles characterizing body position with respect to a stationary system of coordinates $\xi\eta\zeta$ (axes ξ and η are horizontal).

is constructed, where

$$\gamma_* = \frac{1}{2} \arctan \frac{2I_{xy}}{I_{xx} - I_{yy}} \pm k\frac{\pi}{2}, \tag{7.3}$$

and $\alpha_*(t)$ is the solution of the equation

$$(I_{xx} \cos^2 \gamma_* + I_{yy} \sin^2 \gamma_* + I_{xy} \sin 2\gamma_*)\ddot{\alpha} + mgl \sin \alpha = 0. \tag{7.4}$$

From a mechanics viewpoint pendulum oscillations around horizontal axes are like the oscillations of a normal pendulum. Moreover, its undamped oscillations occur around the additional axes 2. Here in the basic oscillation plane there is a middle position of one of the ellipse principal axes obtained by crossing the inertia ellipsoid constructed at the point of intersecting axes ξ and z with a plane xy. This enables one to determine approximately the position of the mentioned principal axis of ellipse section. Thus from an arbitrary system of coordinates xyz one can pass to a new system $x_1 y_1 z_1$, whose axis z_1 coincides with the axis z and the mentioned axis of a section ellipse can be taken as axis x_1. Referring to this new system a centrifugal inertia moment

$$I_{x_1 y_1} = (I_{yy} - I_{xx}) \sin \gamma_* \cos \gamma_* + I_{xy} \cos 2\gamma_*$$

turns out to be close to zero and a system of coordinates $x_1 y_1 z_1$ is supposed to be nearer. Continuing this process one can get closer and closer to the principle axes of body inertia. By numerical experiment, a sufficiently good approach to

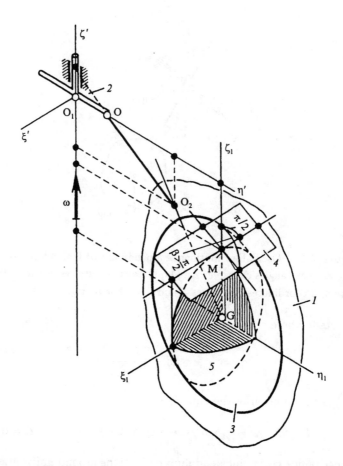

Figure 7.2. Steady-state motions of a rigid body suspended by means of a string and a beam rotating around a vertical axis: 1 is the investigated body; 2 is a beam; 3 is the central ellipsoid of body inertia; 4 is a plane adjacent to the inertia ellipsoid at the point where its surface is crossed by the vertical central axis of body ζ_1; 5 is an ellipse obtained by crossing the central ellipsoid of body inertia with the plane $\xi_1\zeta_1$ (one of the principal axes of this ellipse coincides with the axis ζ_1).

the principal axes of body inertia is obtained after its six consequent oscillatory motions. Thus steady oscillatory motions of a body along with its rotation motions can be used in a problem of experimental determination of principal central inertia axes.

The above mechanical phenomenon is observed in the vibration of a body with one fixed point. Indeed, B.K. Mlodzeevsky established in [46] that if the fixed point is located in a plane containing two principal central axes of inertia of a body (or,

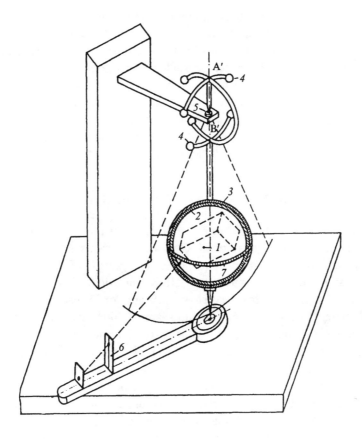

Figure 7.3. Model showing the experimental determination of the principal axes of inertia in rigid bodies of arbitrary configuration: 1 is the tested body; 2, 3 are correspondingly the interior and exterior spheres of body suspension; 4 is the weight for static balancing of the body; 5 is a needle support; 6 is an optical cursor; 7 is the scale for determining the body position with respect to the plane of its oscillations.

in a special case, is located on one of its principal central axes of inertia), then the body will be capable of steady plane-parallel oscillation. In this case, the plane of oscillation will contain two principal central axes of inertia. Similar motions were studied theoretically and experimentally in [47–49]. These investigations showed that when the fixed point is not located on any of the principal central axes of inertia, the body undergoes vibrations that differ little from plane vibration (Figure 7.2). In this case, it becomes possible to fix approximately the location, inside the body, of the principal axes of the ellipse formed by the intersection of the central ellipsoid of inertia and a plane that is perpendicular to the axis that connects the center of mass of the body and the fixed point. This, in its turn, makes it possible to devise a method of determining the principal central axes of inertia in a rigid body of arbitrary form. Analogical motions are observed in a more complicated mechanical system when a body has a ball support, moving along a motionless spherical base [15], that permits us to conduct experimental study of body motion with significant eigen mass. The method just described for successive determination of the directions of principal axes of inertia was realized on a model Figure 7.3. A detailed description of the unit was given in [15].

8 Analysis of Natural Frequencies of Complex Gyroscopic Systems

Identification of the spectrum of natural frequencies is an important point of the analysis of complex mechanical systems. Gyroscopic systems exhibit rather specific distributions of natural vibration frequencies. The motion of the majority of the gyroscopic systems occurs in such a way that the orientation of the gyro axes relative to fixed stars changes very slowly, except for transients at impacts or sharp changes in the forces acting on the system. Such a motion is referred to as the precession motion. When studying precession motions, it suffices to take into account only the variation of proper angular momenta of the gyro rotors. Thus we arrive at the elementary (precession) theory of gyroscopes. In contrast to this, in the transients, the axes of the gyro rotors perform rapid but small conic motions, and the stability analysis requires the angular momenta of all bodies of the system to be taken into account. The corresponding equations of motion are the equations of the nutation theory of gyroscopes. These equations are rather cumbersome and, therefore, approximate methods are needed to investigate them.

Consider, for example, the motion of a gyro compass on a torsional suspension [50], the schematic of which is shown in Figure 8.1.

The linearized equations of motion of this device have the form

$$J_z \frac{d^2\alpha}{dt^2} + H\omega_\eta \alpha + H \frac{d\beta}{dt} + \omega_\eta \omega_\zeta (J_x - I_z)\gamma = 0,$$

$$J_x \frac{d^2\beta}{dt^2} + H\omega_\eta \beta - H \frac{d\alpha}{dt} - \omega_\zeta (J_x + J_y - J_z - C) \frac{d\gamma}{dt}$$
$$+ Ml_2 g(\beta - \lambda) = H\omega_\zeta, \qquad (8.1)$$

$$(J_y - C) \frac{d^2\gamma}{dt^2} + (J_y - C + J_x - J_z)\omega_\zeta \frac{d\beta}{dt} + \omega_\eta \omega_\zeta (J_x - J_z)\alpha$$
$$+ Ml_2 g(\gamma - \mu) = 0,$$

$$l_1 \frac{d^2\lambda}{dt^2} + l_2 \frac{d^2\beta}{dt^2} - 2\omega_\zeta \left(l_1 \frac{d\mu}{dt} + l_2 \frac{d\gamma}{dt} \right) - \omega_\zeta^2 l_2 \beta + g\lambda = 0,$$

$$l_1 \frac{d^2\mu}{dt^2} + l_2 \frac{d^2\gamma}{dt^2} + 2\omega_\varsigma \left(l_1 \frac{d\lambda}{dt} + l_2 \frac{d\beta}{dt} \right) + g\mu - l_2(\omega_\varsigma^2 + \omega_\eta^2)\gamma = 0,$$

$$C\left(\omega_\eta + \omega_\varsigma \beta + \frac{d\gamma}{dt} + \frac{d\delta}{dt} \right) = \text{const} = H,$$

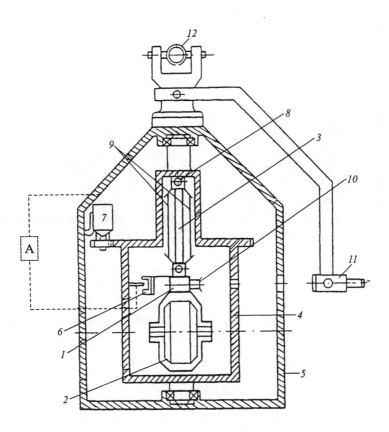

Figure 8.1. A gyro compass on a torsional suspension: 1 is the rod; 2 is the motor of the gyro; 3 is the torsion; 4 is the tracking frame; 5 is the case device; 6 is the data unit of the displacement angle of the tracking hanger and the rod; 7 is the motor; 8 is the torsion clamp; 9 is the current supply; 10 is the speculum; 11 is the autocollimation tube; 12 is the theodolite.

where the angles α, β and γ, determining the position of the gyro compass, together with the angles λ and μ of the deviation of the torsion bar from the vertical and the angle φ of proper rotation of the gyro, form a system of generalized coordinates; ω_ξ, ω_η and ω_ς are the functions characterizing the motion of the suspension point of the device on the Earth spheroid; J_x, J_y and J_z are principal moments of inertia of the system rod-housing-rotor about the center of mass; C is the axial moment of inertia of the rotor; and M is the total mass of the sensor. Equations (8.1) are derived in [50].

The characteristic equation of system (8.1) is too cumbersome to be investigated analytically. Therefore, in [50], this system was substantially simplified. The simplification was based on taking into account realistic relationships between the parameters of the system and the variables characterizing its motion, as a result, the system of equation (8.1) was reduced to the form

$$J_x \frac{d^2\beta}{dt^2} - H \frac{d\alpha}{dt} + H\omega_\eta \beta + Ml_2 g(\beta - \alpha) = H\omega_\varsigma,$$

$$(J_y - C)\frac{d^2\gamma}{dt^2} + Ml_2 g(\gamma - \mu) = 0, \quad J_z \frac{d^2\alpha}{dt^2} + H\omega_\eta \alpha + H \frac{d\beta}{dt} = 0, \quad (8.2)$$

$$l_1 \frac{d^2\lambda}{dt^2} + l_2 \frac{d^2\beta}{dt^2} + g\lambda = 0, \quad l_1 \frac{d^2\mu}{dt^2} + l_2 \frac{d^2\gamma}{dt^2} + g\mu = 0.$$

The analysis of equation of (8.2) permitted one to find analytical expressions for natural vibration frequencies of the system. In particular, an approximate expression for the frequency of the nutation vibration was obtained,

$$\lambda^0 = \frac{1}{2\pi}\left(\frac{H^2}{J_x J_z} + \frac{J_x a + J_z b^*}{J_x J_z} + \frac{g}{l_1}\right)^{\frac{1}{2}}. \quad (8.3)$$

The first term of the radicand in (8.3) considerably exceeds the other terms. In view of this, for practical purposes, formula (8.3) can be approximately replaced by

$$\lambda^0 \approx \frac{1}{2\pi}\left(\frac{H^2}{J_x J_z}\right)^{\frac{1}{2}}. \quad (8.4)$$

This result was highly helpful to the developers of this system. When subjected to the external vibration of a certain frequency, the system failed. It turned out that this dangerous frequency coincided to a high degree of accuracy with the nutation frequency determined by (8.4).

The analysis of natural vibrations in a complex gyroscopic system leads to cumbersome calculations and, moreover, characteristic equations of such systems have high order. In a number of cases, one can replace the input system by a simpler mechanical system that, nevertheless, reproduces all essential features of the processes occurring in the examined system. Approaches making such a reduction possible can be rather helpful. One of these approaches is the method of spring-mass analogy [51].

For a gyro stabilizer (Figure 8.2), the determination of natural vibration frequencies is substantially simplified owing to the specific features of the device. The

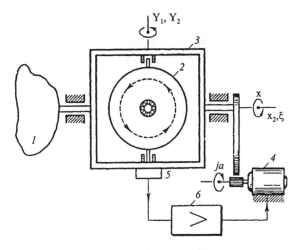

Figure 8.2. The scheme of a one-axis gyroscopic stabilizer: 1 is the stabilized body; 2 is the gyroscope; 3 is the gyroscope external ring; 4 is the electric motor; 5 is the data unit of the gyroscope procession angle; 6 is the booster.

elastic compliance of the bearings of the gyro rotor and the radial deformation of the engaged teeth of the reduction gears substantially exceed the compliance of all other members of the stabilizer. When determining the lower frequencies of small vibrations of this conservative system, one can consider it to consist of four rigid bodies: (1) the rotor of the gyro, (2) the housing of the gyro, (3) the outer ring of the gimbals (together with the body to be stabilized and the driven reduction gear), and (4) the rotor of the electric motor (with the drive gear).

The first two bodies are elastically coupled by the bearing of the rotor axle. We can assume the joint connecting the gyro housing to the outer ring of the suspension to be underformable and the coupling between the outer ring and the rotor of the motor to be elastic. Moreover, we neglect the mass of the housing and friction in the bearings of its axle. Then the equations of small vibrations of this idealized gyroscopic system, when it is placed on in immovable base and the stabilizing motor is not activated, can be represented in the form

$$A\frac{d^2\alpha}{dt^2} + H\frac{d\beta}{dt} = K(\psi - \alpha), \quad A\frac{d^2\beta}{dt^2} - H\frac{d\alpha}{dt} = 0,$$
$$\Psi\frac{d^2\psi}{dt^2} = K(\alpha - \psi) + N(\theta - \psi), \quad \Theta\frac{d^2\theta}{dt^2} = N(\psi - \theta), \tag{8.5}$$

where A is the equatorial moment of inertia of the gyro rotor; H is the proper angular momentum (spin) of the rotor; α and ψ are the angles of rotation of the rotor and housing of the gyro, respectively, about the stabilization axis (these angles are measured with respect to the base); β is the small angle of deviation of the gyro housing from its mean position; K is the stiffness corresponding to the elastic displacement, $\alpha - \psi$ of the gyro rotor with respect to the housing; Ψ is the sum of the moments of inertia of the outer ring, the body to be stabilized, and the driven gear relative to the stabilization axis; θ is the angle of rotation of the motor rotor reduced to the stabilization axis (i.e. the angle of rotation relative to the stator divided by the gear ratio j); N is the stiffness of the reduction gear reduced to the stabilization axis, and Θ is the moment of inertia of the rotor of the motor together with the driven gear, reduced to the stabilization axis.

By integrating the second equation of system (8.5) and subsequently eliminating the angular velocity $\frac{d\beta}{dt}$ from the first equation we arrive at an alternative set of equations governing the motion of the same gyroscopic system. This set of equations has the form

$$A \frac{d^2\alpha}{dt^2} + \frac{H^2}{A} \alpha = K(\psi - \alpha),$$

$$\Psi \frac{d^2\psi}{dt^2} = K(\alpha - \psi) + N(\theta - \psi), \qquad (8.6)$$

$$\Theta \frac{d^2\theta}{dt^2} = N(\psi - \theta).$$

Consider now three point masses A, Ψ and Θ lying on the same straight line. Denote by α, ψ and θ the displacements of the respective masses from the equilibrium positions (see Figure 8.3). The mass A is connected to the mass Ψ by a spring of stiffness K and the mass Θ is attached by means of a spring of stiffness $\frac{H^2}{A}$ to an immovable body of very large mass (to the "inertial" space). The behavior of this system is described by the set of equations coinciding with those of system (8.6). Hence, there exists a mechanical analogy between the frictionless undriven gyro stabilizer and the system of three point masses A, Ψ and Θ elastically connected to each other, with mass A being additionally attached by a "gyroscopic" spring to the "inertial" space.

This and similar analogies permit one to apply methods of the theory of vibrations of lumped-mass elastic systems to the analysis of gyroscopic phenomena. Let, for example,

$$A \ll \Theta < \Psi, \qquad (8.7)$$

which is usually the case for gyro stabilizers. Then one can expect that the mass A virtually does not influence the lower natural vibration frequencies of the system in question. It is apparent that the fundamental frequency ω_1 corresponds to the in-phase motion of all three masses (Figure 8.3). The next frequency corresponds to the vibration in which masses A and Ψ move in phase, whereas mass Θ vibrates

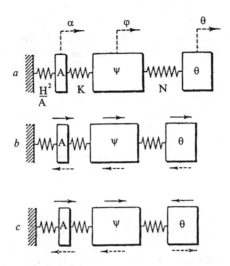

Figure 8.3. A spring-mass analogy of a one-axis gyro stabilizer: (a) the general scheme of "changing" a gyro stabilizer by a system of joint masses; (b) the synphase motion of masses under bottom bounded frequency ω_1; (c) shows the mass motion Θ in antiphase to masses A and Psi under average frequency ω_2.

in anti-phase with respect to A and Ψ. The further simplifications of [51] lead to the approximate relation

$$\omega_3^2 = \frac{1}{A}\left(K + \frac{H^2}{A}\right) = k^2 + \nu^2 \tag{8.8}$$

for the highest vibration frequency of the idealized mechanical system.

References

[1] Malashenko, S.V. (1960). Certain experimental studies pertaining to the rotation of bodies. *Prikl. Mat. Tekh. Fiz.*, **3**, 205–211 (Russian).
[2] Ishlinsky, A.Yu. (1981). Motion of a rigid body on a string. *Usp. Mat. Nauk*, **36**(5), 215–217 (Russian).
[3] Ishlinsky, A.Yu. (1957). Example of bifurcation not leading to unstable modes of steady-state motion. *Dokl. Akad. Nauk SSSR*, **117**(1), 47–49 (Russian).
[4] Ishlinsky, A.Yu., Malashenko, S.V. and Temchenko, M.E. (1958). Bifurcation of stable positions of dynamic equilibrium of one mechanical system. *Izv. Akad. Nauk SSSR, Otd. Tehn. Nauk*, **8**, 53–61 (Russian).
[5] Temchenko, M.E. (1954). Stability of rotation of rigid bodies with an ellipsoidal fluid filler. *Author's Abstract of Physico-Mathematical Sciences Candidate Dissertation*, Kiev (Russian).

[6] Temchenko, M.E. (1969). Study of a criterion of the stability of motion of a rigid body suspended on a string and a gyroscope when both contain an ellipsoidal cavity filled with fluid. *Izv. Akad. Nauk SSSR, Mekh. Tverd. Tela*, **1**, 26–31 (Russian).

[7] Temchenko, M.E. (1983). Stability of gyroscope with the spherical support with moves without friction on spherical surface. In: *Approximate Methods of Studying the Dynamics and Stability of Multidimensional Systems* (Ed.: I.A. Lukovsky). Institute Mat. AN USSR, Kiev, 123–128 (Russian).

[8] Morozova, E.P. (1956). Stability of rotation of a rigid body suspended on a string. *Prikl. Mat. Mekh.*, **30**(5), 621–626 (Russian).

[9] Rubanovsky, V.N. (1983). Analysis of the conditions of stability of uniform vertical rotation of a dynamically symmetric rigid body suspended on a string. In: *Current Problems of Mathematics and Mechanics and Applications*, MFTI, Moscow, 16–21 (Russian).

[10] Temchenko, M.E. (1957). Stability of one of the positions of dynamic equilibrium of one mechanical system. *Dokl. Akad. Nauk SSSR*, **117**(1), 50–52 (Russian).

[11] Chetaev, N.G. (1965). *Stability of Motion*. Izd. AN SSSR, Moscow (Russain).

[12] Vozlinsky, V.I. (1967). Bifurcation of steady-state motions of conservative systems with two cyclic coordinates. *Prikl. Mat. Mekh.*, **31**(5), 841–847 (Russian).

[13] Rumyantsev, V.V. (1966). Stability of steady-state motions. *Prikl. Mat. Mekh.*, **30**(5), 922–933 (Russian).

[14] Skimel', V.N. (1962). Motion of a gyrostat suspended on a string. *Transactions of an Inter-Institute Conference on the Applied Theory of Stability of Motion and Analytical Mechanics*, Kazan', 118–122 (Russian).

[15] Ishlinsky, A.Yu., Storozhenko, V.A. and Temchenko, M.E. (1991). *Rotation of a Rigid Body on a String and Adjacent Problems*. Nauka, Moscow (Russian).

[16] Storozhenko, V.A. and Shuman, B.M. (1991). *Study of the Gyroscopic Stability of One Mode of Steady-State Motion of a Rigid Body on a String*. Preprint 91.32, Institute Mat. AN USSR, Kiev (Russian).

[17] Bugaenko, G.O. and Veligots'ky, G.G. (1963). Modes of dynamics equilibrium of a mechanical system. *Nauk. Zap. Cherk. Derzh. Pedagog. Inst.*, **17**, 21–37 (Russian).

[18] Sobolev, S.L. (1960). Motion of a symmetric gyroscope with a fluid-filled cavity. *Prikl. Mat. Tekh. Fiz.*, **3**, 20–55 (Russian).

[19] Ishlinsky, A.Yu. and Temchenko, M.E. (1960). Small vibrations of the vertical axis of a gyroscope having a cavity filled with an ideal incompressible fluid. *Prikl. Mat. Tekh. Fiz.*, **3**, 65–75 (Russian).

[20] Ishlinsky, A.Yu., Storozhenko, V.A. and Temchenko, M.E. (1994). Dynamics of rigid bodies rotating rapidly on a string and related topics (survey). *Prikl. Mekh.*, **30**(8), 3–30 (Russian).

[21] Malashenko, S.V. and Temchenko, M.E. (1960). Method of experimental studying the stability of motion of a gyroscope having a cavity filled with fluid. *Prikl. Mat. Tekh. Fiz.*, **3**, 76–80 (Russian).

[22] Ishlinsky, A.Yu. and Temchenko, M.E. (1966). Stability of rotation, on a string, of a rigid body with an ellipsoidal cavity filled with an ideal incompressible fluid. *Prikl. Mat. Mekh.*, **30**, 30–41 (Russian).

[23] Gorbachuk, M.L., Sleptsova, G.P. and Temchenko, M.E. (1968). Stability of motion of a rigid body having a liquid filler and suspended on a string. *Ukr. Math. Zh.*, **20**(5), 586–602 (Russian).

[24] Ishlinsky, A.Yu., Gorbachuk, M.L. and Temchenko, M.E. (1986). Stability of motion, on a string, of axisymmetric rigid bodies having cavities filled with fluid. In: *Dynamics of Space Craft and the Study of Outer Space* (Ed.: G.A. Tyulin). Mashinostroenie, Moscow, 234–247 (Russian).

[25] Storozhenko, V.A. (1998). Theory of the stability of vertical motion of systems of connected rigid bodies. *Dokl. RAN.*, **360**(4), 488–490 (Russian).

[26] Storozhenko, V.A. (1999). On studying bifurcation processes in rotating systems of coupled rigid bodies. In: *Problems of Analytical Mechanics and Its Applications* (Ed.: S.M. Onischenko). Institute Mat. NASU, Kiev, 374–386 (Russian).

[27] Ishlinsky, A.Yu., Storozhenko, V.A. and Temchenko, M.E. (1977). Steady-state motions of an axisymmetric rigid body rotation on a string. In: *Dynamics and Stability of Guided Systems* (Ed.: D.G. Korenevsky). Institute Mat. AN USSR, Kiev, 3–20 (Russian).

[28] Ishlinsky, A.Yu., Storozhenko, V.A. and Temchenko, M.E. (1979). Motion of an axisymmetric rigid body suspended on a string. *Izv. Akad. Nauk SSSR, Mekh. Tverd. Tela*, **6**, 3–16 (Russian).

[29] Ishlinsky, A.Yu., Storozhenko, V.A. and Temchenko, M.E (1986). Study of the dynamics of a rigid body with a string suspension at the Institute of Mathematics of the Academy of Sciences of the Ukrainian SSR. In: *Investigations into the Theory of Functions of a Complex Variable, with Applications in Continuum Mechanics* (Ed.: Yu.A. Mitropolsky). Naukova Dumka, Kiev, 16–30 (Russian).

[30] Storozhenko, V.A. and Temchenko, M.E. (1982). Complete pattern of steady-state motions of an axisymmetric rigid body. In: *Analytical Methods of Studying the Dynamics and Stability of Complex Systems* (Ed.: I.A. Lukovsky). Institute Mat. AN USSR, Kiev, 110–136 (Russian).

[31] Ishlinsky, A.Yu., Storozhenko, V.A., Temchenko, M.E. and Shishkin, P.G. (1990). Steady-state motions of an arbitrary rigid body. *Transactions of the All-Union Conference on Analytical Mechanics and the Theory of Stability and Control of Motion (analytical mechanics, rigid-body dynamics)*, VTs AN SSSR, Moscow, 76–94 (Russian).

[32] Storozhenko, V.A. (1985). *Stability of Steady-State Motions of a Solid of Revolution Suspended on a String at an Arbitrary Point*. Preprint 81.13, Institute Mat. AN USSR, Kiev (Russian).

[33] Ishlinsky, A.Yu., Malashenko, S.V., Storozhenko, V.A., Temchenko, M.E and Shishkin, P.G. (1979). Method of balancing rotating bodies on a string drive. *Izv. Akad. Nauk SSSR, Mekh. Tverd. Tela*, **5**, 3–18 (Russian).

[34] Ishlinsky, A.Yu., Storozhenko, V.A. and Temchenko, M.E. (1993). On the study of modes of relative equilibrium of a dynamically asymmetrical solid suspended on a rod. *Izv. Akad. Nauk SSSR, Mekh.Tverd. Tela*, **4**, 60–72 (Russian).

[35] Storozhenko, V.A. (1989). Solution of the equations of relative equilibrium of an arbitrary rigid body suspended on a weightless rod. In: *Stability of Motion of Rigid Bodies and Deformable Systems* (Ed.: I.A. Lukovsky). Institute Mat. AN USSR, Kiev, 31–40 (Russian).

[36] Gernet, M.M. and Ratobyl'sky, V.F. (1969). *Determination of Moments of Inertia*. Mashinostroenie, Moscow (Russian).

[37] Ishlinsky, A.Yu., Malashenko, S.V., Storozhenko, V.A., Temchenko, M.E. and Shishkin, P.G. (1980). Steady-state motions of a rigid body suspended on a string and having one vertical principal central axis of inertia. *Izv. Akad. Nauk SSSR, Mekh. Tverd. Tela*, **2**, 35–45 (Russian).

[38] Storozhenko, V.A. (1986). Determination of the direction of the principal central axes of inertia in a body of arbitrary form by its rotating on a string. In: *Direct Methods in Problems of the Dynamics and Stability of Multi Dimensional System* (Ed.: I.A. Lukovsky). Institute Mat. AN USSR, Kiev, 78–85 (Russian).

[39] Storozhenko, V.A. (1986). *Determination of the Direction of the Principal Central Axes of Inertia in a Body of Arbitrary Form.* Preprint 86.32, Institute Mat. AN USSR, Kiev (Russian).

[40] Ishlinsky, A.Yu., Storozhenko, V.A. and Temchenko, M.E. (1998). On investigation of stability of high-speed centrifuges on electro-magnetic suspension. *Izv. RAN, Mekh. Tverd. Tela*, **3**, 197–208 (Russian).

[41] Ishlinsky, A.Yu., Storozhenko, V.A. and Temchenko, M.E. (1998). Bifurcations and accuracy of the operating of high-speed rotating centrifuge on electro-magnetic suspension. *Izv. RAN, Mekh. Tverd. Tela*, **4**, 69–79 (Russian).

[42] Temchenko, M.E. (1981). Steady-state motions of two coupled bodies. In: *Dynamics and Stability of Complex Systems* (Ed.: I.A. Lukovsky). Institute Mat. AN USSR, Kiev, 95–105 (Russian).

[43] Storozhenko, V.A. (1985). Stability of rotation of a body suspended by a system of serially connected rods. *Izv. Akad. Nauk SSSR, Mekh. Tverd. Tela*, **1**, 45–52 (Russian).

[44] Barnyak, M.Ya. and Storozhenko, V.A. (1988). Study of the stability of vertical rotation of a statically nonequilibrium system of hinged axisymmetric bodies. *Izv. Akad. Nauk SSSR, Mekh. Tverd. Tela*, **1**, 51–58 (Russian).

[45] Storozhenko, V.A. (1983). Stability of motion of a pendulum free to rotate about its principal central axis of inertia. In: *Approximate Methods of Studying the Dynamics and Stability of Multidimensional Systems* (Ed.: I.A. Lukovsky). Institute Mat. AN USSR, Kiev, 115–122 (Russian).

[46] Mlodzeevsky, B.K. (1894). Permanent axes in the motion of a rigid body about a fixed point. *Tr. Otd.-niya Fiz. Nauk O-va Lyubitelei Estestvoznaniya, Antropologii i Etnografii*, **7**(1), 46–48 (Russian).

[47] Ishlinsky, A.Yu., Vasilenko, V.P., Storozhenko, V.A., Temchenko, M.E. and Shishkin, P.G. (1985). Mode of steady-state vibration of a heavy rigid body. *Izv. Akad. Nauk SSSR, Mekh. Tverd. Tela*, **2**, 3–18 (Russian).

[48] Ishlinsky, A.Yu., Storozhenko, V.A., Temchenko, M.E. and Shishkin, P.G. (1985). Certain plane-parallel motions of a rigid body. In: *Problems of Analytical Mechanics and Control of Motion* (Ed.: V.N. Rubanovsky). VTc AN SSSR, Moscow, 67–75 (Russian).

[49] Ishlinsky, A.Yu., Storozhenko, V.A., Temchenko, M.E. and Shishkin, P.G. (1986). Study of the stability of pendulum-like oscillations of a heavy rigid body with one fixed point. *Izv. Akad. Nauk SSSR, Mekh. Tverd. Tela*, **1**, 18–26 (Russian).

[50] Vasilenko, V.P. and Temchenko, M.E. (1966). Theory of a gyro compass on a torsional suspension. *Engeen. Journal, Mekh. Tverd. Tela*, **1**, 6–13 (Russian).

[51] Ishlinsky, A.Yu. (1976). *Orientation, Gyroscopes and Inertial Navigation.* Nauka, Moscow (Russian).

4.4 PROGRESS IN STABILITY OF IMPULSIVE SYSTEMS WITH APPLICATIONS TO POPULATION GROWTH MODELS*

XINZHI LIU[†]

Department of Applied Mathematics, University of Waterloo, Waterloo, Canada

1 Introduction

Many physical systems are characterized by the fact that at certain moments of time they experience a sudden change of their state. For example, when a mass on a spring is given a blow by a hammer, it experiences a sharp change of velocity; and a pendulum in a mechanical clock undergoes a drastic increase of momentum every time when it crosses its equilibrium position. These systems are subject to short-term perturbations which are often assumed to be in the form of impulses in the modelling process. Consequently, impulsive differential equations provide a natural description of such systems [1, 2, 12].

In this paper, we investigate the problem of stability for impulsive differential systems by Lyapunov's direct method. In Section 2, we describe impulsive differential systems and introduce some notations and definitions. We establish, in Section 3, some stability criteria which may be considered as impulsive stabilization of the underlying continuous physical system. It may provide a greater prospect to solving problems that are basically defined by continuous dynamical systems, but on which only discrete-time actions are exercised. As an application, we apply our results, in Section 4, to some population growth models.

---

*Advances in Stability Theory (Ed.: A.A. Martynyuk). Stability and Control: Theory, Methods and Applications, Taylor & Francis, London, **13** (2003) 321–338.

[†]Research partially supported by NSERC Canada.

2 Preliminaries

Consider the impulsive system

$$\begin{cases} x' = f(t,x), & t \neq \tau_k(x), \\ \Delta x = I_k(x), & t = \tau_k(x), \quad k = 1, 2, \ldots, \end{cases} \quad (2.1)$$

where $f: R_+ \times R^n \to R^n$, $\tau_k: R^n \to R_+$, $I_k: R^n \to R^n$ and $\Delta x(t) = x(t^+) - x(t^-)$.

Let $t_0 \in R_+$ and $x_0 \in R^n$. Denote by $x(t, t_0, x_0)$ the solution of (2.1) satisfying the initial condition $x(t_0^+, t_0, x_0) = x_0$. The solutions $x(t) = x(t, t_0, x_0)$ of system (2.1) are, in general, piecewise continuous functions with points of discontinuity of first type at which they are left continuous, i.e. at the moment t_k, when the integral curve of the solution $x(t)$ meets the hypersurface

$$S_k = \{(t, x) \in R_+ \times R^n : t = \tau_k(x)\}$$

the following relations are satisfied:

$$x(t_k^-) = x(t_k) \quad \text{and} \quad \Delta x(t_k) = x(t_k^+) - x(t_k^-) = I_k(x(t_k)).$$

We shall assume that for each $x \in R^n$,

$$0 < \tau_1(x) < \ldots < \tau_k(x) < \ldots \quad \text{and} \quad \lim_{k \to \infty} \tau_k(x) = +\infty,$$

and the integral curve of each solution $x(t) = x(t, t_0, x_0)$ of system (2.1) meets each of the hypersurface $\{S_k\}$ at most once.

Let $\tau_0(x) \equiv 0$ for $x \in R^n$ and introduce the sets

$$G_k = \{(t, x) \in R_+ \times R^n; \tau_{k-1}(x) < t \leq \tau_k(x)\}, \quad k = 1, 2, \ldots,$$

and

$$G = \bigcup_{k=1}^{\infty} G_k.$$

We shall assume that $f(t, x)$ is continuous on each of the sets $\{G_k\}$ and for $(t_k, x) \in S_k$, $k = 1, 2, \ldots$, the limit

$$\lim_{\substack{(t,y) \to (t_k, x) \\ (t,y) \in G_{k+1}}} f(t, y) = f(t_k^+, x)$$

exists. Let $V: R_+ \times R^n \to R_+$. Then V is said to belong to class ν_0 if V is continuous on each of the sets $\{G_k\}$, for $(t_k, x) \in S_k$, $k = 1, 2, \ldots$, the limit

$$\lim_{\substack{(t,y) \to (t_k, x) \\ (t,y) \in G_{k+1}}} V(t, y) = V(t_k^+, x)$$

exists and V is locally Lipshitzian in x on each G_k.

For $(t, x) \in G_k$, we define, as usual, the upper right derivative of V along solutions of (2.1) by

$$D^+V(t,x) = \lim_{\delta \to 0^+} \sup \frac{1}{\delta}[V(t+\delta, x + \delta f(t,x)) - V(t,x)]$$

We denote by K the class of functions $\phi \colon R_+ \to R_+$ which are continuous, strictly increasing and $\phi(0) = 0$, CK the class of continuous functions $\sigma \colon R_+ \times R_+ \to R_+$ such that $\sigma(t, \cdot) \in K$ for each $t \in R_+$, and Γ the class of continuous functions $h \colon R_+ \times R^n \to R_+$ with $\inf h(t,x) = 0$. A function $h \in \Gamma$ is said to be in Γ_0, if $\inf_{x \in R^n} h(t,x) = 0$.

Definition 2.1 Let $h_0, h \in \Gamma$. Then we say that
 (i) h_0 is *finer* than h if $h(t,x) \leq \sigma(t, h_0(t,x))$ whenever $h_0(t,x) < \delta$ for some function $\sigma \in CK$ and constant $\delta > 0$;
 (ii) h_0 is *uniformly finer* than h if the function σ in (i) is independent of t.

Definition 2.2 Let $h_0, h \in \Gamma$. Then we say that $V(t,x)$ is
 (i) *h-positive definite* if $V(t,x) \geq b(h(t,x))$ when $h(t,x) < \rho$ for some function $b \in K$ and constant $\rho > 0$;
 (ii) *weakly h-decrescent* if $V(t,x) \leq a(t, h(t,x))$ when $h(t,x) < \gamma$ for some function $a \in CK$ and constant $\gamma > 0$;
 (iii) *h-decrescent* if a in (ii) is independent of t.

Definition 2.3 Let $h_0, h \in \Gamma$. Then the *system* (2.1) is said to be
 (S_1) (h_0, h)-*stable*, if given $\epsilon > 0$ and $t_0 \in R_+$, there exists $\delta = \delta(t_0, \epsilon) > 0$ such that $h_0(t_0, x_0) < \delta$ implies that $h(t, x(t)) < \epsilon$, $t \geq t_0$, where $x(t) = x(t_1, t_0, x_0)$ is any solution of (2.1);
 (S_2) (h_0, h)-*uniformly stable*, if δ in (S_1) is independent of t_0;
 (S_3) (h_0, h)-*asymptotically stable*, if (S_1) holds and for $t_0 \in R_+$ there exists $\alpha = \alpha(t_0) > 0$ such that $h_0(t_0, \alpha_0) < \alpha$ implies $\lim_{t \to \infty} h(t, x(t)) = 0$;
 (S_4) (h_0, h)-*uniformly asymptotically stable*, if (S_2) holds and there exists $\alpha > 0$ for any given $\epsilon > 0$ there exists $T = T(\epsilon) > 0$ such that $\forall t_0 \in R_+$, $h_0(t_0, x_0) < \alpha$ implies that $h(t, x(t)) < \epsilon$, $t \geq t_0 + T$;
 (S_5) (h_0, h)-*unstable*, if (S_1) fails to hold.

Based on the Definition 2.3 and the usual stability concepts, it is easy to formulate other kinds of stability in terms of two measures (h_0, h). We shall be content in giving a few choices of the two measures (h_0, h) to demonstrate the generality of the Definition 2.3. It is easy to see that Definition 2.3 reduces to
 (1) the well known stability of the trivial solution $x(t) \equiv 0$ of (2.1) if $h(t,x) = h_0(t,x) = \|x\|$;

(2) the stability of the prescribed motion $x_0(t)$ of (2.1) if $h(t,x) = h_0(t,x) = \|x - x_0(t)\|$;
(3) the partial stability of the trivial solution of (2.1) if $h(t,x) = \|x\|_s$, $1 \leq s < n$ and $h_0(t,x) = \|x\|$;
(4) the stability of asymptotically invariant set $\{0\}$, if $h(t,x) = h_0(t,x) = \|x\| + \sigma(t)$, where $\sigma(t)$ is strictly decreasing and $\lim_{t \to \infty} \sigma(t) = 0$;
(5) the stability of the invariant (or an arbitrary) set $A \subset R^n$ if $h(t,x) = h_0(t,x) = d(x,A)$, where $d(x,A)$ is the distance of x from the set A;
(6) the stability of conditionally invariant (or an arbitrary) set B with respect to A, where $A \subset B \subset R^n$, if $h(t,x) = d(x,B)$ and $h_0(t,x)) = d(x,A)$.

Several other combinations of choices are possible for (h_0, h) in addition to those given above. For a detailed discussion of this point, see [5].

We shall assume, in what follows, that each solution $x(t) = x(t, t_0, x_0)$ of (2.1) defined on $[t_0, t_0 + \alpha)$ with $0 < \alpha < \infty$ is continuable to the right if $(t, x(t)) \in S(h, \rho)$ for $t \in [t_0, t_0 + \alpha)$.

3 Main Results

We shall state our main results in this section.

Theorem 3.1 *Assume that*

(i) $h_0 \in \Gamma_0$, $h \in \Gamma$ *and* h_0 *is finer than* h;
(ii) $V \in \nu_0$, $V(t,x)$ *is h-positive definite, weakly h_0-decrescent and*

$$D^+V(t,x) \leq 0, \quad (t,x) \in G \cap S(h,\rho);$$

(iii) $V(t^+, x + I_k(x)) \leq V(t,x)$, $(t,x) \in S_k \cap S(h,\rho)$;
(iv) *there exists* $\rho_0 \in (0, \rho)$, *if* $(t,x) \in S(h, \rho_0)$, *then*

$$h(t^+, x + I_k(x)) < \rho, \quad \forall k \geq 1.$$

Then the system (2.1) is (h_0, h)-stable.

Remark 3.1 We can conclude (h_0, h)-uniform stability if, in Theorem 3.1, h_0 is uniformly finer than h and $V(t,x)$ is h_0-decrescent.

Theorem 3.2 *Assume that conditions (i), (ii) and (iv) of Theorem 3.1 hold. Suppose further that*

(iii*) $V(t^+, x + I_k(x)) \leq V(t,x) - \lambda_k \psi(V(t,x))$, $(t,x) \in S_k \cap S(h,\rho)$, *where* $\lambda_k \geq 0$, $\sum_{k=1}^{\infty} \lambda_k = \infty$, $\psi \in C[R_+, R_+]$, $\psi(s) > 0$ *if* $s > 0$ *and* $\psi(0) = 0$.

Then the system (2.1) is (h_0, h)-asymptotically stable.

Theorem 3.3 *Assume that*

(i) $h_0, h \in \Gamma$, *and there exists* $(t^*, x^*) \in S(h, \rho)$ *such that* $h_0(t^*, x^*) = 0$;
(ii) $V \in v_0$, $V(t,x)$ *is bounded on* $S(h, \rho)$ *and*

$$D^+V(t,x) \geq 0, \quad (t,x) \in G \cap S(h,\rho);$$

(iii) $V(t^+, x + I_k(x)) \geq V(t,x) + \lambda_k \psi(V(t,x))$, $(t,x) \in S_k \cap S(h,\rho)$, *where* $\lambda_k \geq 0$, $\sum_{k=1}^{\infty} \lambda_k = \infty$, $\psi \in C[R_+, R_+]$, $\psi(s) > 0$ *if* $s > 0$ *and* $\psi(0) = 0$.

Then the system (2.1) is (h_0, h)-*unstable.*

Example 3.1 Consider the nonlinear system

$$\begin{cases} x' = y, & t \neq \tau_k(x,y), \\ y' = -x - \gamma x^3, & t \neq \tau_k(x,y), \\ \Delta x = \alpha_k x, & t = \tau_k(x,y), \\ \Delta y = \beta_k y, & t = \tau_k(x,y), \end{cases} \quad (3.1)$$

where $\gamma > 0$, $\beta_k \leq \alpha_k$ $k = 1, 2, \ldots$.

Let $V(x,y) = \frac{1}{2}x^2 + \frac{1}{4}\gamma x^4 + \frac{1}{2}y^2$, $h_0 = h = \sqrt{x^2 + y^2}$. Then V is h-positive definite, h_0-decrescent and for $t \neq \tau(x,y)$

$$D^+V(x,y) = xx' + \gamma x^3 x' + yy' = 0.$$

When $t = \tau_k(x,y)$, we have

$$V(x+\Delta x, y+\Delta y) - V(x,y) = \frac{1}{2}(\alpha_k^2 + 2\alpha_k)x^2 + \frac{1}{4}\gamma[(1+\alpha_k)^4 - 1]x^4 + \frac{1}{2}(\beta_k^2 + 2\beta_k)y^2.$$

Assume that $\lim_{k \to \infty} \beta_k = 0$, $\sum_{k=1}^{\infty} |\beta_k| = \infty$ and $\sum_{k=1}^{\infty} \beta_k^2 < \infty$. If $\alpha_k \leq 0$ and $\beta_k \geq -1$ for all $k = 1, 2, \ldots$, then

$$V(x + \Delta x, y + \Delta y) \leq V(x,y) + (\beta_k^2 + 2\beta_k)V(x,y).$$

Let $\lambda_k = -\beta_k^2 - 2\beta_k$. Then $\lambda_k \geq 0$ and $\sum_{k=1}^{\infty} \lambda_k = \infty$. Thus by Theorem 3.2 the system (3.1) is (h_0, h)-asymptotically stable. On the other hand, if $\beta_k \geq 0$ for all $k = 1, 2, \ldots$, then

$$V(x + \Delta x, y + \Delta x) \geq V(x,y) + (\beta_k^2 + 2\beta_k)V(x,y).$$

Since $\beta_k^2 + 2\beta_k \geq 0$ and $\sum_{k=1}^{\infty}[\beta_k^2 + 2\beta_k] = \infty$, it follows by Theorem 3.3 that the system (3.1) is (h_0, h)-unstable.

Definition 3.1 Let $\lambda\colon R_+ \to R_+$ be a measurable function. Then $\lambda(t)$ is called *integrally positive* if

$$\int_I \lambda(s)ds = \infty \quad \text{wherever} \quad I = \bigcup_{i=1}^{\infty}[\alpha_i, \beta_i],$$

$\alpha_i < \beta_i < \alpha_{i+1}$ and $\beta_i - \alpha_i \geq \delta > 0$.

Theorem 3.4 *Assume that all conditions of Theorem 3.1 hold and suppose further that*
  (i) $D^+V(t,x) \leq -\lambda(t)c(W(t,x))$, $(t,x) \in G \cap S(h,\rho)$, *where* $\lambda(t)$ *is integrally positive, $c \in K$ and $W \in \nu_0$;*
  (ii) $W(t,x)$ *is h-positive definite on* $S(h,\rho)$ *and for any solution* $x(t)$ *of (2.1) such that* $(t, x(t)) \in S(h, \rho)$, *the function*

$$\int_0^t [D^+W(s, x(s))]_+ \, ds \quad \left(\int_0^t [D^+W(s, x(s))]_- \, ds\right)$$

is uniformly continuous on R_+, *where* $[\cdot]_+([\cdot]_-)$ *means that the positive (negative) part is considered for all* $s \in R_+$;
  (iii) $W(t^+, x + I_k(x)) \leq W(t, x)$ $(W(t^+, x + I_k(x)) \geq W(t, x))$ *on* $S_k \cap S(h, \rho)$.

Then system (2.1) is (h_0, h)-*asymptotically stable.*

Remark 3.2 Condition (ii) in Theorem 3.4 can be fulfilled if $D^+W(t,x)$ is bounded from above or from below on $S(h,\rho)$ by a function $q(t)$ which is uniformly continuous on R_+.

Theorem 3.5 *Assume that all conditions in Theorem 3.1 and Remark 3.1 hold and suppose further that*

$$D^+V(t,x) \leq -c(h_0(t,x)), \quad (t,x) \in G \cap S(h,\rho),$$

where $c \in K$. *Then system (2.1) is* (h_0, h)-*uniformly asymptotically stable.*

Theorem 3.6 *Assume that conditions (i) and (iv) of Theorem 3.1 hold and suppose further that*
  (i) $V \in \nu_0$, $V(t,x)$ *is h-positive definite, weakly h_0-decrescent and*

$$D^+V(t,x) \leq -\lambda(t)c(V(t,x)), \quad (t,x) \in G \cap S(h,\rho),$$

where $\lambda\colon R_+ \to R_+$ *is locally integrable and* $c \in K$;
  (ii) $V(t,x) \leq V(t^+, x + I_k(x)) \leq \psi_k(V(t,x))$, $(t,x) \in S_k \cap S(h,\rho)$, *where* $\psi_k \in C[R_+, R_+]$, $\psi_k(0) = 0$ *and* $\psi_k(s) > 0$ *if* $s > 0$;

(iii) there exists $\sigma > 0$ such that for any solution $x(t)$ of (2.1) which meets each hypersurface S_k at t_k and $V(t_k, x(t_k)) \in (0, \sigma)$,

$$-\int_{t_{k-1}}^{t_k} \lambda(s)\,ds + \int_{V(t_k,x(t_k))}^{\psi_k(V(t_k,x(t_k)))} \frac{ds}{c(s)} \leq -\gamma_k,$$

where $\gamma_k \geq 0$ and $\sum_{k=1}^{\infty} \gamma_k = \infty$.

Then system (2.1) is (h_0, h)-asymptotically stable.

Remark 3.3 Condition (iii) of Theorem 3.6 is satisfied if

$$\inf_k \left[\inf_{(\tau_k(x),x) \in S(h,\rho)} \tau_k(x) - \sup_{(t_k(x),x) \in S(h,\rho)} \tau_k(x) \right] = \beta > 0,$$

$$\inf_{t \in R_+} \int_t^{t+\beta} \lambda(s)\,ds = \alpha > 0 \quad \text{and} \quad -\alpha + \int_\eta^{\psi_k(\eta)} \frac{ds}{c(s)} = -\gamma < 0, \quad \forall \eta \in (0,\sigma).$$

The next result is on (h_0, h)-instability.

Theorem 3.7 *Assume that*

(i) $h, h_0 \in \Gamma$ and there exists $(t^*, x^*) \in S(h, \rho)$ such that $h_0(t^*, x^*) = 0$;
(ii) $V \in \nu_0$, $V(t, x)$ is bounded on $S(h, \rho)$ and

$$-\lambda(t)c(V(t,x)) \leq D^+V(t,x) \leq 0, \quad (t,x) \in G \cap S(h,\rho),$$

where $\lambda: R_+ \to R_+$ is locally integrable and $c \in K$;
(iii) $V(t^+, x + I_k(x)) \leq \psi_k(V(t,x))$, $(t,x) \in S_k \cap S(h,\rho)$, where $\psi_k \in C[R_+, R_+]$, $\psi_k(0) = 0$ and $\psi_k(s) > 0$ if $s > 0$;
(iv) for any solution $x(t)$ of (2.1) which meets S_k at t_k and $(t, x(t)) \in s(h, \rho)$,

$$-\int_{t_{k-1}}^{t_k} \lambda(s)\,ds + \int_{V(t_k,x(t_k))}^{\psi_k(V(t_k,x(t_k)))} \frac{ds}{c(s)} \geq \gamma_k,$$

where $\gamma_k \geq 0$ and $\sum_{k=1}^{\infty} \gamma_k = \infty$.

Then system (2.1) is (h_0, h)-unstable.

Example 3.2 Consider the nonlinear system

$$\begin{cases} x_1' = -2x_2 + x_2x_3 - x_1, & t \neq \tau_k(x), \\ x_2' = x_1 - x_1x_3 - x_2, & t \neq \tau_k(x), \\ x_3' = x_1x_2 - x_3, & t \neq \tau_k(x), \\ \Delta x_i = d_i x_i, \; i = 1,2,3, & t = \tau_k(x), \end{cases} \quad (3.2)$$

where $\tau_k(x) \equiv kT$, $T > 0$, $k = 1, 2, \ldots$, and $d_i > 0$, $i = 1, 2, 3$.

Let $V(x) = x_1^2 + 2x_2^2 + x_3^2$, $h_0 = h = \sqrt{x_1^2 + x_2^2 + x_3^2}$. Then V is h-positive definite, h-decrescent and for $t \neq kT$,

$$D^+V(x) = 2x_1(-2x_2 + x_2x_3 - x_1) + 4x_2(x_1 - x_1x_3 - x_2)$$
$$+ 2x_3(x_1x_2 - x_3) = -2V(x).$$

When $t = kT$ we have

$$V(x + \Delta x) = (1 + d_1)^2 x_1^2 + 2(1 + d_2)^2 x_2^2 + (1 + d_3)^3 x_3^2.$$

Let $m = \min\{d_1, d_2, d_3\}$ and $M = \max\{d_1, d_2, d_3\}$. Then

$$(1+m)^2 V(x) \leq V(x + \Delta x) \leq (1+M)^2 V(x).$$

It is easy to see that in this example condition (iv) of Theorem 3.6 reduces to

$$-T + \ell n(1 + M) < 0.$$

Thus we conclude that the system (3.6) is (h_0, h)-asymptotically stable if $M < e^T - 1$. On the other hand, application of condition (iv) of Theorem 3.4 yields

$$-T + \ell n(1 + m) > 0 \quad \text{or} \quad m > e^T - 1,$$

which implies that the system (3.7) is (h_0, h)-unstable if $m > e^T - 1$. In case $d_1 = d_2 = d_3$, then $m = M$ and this shows the sharpness of the assumptions given in our theorems.

Theorem 3.8 *Assume that conditions (i) and (iv) of Theorem 3.1 hold and suppose further that*

(i) $V_1 \in \nu_0$, $V_1(t, x)$ *is h-positive definite, weakly h_0-decrescent and*

$$D^+V_1(t, x) \leq -\lambda(t)c_1(W(t, x)), \quad (t, x) \in G \cap S(h, \rho),$$

where $\lambda(t)$ is integrally positive, $c_1 \in K$ and $W \in \nu_0$;

(ii) *conditions (ii) and (iii) of Theorem 3.4 hold and*

$$V_1(t^+, x + I_k(x)) \leq V_1(t, x), \quad (t, x) \in S_k \cap S(h, \rho);$$

(iii) $V_2 \in \nu_0$, $V_2(t, x) + W(t, x)$ *is h-positive definite and*

$$D^+V_2(t, x) \leq -c_2(V_2(t, x)) + \psi(W(t, x)), \quad (t, x) \in G \cap S(h, \rho),$$

where $c_2 \in K$, $\psi \in C[R_+, R_+]$ and $\psi(0) = 0$;

(iv) $V_2(t^+, x + I_k(x)) \leq V_2(t, x)$, $(t, x) \in S_k \cap S(h, \rho)$.

Then system (2.1) is (h_0, h)-asymptotically stable.

Example 3.3 Consider the nonlinear system

$$\begin{cases} x_1' = -\frac{1}{2}x_1 - (1+\cos^2 t)x_2, & t \neq \tau_k(x), \\ x_2' = x_1 - 2x_1x_2^2 - (5+4\cos^2 t)x_2^3, & t \neq \tau_k(x) \\ \Delta x_1 = -\frac{1}{1+1/2k}x_1, & t = \tau_k(x) \\ \Delta x_2 = -\frac{1}{1+1/k}x_2, & t = \tau_k(x) \end{cases} \quad (3.3)$$

where $\tau_k(x) \equiv t_k$, $0 < t_1 < t_2 < \cdots < t_k < \cdots$ and $t_k \to \infty$ as $k \to \infty$. Choose $V_1(t,x) = 2x_1^2 + 2x_1x_2 + (\frac{5}{2} + 2\cos^2 t)x_2^2$ and $h_0 = h = \sqrt{x_1^2 + x_2^2}$. Then

$$D^+V_1(t,x) \leq -2\cos^2 t\, x_2^2, \quad t \neq t_k.$$

where $\cos^2 t$ is integrally positive and

$$V_1(t, x+\Delta x) \leq V_1(t,x), \quad t = t_k.$$

Let $V_2(t,x) = x_1^2$. Then

$$D^+V_2(t,x) \leq -V_2(t,x) + Mx_2^2, \quad t \neq \tau_k, \quad M > 0,$$
$$V_2(t, x+\Delta x) \leq V_2(t,x), \quad t = t_k.$$

Thus by Theorem 3.8 system (3.3) is (h_0, h)-asymptotically stable.

We shall next consider the impulsive system with fixed moments of impulses

$$\begin{cases} x' = f(t,x), & t \neq t_k, \\ \Delta x = I_k(x), & t = t_k, \end{cases} \quad (3.4)$$

and the scalar impulsive equation

$$\begin{cases} u' = g(t,u), & t \neq t_k, \\ \Delta u = J_k(u) - u, & t = t_k, \end{cases} \quad (3.5)$$

where $g\colon R_+ \times R \to R$ is continuous on $(t_{k-1}, t_k] \times R$ and $\lim_{(t,\nu)\to(t_k^+,u)} g(t,\nu) = g(t_k^+, u)$ exists and $J_k\colon R_+ \to R_+$ is nondecreasing.

We have the following comparison result.

Theorem 3.9 *Assume that conditions in Remark 2.1 hold and suppose that*
(i) $V \in \nu_0$, $V(t,x)$ *is h-positive definite, h_0-decrescent and*

$$D^+V(t,x) \leq g(t, V(t,x)), \quad t \neq t_k, \quad (t,x) \in S(h, \rho),$$

where g is given in (3.5) and $g(t,0) \equiv 0$;
(ii) $V(t_k^+, x + I_k(x)) \leq I_k(V(t_k, x))$, $(t_k, x) \in S(h, \rho)$.

Then the stability properties of the trivial solution of (3.5) imply the corresponding (h_0, h)-stability properties of (3.4).

The next result shows the advantage of utilizing a family of Lyapunov functions in proving uniform stability.

Theorem 3.10 *Assume that condition (iv) of Theorem 3.1 holds. Suppose further that*

(i) $h_0, h \in \Gamma$ and h_0 is uniformly finer than h;
(ii) for every $\eta > 0$, there exists a $V_\eta \in \nu_0$ such that

$$b(h(t,x)) \leq V_\eta(t,x) \leq a(h_0(t,x)), \quad (t,x) \in S(h,\rho) \cap S^C(h_0,\eta),$$

$$D^+V_\eta(t,x) \leq g(t,V_\eta(t,x)), \quad (t,x) \in S(h,\rho) \cap S^C(h_0,\eta), \ t \neq t_k,$$

$$V_\eta(t_k^+, x + I_k(x)) \leq \psi_k(V_\eta(t_k,x)), \quad k = 1,2,\ldots,$$

where $a, b \in K$, $g: R_+ \times R_+ \to R$ is continuous on $(t_{k-1}, t_k]$, $g(t,0) \equiv 0$ and $\lim_{\substack{(t,v) \to (t_k,u) \\ t > t_k}} g(t,v) = g(t_k^+, u)$ exists, $\psi_k: R_+ \to R_+$ is nondecreasing and $\psi_k^{(0)} = 0$ for all $k = 1,2,\ldots$;

(iii) *the trivial solution of*

$$\begin{cases} u' = g(t,u), & t \neq t_k, \\ u(t_k^+) = \psi_k(u(t_k)), & k = 1,2,\ldots, \\ u(t_0^+) = u_0 \geq 0 \end{cases}$$

is uniformly stable.

Then the system (3.4) is (h_0, h)-uniformly stable.

In the following result, the functions g and ψ_k are given so that (h_0, h)-uniform asymptotic stability is obtained.

Theorem 3.11 *Assume that condition (ii)–(iv) of Theorem 3.1 holds. Suppose further that*

(i) $h_0, h \in \Gamma$ and h_0 is finer than h;
(ii) for every $\eta > 0$, there exist an integrally positive function $\lambda_\eta(t)$ and $V_\eta \in \nu_0$ such that for $a, b \in K$,

$$b(h(t,x)) \leq V_\eta(t,x) \leq a(h_0(t,x)), \quad (t,x) \in S(h,\rho) \cap S^C(h_0,\eta),$$

$$D^+V_\eta(t,x) \leq -\lambda_\eta(t), \quad (t,x) \in S(h,\rho) \cap S^C(h_0,\eta), \ t \neq t_k,$$

$$V_\eta(t_k^+, x + I_k(x)) \leq V_\eta(t_k,x), \quad k = 1,2,\ldots.$$

Then the system (3.4) is (h_0, h)-uniformly asymptotically stable.

As an application of Theorem 3.1, we consider the following example.

Example 3.4 Consider the impulsive differential equation

$$\begin{cases} x'' + e(t)x' + \beta x + \gamma x^3 = 0, & t \neq t_k, \\ \Delta x = -b_k x, & t = t_k, \\ \Delta x = 0, & t = t_k, \ k = 1, 2, \ldots. \end{cases} \quad (3.6)$$

where $h \in C[R_+, R_+]$, $0 < e_1 \leq e(t) \leq e_2$, $B > 0$, $\gamma > 0$, $0 < b_k < 1$ for all $k = 1, 2, \ldots$.

Setting $y = x'$, (3.6) becomes

$$\begin{cases} x' = y, & t \neq t_k, \\ y' = -e(t)y - \beta x - \gamma x^3, & t \neq t_k, \\ \Delta x = -b_k x, & t = t_k, \\ \Delta y = 0, & t = t_k, \ k = 1, 2, \ldots. \end{cases} \quad (3.7)$$

Now, for any $0 < \eta < 1$, choose a function $\phi_\eta \in C^1[R_+, R_+]$ such that

$$\phi_\eta(s) = \begin{cases} 1, & 0 \leq s \leq \frac{\eta}{2}, \\ 0, & s \geq \eta. \end{cases} \quad (3.8)$$

Next, we define functions ψ_η^+ and ψ_η^- by

$$\psi_\eta^+(x, y) = \begin{cases} \phi_\eta(|y|), & \text{if } \eta \leq \sqrt{x^2 + y^2} \leq 1 \text{ and } e(t)y + \beta x + \gamma x^3 > 0, \\ 0, & \text{otherwise} \end{cases} \quad (3.9)$$

$$\psi_\eta^-(x, y) = \begin{cases} \phi_\eta(|y|), & \text{if } \eta \leq \sqrt{x^2 + y^2} \leq 1, \ e(t)y + \beta x + \gamma x^3 < 0, \\ 0, & \text{otherwise.} \end{cases} \quad (3.10)$$

Construct a Lyapunov function $V_\eta(x, y)$ by

$$V_\eta(x, y) = \frac{y^2}{2} + \frac{1}{2}\beta x^2 + \frac{1}{4}\gamma x^4 + u(\psi_\eta^+(x, y) - \psi_\eta^-(x, y))y, \quad (3.11)$$

where $u > 0$ is a constant to be determined later.

Let $\beta_0 = \min\{\frac{1}{4}, \frac{1}{4}\beta\}$ and $\alpha_0 = 2\max\{\frac{1}{2}, \frac{1}{2}\beta + \frac{1}{4}\gamma, \frac{\beta_0}{2}\}$. Then, it is not difficult to verify that for $u \leq \frac{\beta_0 \eta}{2}$

$$\beta_0(x^2 + y^2) \leq V_\eta(x, y) \leq \alpha_0(x^2 + y^2), \quad \eta \leq \sqrt{x^2 + y^2} \leq 1. \quad (3.12)$$

For $t \neq t_k$ we obtain from (3.11) that

$$\begin{aligned} D^+ V_\eta(x, y) \leq &-e(t)y^2 - u(\psi_\eta^+(x, y) - \psi_\eta^-(x, y))(e(t)y + \beta x + \gamma x^3) \\ &+ u(D^+ \psi_\eta^+(x, y) - D_+ \psi_\eta^-(x, y))y, \end{aligned} \quad (3.13)$$

1. If $|y| > \eta$, then it follows from (3.13) that

$$D^+V_\eta(x,y) \leq -e_1\eta^2, \quad \eta \leq \sqrt{x^2+y^2} \leq 1. \tag{3.14}$$

2. Suppose that $\frac{\eta}{2} \leq |y| \leq \eta$. Then if $u \leq \frac{e_1\eta^2}{\delta M}$, where

$$M = \max_{(x,y)\in\Omega} |D^+\psi_\eta^+(x,y)| + \max|D_+\psi_\eta^-(x,y)|,$$

$$\Omega = \{(x,y) \in R^2;\ \eta \leq \sqrt{x^2+y^2} \leq 1,\ \frac{\eta}{2} \leq |y| \leq \eta\},$$

we have from (3.14)

$$D^+V_\eta(x,y) \leq -\frac{e_1}{\delta}\eta^2, \quad \eta \leq \sqrt{x^2+y^2} \leq 1. \tag{3.15}$$

3. In case $|y| < \frac{\eta}{2}$ holds, then $D^+\psi_\eta^+(x,y) = D_+\psi_\eta^-(x,y) \equiv 0$ and

$$D^+V_\eta(x,y) \leq -e(t)\eta^2 - u|e(t)y + \beta x + \gamma x^3|$$
$$\leq -e_1 y^2 - u\min\{\max(0, e_1|y| - |\beta x + \gamma x^3|), \max(0, |\beta x + \gamma x^3| - e_2|y|)\}.$$

It is easy to see that the function

$$F(x,y) = e_1 y^2 + u\min\{\max(0, e_1|y| - |\beta x + \gamma x^3|), \max(0, |\beta x + \gamma x^3| - e_2|y|)\}$$

is continuous and positive on the set $\eta \leq \sqrt{x^2+y^2} \leq 1$. Set $u = \min\{\frac{\beta_0}{2}\eta, \frac{e_1\eta^2}{\delta M}\}$. Then there exists $N > 0$ such that

$$N = \max F(x,y),$$
$$\eta \leq \sqrt{x^2+y^2} \leq 1.$$

Thus we have

$$D^+V_\eta(x,y) \leq -N, \quad \eta \leq \sqrt{x^2+y^2} \leq 1. \tag{3.16}$$

Let $\lambda_\eta = \min\{e_1\eta^2, \frac{e_1}{\delta}\eta^2, N\}$. Then it follows from (3.14), (3.15) and (3.16) that

$$D^+V_\eta(x,y) \leq -\lambda_\eta, \quad \eta \leq \sqrt{x^2+y^2} \leq 1,\ t \neq t_k. \tag{3.17}$$

Since $0 < b_K < 1$ for all $k = 1,2,\ldots$, a direct calculation from (3.11) gives

$$V_\eta(x^+,y^+) \leq V_\eta(x,y), \quad t = t_k,\ k = 1,2,\ldots. \tag{3.18}$$

Setting $h_0 = h = \sqrt{x^2+y^2}$, $b(s) = \beta_0 s^2$ and $a(s) = \alpha_0 s^2$, we see that all conditions of Theorem 3.11 are met and therefore we conclude that the trivial solution $(x,y) = (0,0)$ of (3.7) is uniformly asymptotically stable.

To motivate our next theorem, let us consider a simple example.

Example 3.5 Consider the linear impulsive system

$$\begin{cases} x' = Ax, & t \neq k, \\ \Delta x = Bx, & t = k, \\ x(0) = x_0, & k = 1, 2, \ldots, \end{cases} \quad (3.19)$$

where $x = (x_1, x_2)^T$, $A = \begin{bmatrix} 0 & 1 \\ 1 & 0 \end{bmatrix}$, $B = \begin{bmatrix} -b_1 & 0 \\ 0 & -b_2 \end{bmatrix}$, $0 < b_i < 1$ and $(1-b_i)e < 1$. It should be noted that $x = 0$ is an unstable saddle point of the underlying system $x' = Ax$. But it is asymptotically stable with respect to system (3.19). In fact, letting $V(x) = \frac{1}{2}(x_1^2 + x_2^2)$, we have

$$D^+V(x) \leq 2V(x), \quad t \neq k,$$
$$V(x(k^+)) \leq (1-m)^2 V(x(k)), \quad m = \min_{1 \leq i \leq 2} b_i,$$

which implies

$$V(x(t)) \leq \begin{cases} V(x(k^+))e^{(t-k)}, & k < t \leq k+1, \ k = 0, 1, 2, \ldots, \\ V(x_0), & t = 0. \end{cases}$$

Thus the conclusion follows from the fact

$$V(x(k^+)) \leq V(x_0)\left[(1-m)e\right]^{2k} \to 0 \quad \text{as} \quad k \to \infty.$$

This example shows that an unstable system may be stabilized by impulses. We have the following general result.

Theorem 3.12 *Assume that*

(i) $h_0, h \in \Gamma$, h_0 *is finer than* h, *and there exist constants,* ρ, ρ_0, *with* $0 < \rho_0 < \rho$ *such that* $(t_k, x) \in S(h, \rho_0)$ *implies* $(t_k, x + I_k(x)) \in S(h, \rho)$ *for all* $k = 1, 2, \ldots;$

(ii) $V \in v_0$, $V(t, x)$ *is* h-*positive definite,* h_0-*decrescent and*

$$V(t_k^+, x + I_k(x)) \leq \psi_k(V(t_k, x)), \quad k = 1, 2, \ldots,$$

where $\psi_k \in C[R_+, R_+]$, $\psi_k(s) > 0$ *if* $s > 0$ *and* $\psi_k(0) = 0;$

(iii) *there exist* $c \in \mathcal{K}$ *and* $p \in PC$ *such that*

$$D^+V(t, x) \leq p(t)c(V(t, x)), \quad (t, x) \in S(h, \rho), \ t \neq t_k;$$

(iv) *there exists a constant* $\sigma > 0$ *such that for all* $z \in (0, \sigma)$

$$\int_{t_k}^{t_{k+1}} p(s)\,ds + \int_z^{\psi_k(z)} \frac{ds}{c(s)} \leq -\gamma_k,$$

for some constant γ_k *and* $k = 1, 2, \ldots$.

Then the system (3.4) is (h_0, h)-stable if $\gamma_k \geq 0$ for all $k = 1, 2, \ldots$, and (h_0, h)-asymptotically stable if in addition $\sum_{k=1}^{\infty} \gamma_k = \infty$.

Remark 3.3 It should be noted that assumption (iii) includes the case when

$$D^+V(t,x) > 0, \quad (t,x) \in S(h,\rho), \quad t \neq t_k,$$

which indicates unstable behaviour of the underlying system. That is, an unstable system has been stabilized by impulses.

Theorem 3.13 *Assume that*

(i) $h_0, h \in \Gamma$, h_0 is finer than h, and for $\rho > 0$

$$\inf h_0(t,x) = 0, \quad (t,x) \in S(h,\rho);$$

(ii) $V \in v_0$, $V(t,x)$ is bounded on $S(h,\rho)$ and

$$V(t_k^+, x + I_k(x)) \geq \psi_k(V(t_k, x)), \quad k = 1, 2, \ldots,$$

where $\psi_k \in C[R_+, R_+]$, $\psi_k(s) > 0$ if $s > 0$ and $\psi_k(0) = 0$;

(iii) there exist $c \in K$ and $\lambda \in PC$ such that

$$-\lambda(t)c(V(t,x)) \leq D^+V(t,x), \quad (t,x) \in S(h,\rho), \quad t \neq t_k;$$

(iv) there exists a sequence $\{\gamma_k\}$ with $\gamma_k \geq 0$ for $k = 1, 2, \ldots$, and $\sum_{k=1}^{\infty} \gamma_k = \infty$ such that

$$-\int_{t_{k-1}}^{t_k} \lambda(s)\, ds + \int_z^{\psi_k(z)} \frac{ds}{c(s)} \geq \gamma_k, \quad z \in (0, \infty).$$

Then the system (3.4) is (h_0, h)-unstable.

4 Applications

In this section, we present a fish population model which exhibits impulsive behaviour in its state variable and may have applications in fisheries management.

Consider a fish population in a lake which connects the upper and down streams of a creek. Suppose that all members of the fish population have identical ecological properties. This means that age differences among members of the population are not important. Under this assumption the population can be modelled by the nonlinear differential equation

$$N' = NF(N) + u, \qquad (4.1)$$

where $N(t)$ is the population at time t, $N(t)F(N(t))$ is the natural growth rate of the fish population and $u \geq 0$ represents a constant influx rate of the population into the lake from the creek.

Suppose that the natural growth of the fish population is disturbed by making catches and adding fish brood, i.e. at times $t_1, t_2, \ldots, t_k$, a part of the fish population with amount $E_1(N), F_2(N)$, are removed from the lake and simultaneously a new brood of fish with amount $D_1(N), D_2(N), \ldots, D_K(N)$, are released. Then the growth of the fish population is impulsive, and can be described by the following impulsive differential equation

$$\begin{cases} N' = NF(N) + u, & t \neq t_k, \\ \Delta N = I_k(N), & t = t_k, \ k = 1, 2, \ldots. \end{cases}$$

We shall first consider the case $u > 0$ and obtain some results on stability properties of system (4.1). To motivate appropriate assumptions, let us consider the logistic equation

$$N' = \alpha N(C - N) + u, \qquad (4.2)$$

where $\alpha, C > 0$ are constants. It is easy to verify that

$$\alpha N(C - N) + u = -\alpha(N - L)(N + M), \qquad (4.3)$$

where $L = \frac{1}{2}\left(\sqrt{C^2 + 4u/\alpha} + C\right) > 0$ and $M = \frac{1}{2}\left(\sqrt{C^2 + 4u/\alpha} - C\right) > 0$. Thus we have

$$(N - L)\left[\alpha N(C - N) + u\right] \leq -\alpha M(N - L)^2. \qquad (4.4)$$

Motivated by this observation, we make the assumption on the right-hand side of (4.1) that there exist constants $L, \sigma > 0$ such that

$$(N - L)\left[NF(N) + u\right] \leq -\sigma(N - L)^2, \quad N \geq 0. \qquad (4.5)$$

Under this assumption and applying Theorem 3.6 to system (4.1), we get the following result.

Theorem 4.1 *Assume that*

(i) $F(N)$ *is continuously differentiable and there exist constants $L, \sigma > 0$ such that (4.5) holds;*

(ii) $I_k(N)$ *is continuous and $I_k(L) = 0$ for all $k = 1, 2, \ldots$;*

(iii) *for any* $z \in (0, L)$, *there exists a sequence* $\{\gamma_k\}$, $\gamma_k \geq 0$ *for all* $k = 1, 2, \ldots$, *and* $\sum_{k=1}^{\infty} \gamma_k = \infty$, *such that*

$$-\sigma(t_k - t_{k-1}) + \max\left(\ln\left[1 + \frac{I_k(L+z)}{z}\right], \ln\left[1 - \frac{I_k(L-z)}{z}\right]\right) \leq -\gamma_k. \quad (4.6)$$

Then the steady state $N = L$ *of (4.1) is asymptotically stable.*

Proof Condition (4.5) implies $LF(L) + u = 0$. This, together with condition (ii), shows that $N = L$ is a solution of (4.1). Let $h = h_0 = |N - L|$ and $V(N) = (N - L)^2$. Then it follows from (4.1) and (4.5)

$$D^+V(N) \leq -2\sigma V(N), \quad t \neq t_k, \quad (4.7)$$

$$V(N(t_k^+)) = \begin{cases} \left[\sqrt{V(N(t_k))} + I_k\left(L + \sqrt{V(N(t_k))}\right)\right]^2, & \text{if } N \geq L, \\ \left[\sqrt{V(N(t_k))} - I_k\left(L - \sqrt{V(N(t_k))}\right)\right]^2, & \text{if } N < L, \end{cases} \quad (4.8)$$

for all $k = 1, 2, \ldots$.

Set $\lambda(t) = \sigma$, $c(V) = 2V$ and $\psi_k(V) = \left(\sqrt{V} + I_k(L + \sqrt{V})\right)^2$ $\left(\psi_k(V) = (\sqrt{V} - I_k(L - \sqrt{V}))^2\right)$. Then application of (3.17) yields (4.6). Thus all condition of Theorem 3.6 are satisfied and the conclusion of Theorem 4.1 follows.

In case the inequality (4.5) is reversed or equality holds, then we get, in view of (4.8) and Theorem 3.7, the following result on instability.

Theorem 4.2 *Assume that conditions (i) and (ii) of Theorem 4.1 hold except that the equality (4.5) is reversed or equality holds. Suppose further that*

(iii*) *for any* $z \in (0, L)$, *there exists a sequence* $\{\gamma_k\}$, $\gamma_k \geq 0$ *for all* $k = 1, 2, \ldots$, *and* $\sum_{k=1}^{\infty} \gamma_k = \infty$, *such that*

$$-\sigma(t_k - t_{k-1}) + \min\left(\ln\left[1 + \frac{I_k(L+z)}{z}\right], \ln\left[1 - \frac{I_k(L-z)}{z}\right]\right) \geq \gamma_k. \quad (4.9)$$

Then the solution $N = L$ *of (4.1) is unstable.*

In the case when $u = 0$, which means biologically, that the environment is closed, it can be seen from (4.2) that we will have, instead of (4.4),

$$(N - C)\alpha N(C - N) = -\alpha N(N - C)^2 \quad (4.10)$$

since $L = C$ and $M = 0$.

Thus assumption (4.5) has to be modified. One possibility is that we set a threshold $N_0 > 0$ such that

$$E_k(N) = 0 \quad \text{if} \quad N \leq N_0, \tag{4.11}$$

for all $k = 1, 2, \ldots$, which means that harvesting is not allowed when the fish population stays below the threshold value N_0. Under assumption (4.11), we can obtain the same conclusions as in Theorem 4.1 and Theorem 4.2 if we revise (4.5) to

$$(N - C)NF(N) \leq \begin{cases} -\sigma(N - C)^2, & \text{if } N > N_0, \\ 0, & \text{if } 0 < N \leq N_0. \end{cases} \tag{4.12}$$

The details for this case are left to the interested reader.

Finally, we consider a special case for the logistic equation (4.2) with $u = 0$. We set $t_k = kd$, $d > 0$ and

$$I_k(N) = \begin{cases} \delta(N - C), & \text{if } N > C, \\ 0, & \text{otherwise}, \end{cases} \tag{4.13}$$

where $\delta \in (0, 1)$ is a constant. For this case, the condition (3.17) in Theorem 3.6 reduces to, in view of (4.12),

$$e^{\sigma d} > 1 + \delta. \tag{4.14}$$

If $e^{\sigma d} < 1 + \delta$, then it can be verified by direct computation that

$$N(t) = \begin{cases} \dfrac{C\beta}{(C - \beta)e^{-\alpha C(t - kd)} + \beta}, & t \in (kd, (k+1)d], \\ & k = 0, 1, 2, \ldots, \\ \beta = \dfrac{\alpha C \delta}{(1 + \delta)(1 - e^{\alpha C d})}, & t = 0, \end{cases} \tag{4.15}$$

is a periodic solution of the impulsive system

$$\begin{cases} N' = \alpha N(C - N), & t \neq kd, \\ \Delta N = I_k(N), & t = kd. \end{cases} \tag{4.16}$$

Thus $N = C$ is unstable which shows the sharpness of condition (3.27).

References

[1] Bainov, D.D. and Simeonov, P.S. (1989). *Systems with Impulsive Effect.* Ellis Horwood.
[2] Lakshmikantham, V., Bainov D.D. and Simeonov, P.S. (1989). *Theory of Impulsive Differential Equations.* World Scientific, Singapore.
[3] Lakshmikantham, V. and Xinzhi Liu (1989). Stability for impulsive differential systems in terms of two measures. *J. Appl. Math. and Comp.*, **29**, 591–604.
[4] Lakshmikantham, V. and Xinzhi Liu (1989). Stability criteria for impulsive differential equations in terms of two measures, *J. Math. Anal. Applic.*, **137**, 591–604.
[5] Lakshmikantham, V. and Xinzhi Liu (1993). *Stability Analysis in Terms of Two Measures.* World Scientific, Singapore.
[6] Xinzhi Liu (1990). Stability theory for impulsive differential equations in terms of two measures. In: *Diff. Eqns. Stability and Control* (Ed.: S. Elaydi). Marcel Dekker, New York, 61–70.
[7] Xinzhi Liu (1993). Stability analysis of impulsive systems via perturbing families of Lyapunov functions. *Rocky Mountain J. Math.*, **23**, 651–670.
[8] Xinzhi Liu (1993). Impulsive stabilization of nonlinear systems, *IMA J. Math. Control Information*, **10**, 11–19.
[9] Xinzhi Liu and Pirapakaran, R. (1989). Asymptotic stability for impulsive differential systems, *J. Appl. Math. Comp.*, **30**, 149–163.
[10] Matrosov, V.M. (1962). On the stability of motion. *Prikl. Mat. Mekh.*, **26**, 885–895 (Russian).
[11] Movchan, A.A. (1960). Stability of processes with respect to two matrices, *Prikl. Mat. Mekh.*, **24**, 998–1001 (Russian).
[12] Samoilenko, A.M. and Perestyuk, N.A. (1987). *Differential Equations with Impulsive Effect.* Vysshaya Shkola, Kiev (Russian).

Index

A
Agriculture-environment
  submodel   257, 261
Agriculture-industry model   256, 258

C
Centrifuges on a string suspension   304
Chaotization of solutions   281
Class $G_m(A)$ 33
Cocycle   2
Comparison system   50, 53
Complex gyroscopic systems   312
Condition
— Bolshakov – Prokhorova's   28
— Grobman's   28
— Lyaponov's   27
— Malkin's   28
— Massera's   28
— Vinograd's   28
Constructive m-exponent   34

D
Domain of stability   65
— — w.r.t. $\{t_0\}$ 65
— — w.r.t. $\mathcal{T}$ 65
Domains of uniform stability   65

E
E-function   112
— global   112
Equation
— differential of perturbed motion   3
— functional differential   52
Equilibrium
— stable hyperbolic   221, 225
Exponential indices
— — higher   29
— — lower   29

F
Family of function
— — definitely negative   77

Function
— γ^+-quasibounded   2
— γ^--quasibounded   2
— $\gamma^\pm$-quasibounded   2
— h-decrescent   323
— h-positive definite   323
— belongs to class ν_0 156
— belongs to class PIM 157
— belongs to class $\Re$ 157
— belongs to the functional family   110
— integrally positive   177, 326
— vector Lyapunov   51, 57
— — — radially unbounded   57
— weakly h-decrescent   323
Functional
— belongs to class $\nu_0(\cdot)$ 156
— belongs to class $\nu_0^*(\cdot)$ 156
— mixed quasi-monotone non-decreasing
  in x, non-decreasing in y 54
— non-decreasing in y 54
— quasi-monotone non-decreasing
  in x 54

H
Higher sigma-exponent   31

L
Lyapunov methodology
— — classical   107
— — consistent   107

M
Mapping
— horizontal fiber   6
— horizontal foliation   6
— vertical fiber   6
— vertical foliation   6
Measure
— finer   323
— uniformly finer   323

O
Orbit
— positive   218
— periodic   218

INDEX

P
P-comparison system   50, 53
Physical Continuity and Uniqueness
  Principle   109
Point
— ω-periodic   218
— equilibrium   218
Problem
— general Lyapunov (linear problem)   27
— — — simplified form   27
— plane elliptical of three bodies   273
— restricted of three bodies   268
— special Lyapunov (linear problem)   26

R
Recurrent event
— — persistent   93
— — transient   93
Rigid bodies rotating on a string   290

S
Semiflow   218
— asymptotic   219
— continuous   218
— discrete   218
— Lagrange stable   218
— Lyapunov stable   219
— phase asymptotic   219
Sequence of period-doubling
  bifurcations   281
Set
— asymptotically invariant   128
— asymptotically stable   65
— asymptotically stable w.r.t. $\{t_0\}$ 65
— asymptotically stable w.r.t. $\mathcal{T}$ 65
— attractive   64
— attractive w.r.t. $\{t_0\}$ 64
— attractive w.r.t. $\mathcal{T}$ 64
— conditionally invariant   128
— conditionally quasi-invariant   128
— exponentially stable   64
— exponentially stable w.r.t. $\{t_0\}$ 64
— exponentially stable w.r.t. $\mathcal{T}$ 64
— invariant   129
— omega limit   218
— positively invariant   68
— stable   64

— stable w.r.t. $\{t_0\}$ 64
— stable w.r.t. $\mathcal{T}$ 64
— total-asymptotically stable   77
— total-attractor   77
— total-stable   77
— uniform   64, 65
— uniformly asymptotically
  stable   128, 129
Solution
— asymptotic   223
— collectively uniformly asymptotically
  stable   99
— collectively uniformly stable   99
— exponentially stable   26
— harmonic   223
— initial value problem   155
— strictly stable   126
— totally stable   99
— uniformly asymptotically
  stable   156, 223
— uniformly Lyapunov stable   223
— uniformly stable   156
Stability   16
— asymptotic   16
— criterion   194
— exponential asymptotic   16
— triangular points of libration   276
— vertical rotations   294
— uniform   16
— uniform asymptotic   16
System
— (h_0, h)-asymptotically stable   323
— (h_0, h)-stable   124, 323
— (h_0, h)-uniformly asymptotically
  stable   323
— (h_0, h)-uniformly stable   323
— (h_0, h)-unstable   323
— eventually stable   124
— perturbed   76
— practically stable   124
— unperturbed   76

T
τ-stable motion   132

V
Vector norm   51, 58
— — regular   58